CW00519064

Back Cover: Top Left: 16in (40.6cm) XI. See Illustration 32 on page 27 for detailed description. *Maxine Salaman Collection.* **Top Right:** 14in (35.6cm) mold 221, original outfit dressed as a nurse, a popular World War I style. *Ruth Noden Collection.* **Bottom Left:** 9in (22.9cm) ummarked all-bisque girl with swivel neck, all original. *Ruth Noden Collection.*

Above: 14in (35.6cm) 243 baby. For detailed description see page 147, Illustration 312. *Mary Lou Rubright Collection.*

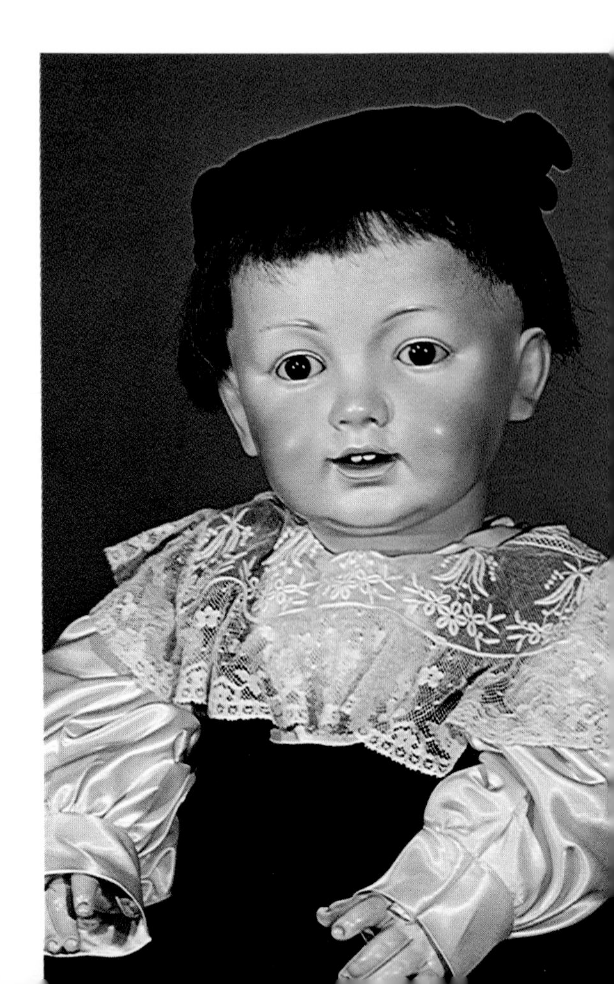

Right: 27in (68.6cm) 220 boy. For detailed description see page 141, Illustrations 297-301. *Jan Foulke Collection.*

Front Cover and Title Page: 19in (48.3cm) pair of 143 children. *Mary Lou Rubright Collection.*

KESTNER

King of Dollmakers

Jan Foulke

photographs by Howard Foulke

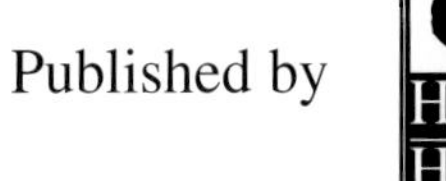
Published by

Hobby House Press, Inc.
Grantsville, MD 21536

Acknowledgements

Many collectors helped make this book possible by allowing photographs of their dolls to be used. Thanks to Jane Alton, Sue Bear, Edna Black, Elba Buehler, John and Janet Clendenien, Dorothy, Jane and Ann Coleman, Rosemary Dent, Rosemary Hanline, Gail Hiatt, Maxine Look, Becky Lowe, Elizabeth McIntyre, Pearl Morley, Kathleen Moyne, Sheila Needle, Ruth Noden, Joanna Ott, Maurine Popp, Jimmy and Faye Rodolfos, Mary Lou Rubright, Maxine Salaman, Richard Saxman, Esther Schwartz, Irene Smith, Helen Teske, Emma Wedmore, May Wenzel, Richard Wright, Virginia Yeatman, and Dr. Carole Zvonar, as well as several collectors who wished to remain anonymous.

Thanks also to collectors who sent information and photographs for the revised edition: Mary Thomasson, Hazel Coons, Joanne Knoedler, Nancy Forester, Joan Williets, Karen Potthoff, Dolly Valk, Jane Walker, Jackie Kaner, LaRue Armstrong, Helen F. Smith, Sarah Doherty and Betty Harms.

Authors whose books and articles were helpful are listed in the Bibliography. Back issues of *Playthings* magazine are available for reference use in the Library of Congress, Washington, D.C.

Additional copies of this book may be purchased at $39.95 (plus postage and handling) from

HOBBY HOUSE PRESS, INC.

1 Corporate Drive

Grantsville, Maryland 21536

1-800-554-1447

or from your favorite bookstore or dealer.

Printed in the United States of America

ISBN: 0-87588-538-1

Table of Contents

27in (68.6cm) turned shoulder head incised "Made in Germany" mounted on a gusseted kid body. *Mary Lou Rubright Collection.*

19in (48.3cm) 187 character child. *Richard Wright Collection.*

19in (48.3cm) XI. For detailed description see page 28, Illustration 36. *Joanna Ott Collection.*

16in (40.6cm) incised "10." For detailed description see page 38, Illustration 58. *Edna Black Collection.*

19in (48.3cm) girl doll incised "13". For detailed description see page 38, Illustration 60.

Redressed in attractive sailor suit style of the 1880s and 1890s. For detailed description see page 39, Illustration 63. *Mary Lou Rubright Collection.*

25in (63.5cm) child incised with letter "O." For detailed description see page 54, Illustrations 96 and 97. *Edna Black Collection.*

Totally original 18in (45.7cm) mold 129 with head size G 11. For detailed description see page 62, Illustrations 112 and 113. *Mary Lou Rubright Collection.*

Introduction

I have been fascinated with the dolls of J. D. Kestner since the very first of my "dolling" days. I have always admired the high quality of workmanship in a Kestner doll and the tremendous variety produced by that company over its long period. I am particularly interested in the bisque dolls — not the earlier papier-mâché ones or the later celluloids. I have spent many hours with friends speculating about the early Kestner bisque dolls which preceded to 100 series, identifying closed-mouth dolls which were possible Kestner products, eliminating, wondering, and hoping for some definite evidence. Kestner all-bisque dolls also have always been my favorites with their pert faces, tiny sleep eyes, and wide range of molded footwear. The factory certainly put a lot of work and care into such a tiny product. I have spent a lot of time studying these little dolls. And so from personal interest and with the encouragement of many friends this book *Kestner, King of Dollmakers,* has evolved, focusing on the bisque dolls produced by this prolific company.

In the 100 and 200 series dolls, I have listed all of the mold numbers which I could find. All were hunted down for photographing except three. We would appreciate hearing about any additional numbers which readers have in their collections.

Jan Foulke
1982

I knew soon after this book came out in 1982 that there were thousands of Kestner collectors and fans in the doll world who shared my enthusiasm and love for Kestner dolls. Many of you wrote to us and we appreciate all of your letters. We tried to respond to all of them, but may have missed a few during particularly busy times; even so we are grateful for all of them. Some of you sent us missing Kestner numbers and numbers that we did not even know existed which we were able to use in this revised edition. Many of you sent photographs of the Kestner kid body with composition arms. Thank you. We have used one of them.

The publication of the Ciesliks' *German Doll Encyclopedia, 1800-1939* in 1984 in German and 1985 in English has added tremendously to our knowledge of Kestner dolls. It is an invaluable source regarding the history of Kestner dolls. Also the Ciesliks have determined that some dolls which collectors long attributed to Kestner were actually made by Alt, Beck & Gottschalck (molds 639, 698 and 911) and Hertel, Schwab & Co. (molds 150-152, 142, 163, 165 and 173), so these have been omitted from this revised edition.

Again, our research on Kestner dolls is continuing and we would appreciate hearing about any additional numbers.

Jan Foulke
1989

OPPOSITE PAGE: Two J.D.K. *Hildas* and a 243 oriental baby. *Mary Lou Rubright Collection.*

All original and fancy costumes for the pair of 12in (30.5cm) dolls incised with "178." For detailed description see page 120, Illustrations 242-244. *Richard Wright Collection.*

15in (38.1cm) 182 mold number. For a detailed description see page 124, Illustration 254. *Richard Wright Collection.*

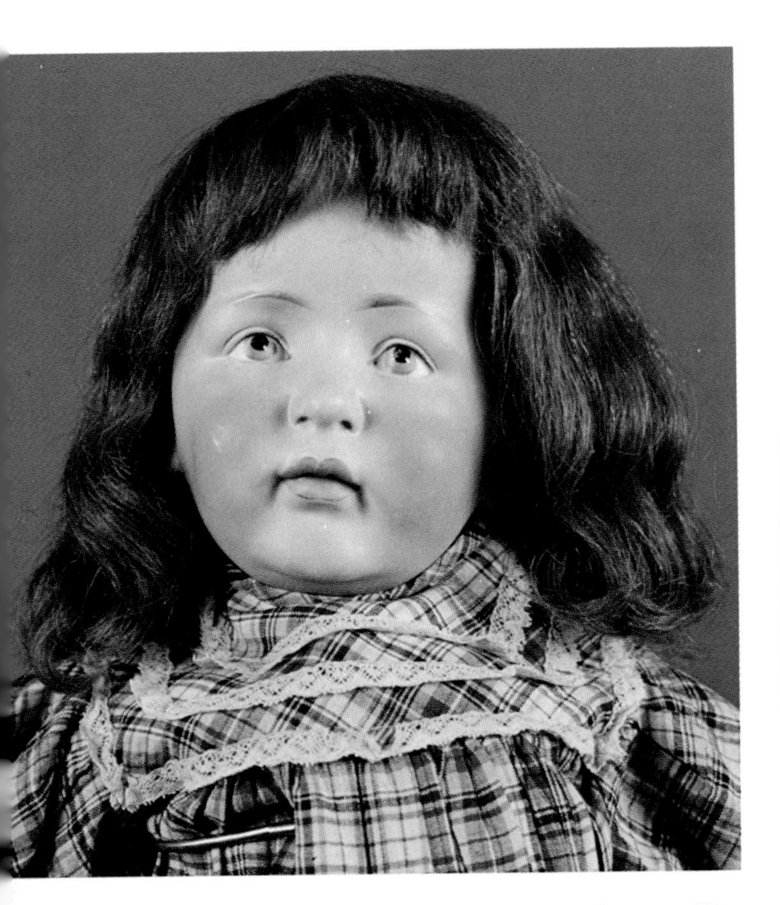

The largest of the Kestner character children at 23in (58.4cm) mold 208. For detailed description see page 132, Illustration 276. *Richard Wright Collection.*

23in (58.4cm) 241 girl. For detailed description see page 135, Illustration 283. *Mary Lou Rubright Collection.*

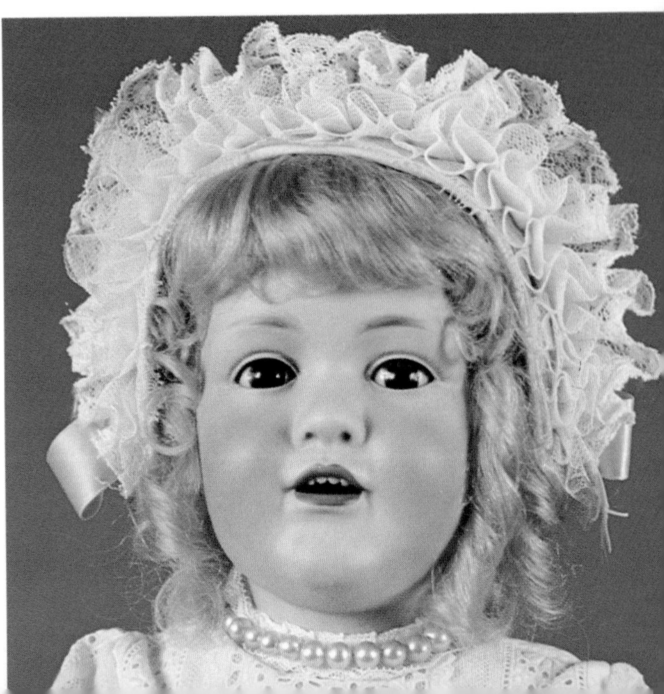

Front row: Left is a 14in (35.6cm) 243 baby. See Illustration 312, page 147. Right: 22in (55.9cm) *Hilda.* See Illustration 327, page 153. Back Row: 21in (53.3cm) *Hilda.* See Illustrations 321 and 322 on page 151. All dolls from *Mary Lou Rubright Collection.*

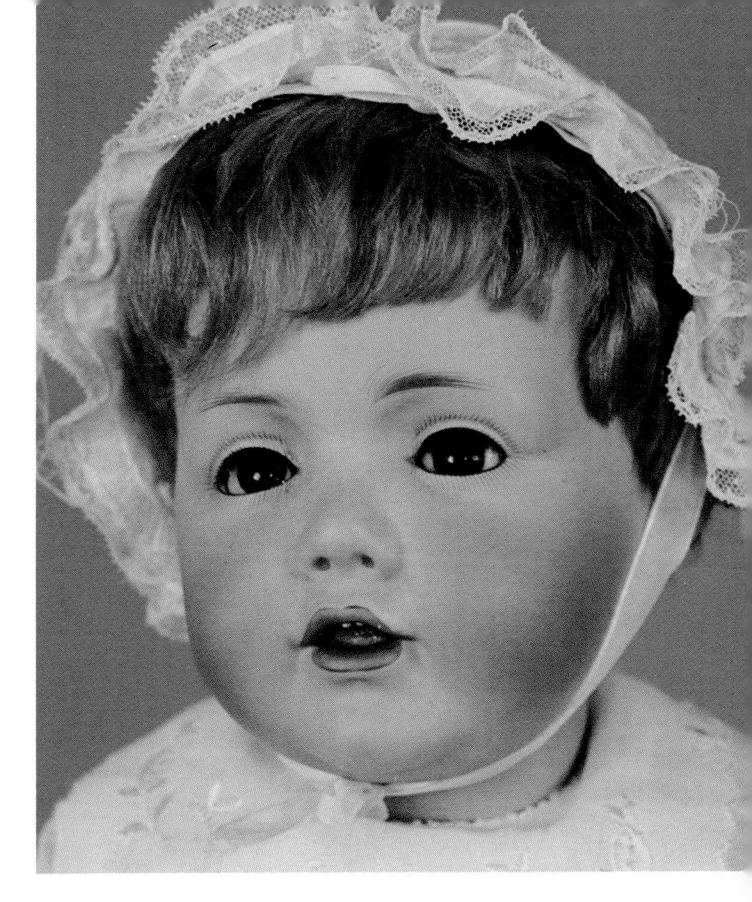

20in (50.8cm) *Hilda,* number 245. See Illustration 318, page 150. *Jan Foulke Collection.*

A large 28in (71.1cm) 260 girl. For detailed description see page 159, Illustration 345. *Mary Lou Rubright Collection.*

24in (61.0cm) boy incised "J.D.K." For detailed description see page 166, Illustration 366. *Mary Lou Rubright Collection.*

The Kestner Company

Johannes Daniel Kestner, Jr., did not start out in doll manufacturing. In 1805 when he started his business, his first products were papier-mâché slates and wooden buttons. It seemed a natural extension to use the lathe on which the buttons were turned for other wooden objects, and by 1816 he was making wooden dolls and toys. How exciting it would be for collectors to find one of these early wooden dolls with a Kestner label! (It is also very unlikely and improbable.)

Early material on the German doll industry is quite sketchy indeed, but Jürgen and Marianne Cieslik are doing fascinating work in this field. According to them Kestner advertised in 1823 that he was making doll heads and bodies of papier-mâché as well as bodies in white leather; the factory was also making buttons, chemises, and ladies' clothing. These dolls must have been the type referred to as milliners' models. Again it would be a collector's dream to be able to identify one of these as a Kestner doll. In 1824 Kestner built a new factory. Apparently Kestner was already using the crown as his trademark as the Ciesliks mention that a crown was in its gable. His business flourished, but was still diversified as he produced not only dolls, but a wide variety of toys and other small objects. The factory provided employment for many Waltershausen families and was the beginning of the great Waltershausen doll industry. Sonneberg, also in Thüringia just over the mountains, had long been known for its dolls and toys, but now Kestner from Waltershausen entered the market. His products soon became known for their fine quality, even better than those from Sonneberg.

Kestner added porcelain heads to his line in the 1850s and greatly increased his sales. These would have been the molded-hair china head dolls with china limbs. He also made waxed-over papier-mâché dolls at this time, the so-called Sonneberg Täufling. Again, and unfortunately, it has been impossible to identify any Kestner doll heads from this period, as they were apparently unmarked.

Illustration 2. Two J.D.K. *Hildas* and a 243 oriental baby *Mary Lou Rubright Collection.*

In 1858, J. D. Kestner, Jr., died and the business was run by others until his grandson, Adolf Kestner, could take over in 1863.

The year 1860 was important for the Kestner firm, for at this time they bought a porcelain factory in nearby Ohrdruf. Now they could manufacture their own china heads and also the new style bisque heads with the lovely molded hairdos. The porcelain works was known as Kestner & Co., and again it is unfortunate that these early dolls were not marked, although collectors enjoy speculating about which dolls of this period could have been made by Kestner. In Janet Johl's *Still More About Dolls* on page 75 is a beautiful doll with molded and braided hair which could be a Kestner, but this is pure speculation. During this period also the Kestner factory in Waltershausen made dolls with waxed-over papier-mâché heads as well as kid bodies.

Illustration 1. 25in (63.5cm) shoulder head 0. *Edna Black Collection.*

In the early 1880s with the introduction of the bisque-headed bébé or child doll on the jointed composition body developed in France, the Waltershausen doll industry, with Kestner as the leader, came into its own. The Ciesliks report that in 1890 the J. D. Kestner firm employed 100 workers in the factory and 500 to 600 home workers, making not only dolls and clothes, but Christmas garden items as well. In 1900 the porcelain factory in Ohrdruf employed 300 workers. Apparently, the Waltershausen factory continued to prosper because the large complex shown on the cover of Kestner's catalog of the 1930s would have housed a good many more than 100 workers. Sometime after 1900 celluloid dolls were introduced to the Kestner line. Made by Rheinische Gummi und Celluloid Fabrik Co., these dolls from Kestner's own molds carry the J.D.K. initials.

Adolph Kestner died in 1918, but the business was continued by several of his employees. The first world war did stop the German toy industry for quite a few years, but the Kestner factory recovered and resumed its doll production. However, during the war years, the American doll industry seized its chance to expand, and even after the war German dolls did not regain their former overwhelming portion of the market.

The focus of this book is on the bisque dolls which Kestner produced from the early 1880s until he went bankrupt, sometime between 1935 and 1938.

Certainly, J. D. Kestner's factory was the first in Waltershausen, and he should be regarded as the founder of that city's famous and prosperous doll industry. The Waltershausen dolls were recognized, advertised, and bought for their fine quality, much better than those from Sonneberg, the other large doll manufacturing area just over the mountains. Both of these areas are in the part of Germany referred to as Thüringia, which is now just over the border in East Germany.

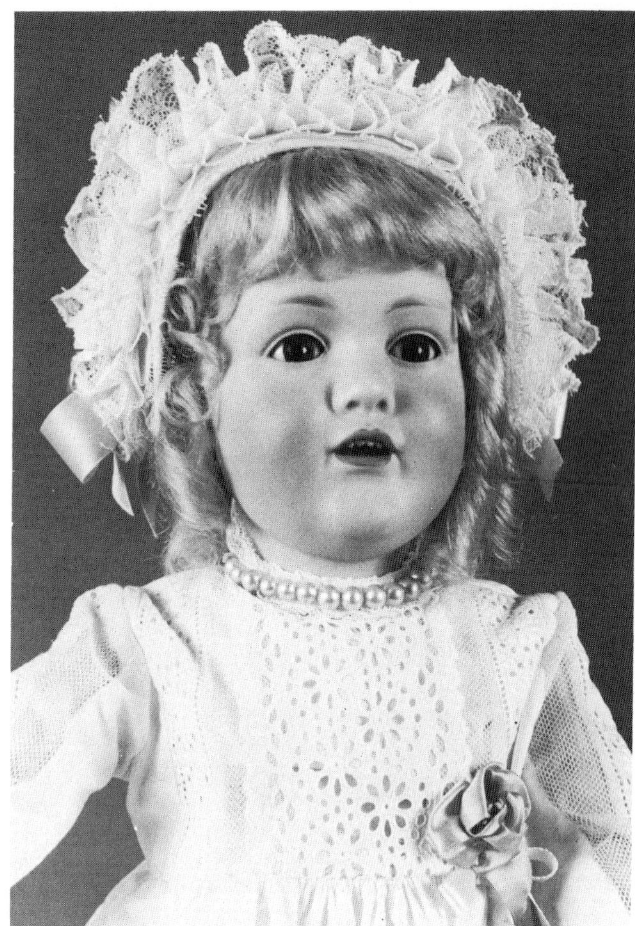

Illustration 4. 23in (58.4cm) J.D.K. 241. *Mary Lou Rubright Collection.*

Illustration 3. 11in (27.9cm) J.D.K. O'Neill *Kewpie. Richard Wright Collection.*

The J. D. Kestner firm was one of the few German dollmakers who made both heads and bodies, thus producing its own complete doll. Most of the other doll companies ordered their heads from one of the porcelain factories, many of which were apparently small and remain unidentified today, judging from the fact that Borgfeldt claimed that 21 porcelain factories were making *Kewpie* dolls in 1914. Well-known porcelain factories in the Waltershausen area were Kestner, Simon & Halbig, Kling & Co., Bähr & Pröschild, Alt, Beck & Gottschalck and later Hertel, Schwab & Co. Well-known doll producers in the area who did not make their own heads were Franz Schmidt, Bruno Schmidt, C. M. Bergmann, Kämmer & Reinhardt, Heinrich Handwerck, Catterfelder Puppenfabrik, Konig & Wernicke, and Kley & Hahn. There were many other smaller and less well-known doll producers in the area as well.

Obviously, with all of these companies located in such a small area, there must have been a lot of give-and-take and going back and forth among the producers. Sculptors of heads, artists, mold makers — many of these probably worked for more than one factory at one time or another. A large factory like Kestner doubtless provided heads for

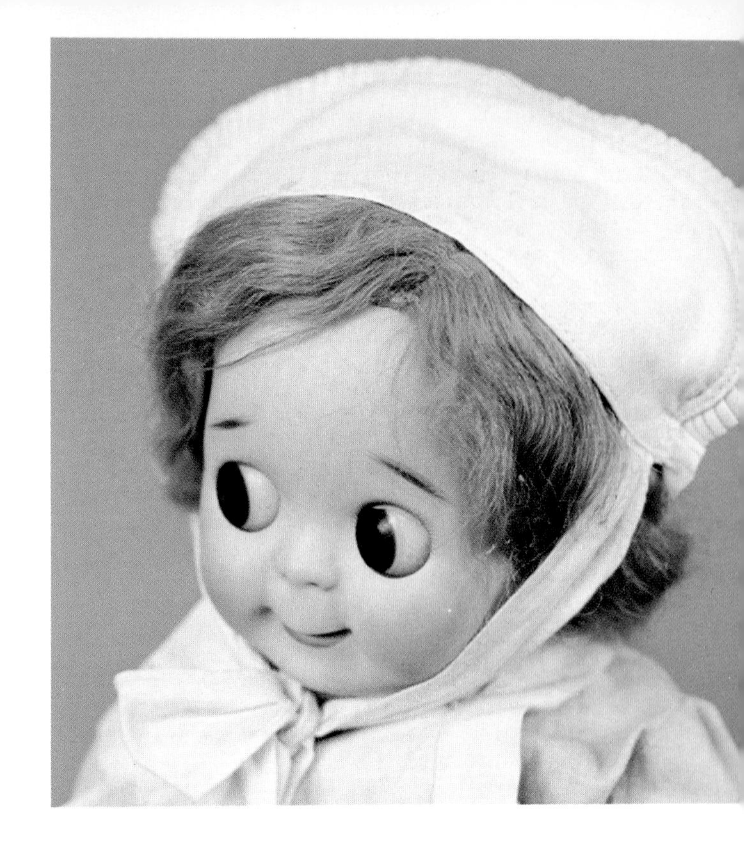

Above: 14in (35.6cm) mold 221, shown in original outfit. For detailed description see page 178. *Ruth Noden Collection.*

Top Left: 16in (40.6cm) XI. For detailed description see page 27. *Maxine Salaman Collection.*

Bottom Left: 9in (22.9cm) unmarked all-bisque girl. *Ruth Noden Collection.*

ABOVE LEFT: 11in (27.9cm) *Kewpie*. For a detailed description see page 187, Illustration 421. *Richard Wright Collection.*

ABOVE RIGHT: Left to right: 4¾in (12.2cm) little girl is incised "111, 0" (Illustration 412); 5in (12.7cm) girl is incised "112 1" (Illustration 413); 5in (12.7cm) boy is incised "112 1" (Illustration 411). See pages 183 and 184. *Jan Foulke Collection.*

15in (38.1cm) girl is a 165 googly, size 5. For a detailed description see page 179, Illustration 398. *Ruth Noden Collection.*

Completely original outfit on a 20in (50.8cm) baby, mold number 211. For a detailed description see page 137, Illustration 286. *Richard Wright Collection.*

many of these producers. It is also likely that producers owned their own molds and simply had the heads made in whatever factory had the time or gave the lowest bid on the price. As doll historians we must realize that there was much going back and forth for a profit and probably at this late date we will never figure out all of the relationships between these companies. It would have been much simpler if all porcelain factories had marked all of their heads and all producers had marked all of their bodies, but unfortunately these companies were not thinking of the doll collector 100 years later who might like to know these things!

Mr. Fred Kolb, manager of the doll department and later president of George Borgfeldt & Co., wrote about Kestner in a letter to Janet Johl quoted in her book *More About Dolls,* page 278. He said that he considered "Kestner to be the most prominent and most important [doll] manufacturer in the world . . . . Kestner dolls had a fine name and reputation with the buying public." He further stated that in his opinion no other manufacturer made a better doll.

Kestner dolls were very popular and respected in the United States. During the period of their manufacture of bisque dolls, Kestner's United States representative was George Borgfeldt & Co. of New York who distributed the Kestner dolls in this country and Canada. The Kestner dolls were not the cheapest ones on the market; although Kestner did have a less expensive, more competitive line, the company relied upon their reputation of supplying good value for the money. As a comparison, in the 1899 Butler Brothers wholesale catalog, Kestner bisque shoulder head dolls with gusseted kid bodies, open mouths, curly hair, and sleep eyes sold for $8.35 per dozen for 17½in (44.5cm) dolls and $12.00 per dozen for 21¼in (61.7cm) dolls. Dolls comparable in size and description made by unidentified manufacturers sold for $3.90 and $8.35 per dozen, respectively. Again, bisque *Florodora* shoulder heads (by Armand Marseille of Sonneberg) in a 1904 wholesale catalog sold for $3.00 to $6.00 per dozen, whereas Kestner heads on the same page sold for $4.20 to $10.50. Similarly, in the same catalog, a beautifully dressed *Florodora* doll, 21in (53.3cm) tall sold for $2.50; a comparable Kestner doll for $3.50. These were wholesale prices, and the retail would have been about twice that amount.

The Kestner name was known to retail as well as wholesale customers. The November 1911 *Playthings* carried a sample ad from the MacDougall & Southwick Co. Toy Department which featured Kestner dolls. A special at $5.00 was "a little blonde beauty exactly 2 feet tall - made in the Kestner way with 'ball & socket' joints, sewn wig -- real eyelashes -- the pretty lawn dress lace-edged, a complete set of under muslins, stockings, pumps." They also offered a Kestner doll at $100.00: "Wears clothes made for a five-year-old girl -- that shows you how big she is! The undergarments are of silk and fine flannel -- her gown is most elaborate. You must see her -- she is b-e-a-u-t-i-f-u-l!" Dressed Kestner dolls were priced at $3.25 on up to $100.00. Undressed ones were $1.25 and 20 other prices up to $16.00.

Illustration 5. 18in (45.7cm) 129 dolly, all original. *Mary Lou Rubright Collection.*

The Heads

The bisque of the Kestner dolls is consistently of excellent quality; seldom is a poor one found. The making of the heads is a very tedious process involving many complicated steps, at any of which something could go wrong and cause an inferior product. When the heads first come out of the mold before they are entirely dry, the eye sockets and the mouth are cut and any other openings which are necessary. (Kestner dolls do not usually have pierced ears, but they would have been done at this time.) After the head is dry, it is completely smoothed to remove any mold debris, marks, or extra clay. Then the piece is ready for its first high firing; it is the high degree of heat which makes the clay hard. It is then sanded; this is when the quality of the bisque shows up and sanding is important to make it smooth. Next the heads are given the pink complexion coat. When this is dry, they are ready for the features to be painted: eyebrows, eyelashes, lips, cheeks, and on the all-bisque dolls, any parts of the body which were to be blushed as on the *Kewpie* in *Illustration 422* and the shoes and socks. Then comes the last firing. This time the completed doll head comes out.

It is easy to see that the porcelain factories had to employ skilled people to carry out these tasks. The process of making heads was long and involved, and it seems incredible to us that they could have originally sold for so little. The retail price in 1904 for an 11in (27.9cm) Kestner with a shoulder head was as little as 35¢. (Sometimes these could be 25¢ if the merchant wanted to use it as a leader item.) One wonders about the cost when it is considered that Kestner made a profit, Borgfeldt made a profit, Butler Brothers made a profit, and the store made a profit! Just how much did the doll cost Kestner to produce? However, here it must be admitted when talking about these small, inexpensive Kestners that sometimes the decoration on them is not quite as well-done as on the larger dolls, but still they are very good when compared to other inexpensive dolls.

The decoration on the Kestner dolls is extremely well-done. Apparently, the factory decided upon a style and all of the artists painted according to the model. The heavy eyebrows are an outstanding Kestner characteristic which could perhaps have been adopted from the Jumeau dolls also noted for this trait, as a Jumeau doll is on exhibit in Waltershausen and Adolf Wislizenus is supposed to have brought one back from Paris, according to the Ciesliks. The Kestner mouth is always done the same way, with a nice high bow on the top lip and upturned ends. There are usually shading strokes at the top edge of the upper lip and the bottom edge of the lower lip; on the closed-mouth dolls a darker red line divides the lips. The complexion coat of the face is usually a light flesh color and the cheeks are nicely rouged. On the dolls after World War I, the complexion coat and the rouge are both quite a bit heavier. The eyelashes are long and carefully painted black with one quick brush stroke each. Each nostril opening is shaded with a red dot, also placed at the inner corner of each eye; the red color usually matches the lip color. The color is probably more accurately described as soft orange than red.

The fine wigs of the Kestner dolls were a good selling point. In many advertisements it was particularly noted that the dolls had curly hair and sewn wigs which meant that the hair was sewn onto a cloth cap, not merely a clump of hair glued onto the dome or inserted in a hole. The wigs were usually of mohair, also called angora fleece, and were predominantly blonde, although some were brown. An interesting sidelight on the blonde hair came up in *Playthings* in April 1913, as someone was decrying the increasing number of brunette dolls. The call was for a return to only the blonde-haired doll as she was "a golden fairy creature for the children" and not only that but she represented "the German ideals of feminine beauty." In the late 1870s and into the 1880s, the *Harper's Bazar* doll column often mentions the blonde-haired dolls with the curly angora fleece which looks like natural hair. In 1877 the wigs themselves cost 40¢ to $2.00 each. In 1904 a comparison of two Kestner dolls exactly the same except that one had fine flowing hair down below her waist, showed that that latter doll wholesaled for $2.25 each, whereas the other doll with a shoulder length wig wholesaled for $1.50. While most Kestner dolls did have mohair wigs, some were

Illustration 6. 22in (55.9cm) 192, all original. *J. C. Collection.*

provided with human hair. In the same 1904 catalog, the Kestner doll just mentioned but with a human hair wig with long ringlets falling to below her waist nearly 12in (30.5cm) long, wholesaled for $2.50.

Of course, we know that Kestner made both bisque and china heads with molded hair, but we have not identified any of these heads for certain as yet. In a 1904 wholesale catalog, Kestner china head dolls are advertised by name. Two line drawings are shown. One had short curly hair with several locks combed down on the forehead, perhaps intended to be a boy doll. The other is a girl with a more conventional hairdo with a center curl dipping low on the forehead. The first listing is for a fine china head with a bisque finish at 75¢ a dozen, but no information is given about hair color or size. Fine china doll heads, handsome in a large size, are listed with assorted light and dark hair; no size is given, but these heads were $1.75 per dozen. Four more entries are given which are larger sizes of the same head. A 5½in (14.0cm) high size sold for $2.50 per dozen; a 6½in (16.5cm) size was $4.20 per dozen.

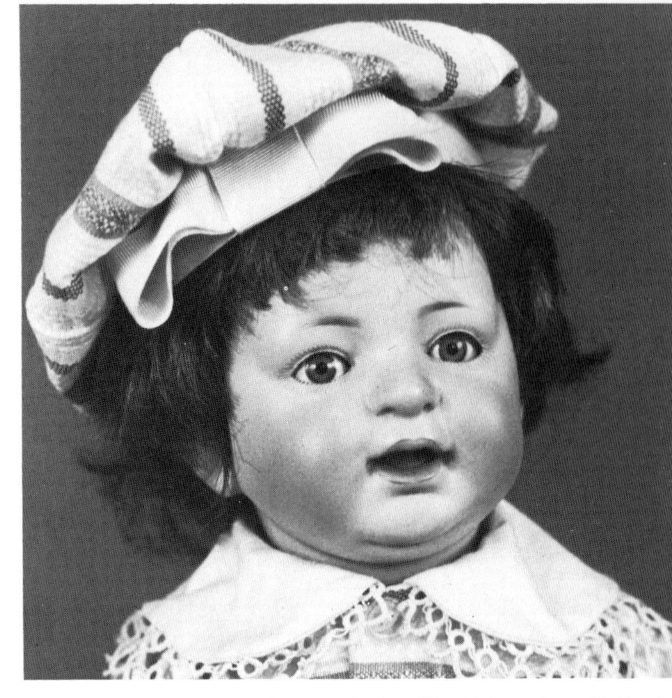

Illustration 7. 17in (43.2cm) J.D.K. 226 baby. *H & J Foulke.*

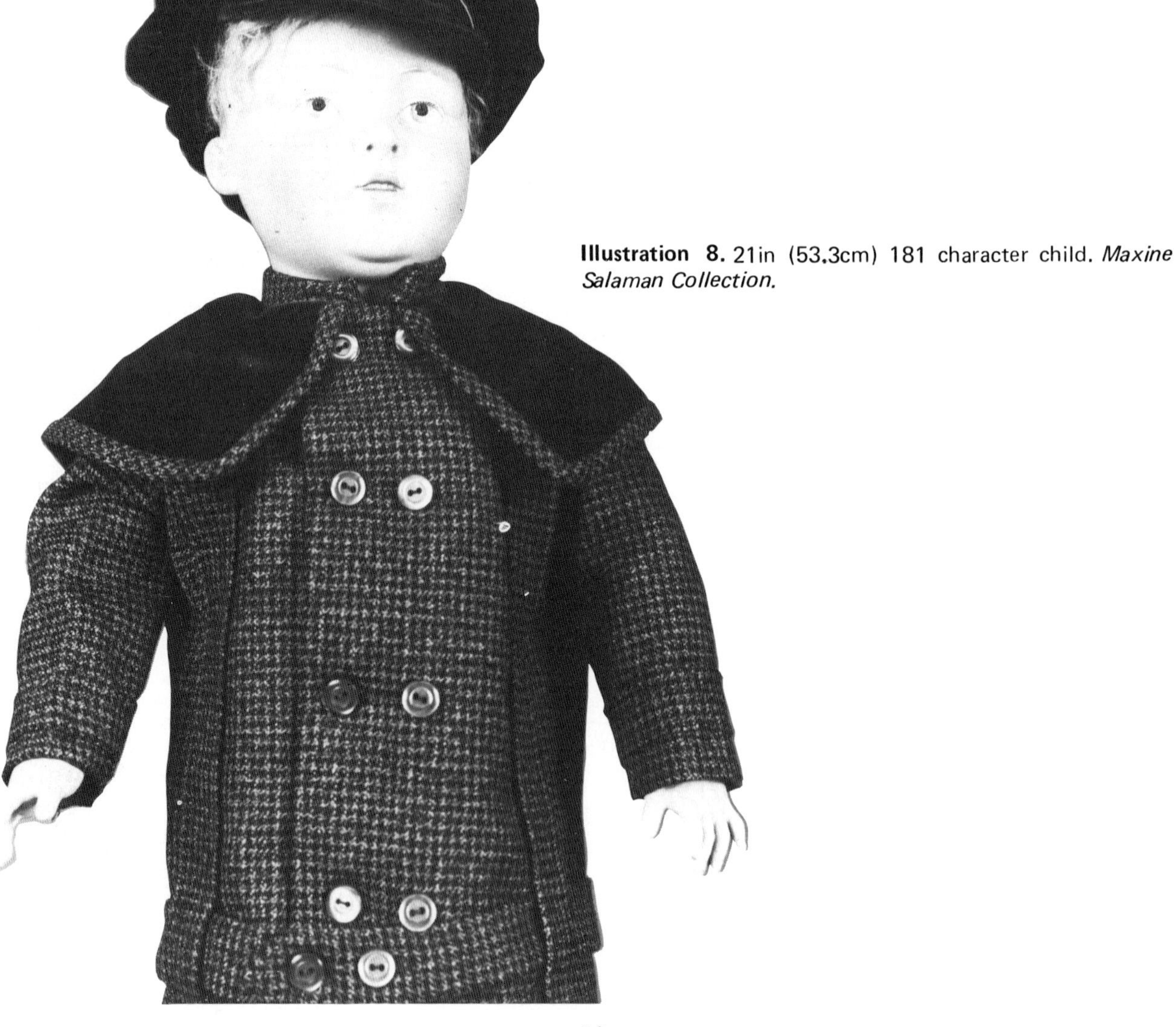

Illustration 8. 21in (53.3cm) 181 character child. *Maxine Salaman Collection.*

Some of the Kestner character babies were made with solid heads and painted hair. Most of these are nearly bald with just a molded curly forelock and the remainder of the hair indicated by brush strokes painted naturally, the way the hair would grow.

The crown opening in the Kestner dolls, which was provided so that the glass sleep eyes could easily be set, is usually covered by a plaster pate. This is often an identifying factor of Kestner dolls as no other specific factory is known to have used this method. However, our study has turned up a few dolls which we feel are not Kestners which do have plaster domes. So it should not be discounted that another company might have used them. Or perhaps we might discover that the dolls we did not think were Kestners are indeed by him after all. This is what makes the work of the doll historian so intensely interesting.

The weighted glass sleep eye was nearly always used in the Kestner heads, even the ones which we consider to be the earlier types from the 1880s. Kestner claims to have invented the sleeping eye in 1860, but other doll sources credit Heinrich Stier of Sonneberg in about 1880. This was definitely a German invention which the French did not pick up until nearly 1900. The Kestner eyes are always of good quality blown glass with threading in the irises to make the eye look natural. Favorite colors are, of course, the blue ones, and the unusual gray ones, a color seldom used by other makers, and sometimes a very deep brown. Kestner apparently agreed with Manfred Bachmann who emphasized the importance of a natural expression in the doll's eyes because the eyes greatly influence the child's relationship to its doll. They are a very important factor in making a doll look real. Otto Gans, inventor and doll manufacturer from Waltershausen introduced the flirty-eyed doll. His patent DRP 135 513 dated from 1901. A mechanism allowed the eyes to move from side to side as well as sleep. Kestner used this flirty-eyed device only rarely in his dolls.

A few of the Kestner character babies have painted eyes. Of course, it was less expensive to produce dolls without glass eyes, as it reduced the expense of the glass eyes and plaster to cement them and the time involved in cutting the eye sockets and placing them in. Some of the Kestner character children have painted eyes which are executed in quite a unique way with the tiny little lashes painted over the colored part of the eye. I do not recall any other company which used this particular little decorating technique. Some of the character childern came with either painted or glass eyes.

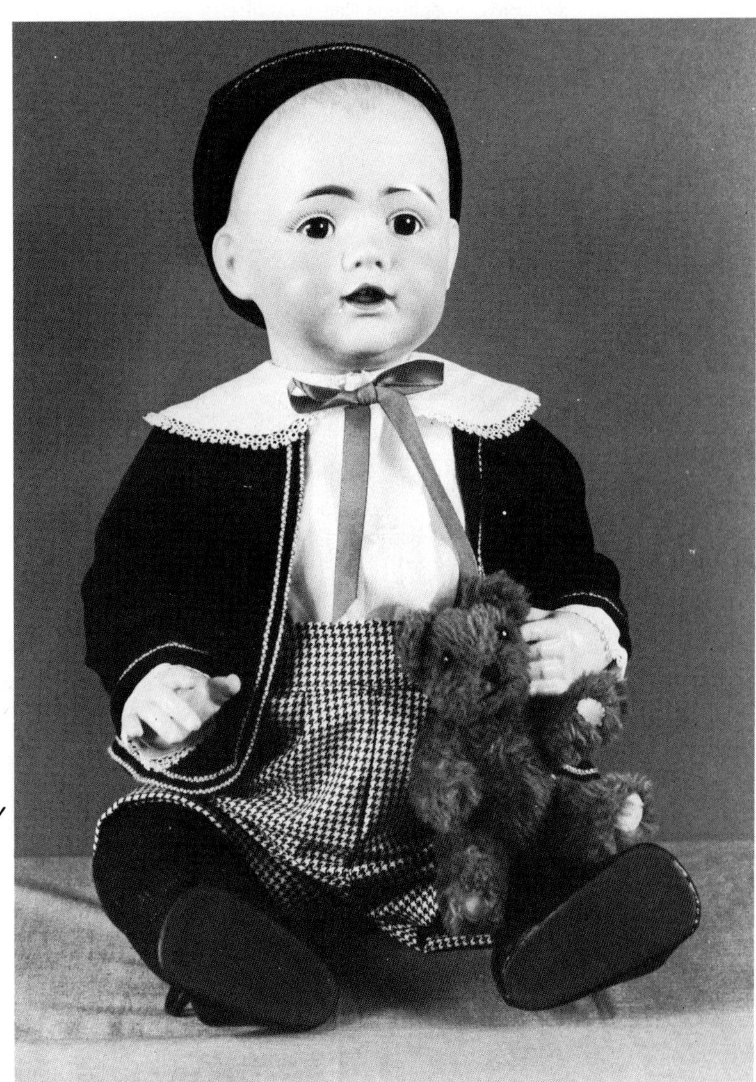

Illustration 9. 24in (61.0cm) J.D.K. baby. *Mary Lou Rubright Collection.*

The Bodies

The Kestner shoulder heads were designed to be used on muslin or kid bodies. The dolls with the muslin bodies would have been their least expensive line. The kid bodies with the gusseted joints would be next. These bodies were made of very fine quality kid with good strong stitching. Most of them were entirely of kid, not with the cloth lower legs, although some did have this feature because after all that part would be covered by stockings and it would be a little cheaper than the all-kid body. The lower arms were of bisque, nicely modeled usually with tinting on the tops of the hands and fingers well-done with free thumbs. A more expensive type of kid body was the one with the hinged joints. This gave much more flexibility to the doll as she could move her limbs more easily; also the hinged joint was less likely to tear open as the gusseted one was apt to do. The *Ne Plus Ultra* joint was used after 1883 for the hip, knee, and elbow. After 1895, the *Universal* joint was used for the knee and elbow but the *Ne Plus Ultra* joint was still used for the hip. A comparison of a 1904 price list shows that with comparable Kestner dolls having kid bodies, the one with the hinged joint was 50 percent higher in price than the one with gusseted joints. Kestner also used the rivet-jointed kid body with shoulder head character baby dolls. In 1895, the Kestner firm began to use its crown and streamers trademark which is often found on its kid-bodied dolls.

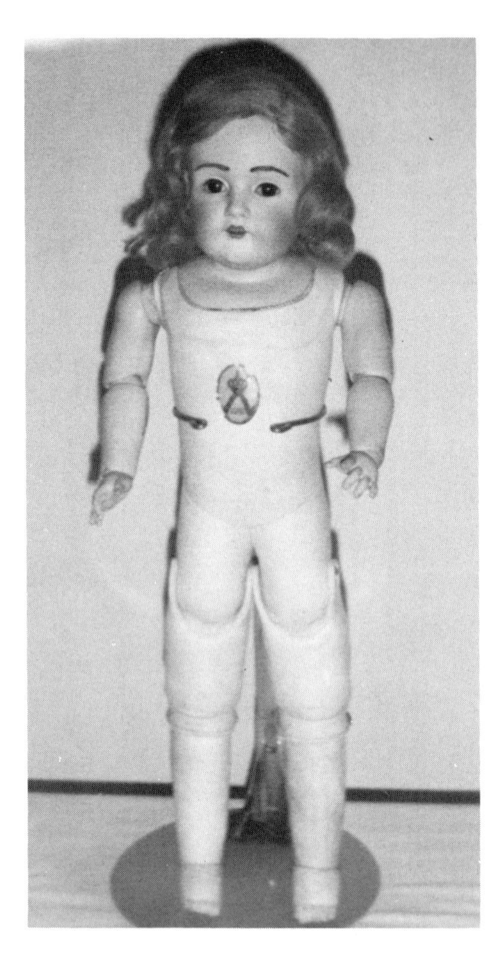

Illustration 10. Kestner 195 shoulder head child on marked Kestner kid body with rivet joints at hips and knees and ball-jointed composition arms. *Courtesy of LaRue Armstrong.*

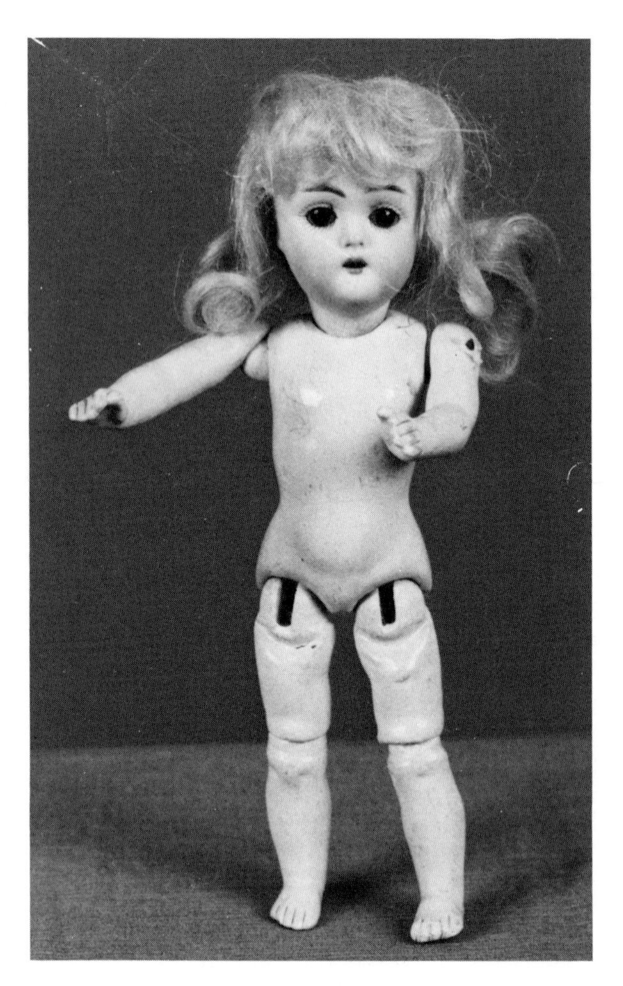

Illustration 11. Small Kestner jointed composition body with ball joint at shoulders, but no elbow joint; mold 155 head. *Edna Black Collection.*

The jointed wood and composition child body was a French contribution to the doll world. Bachmann says that Heinrich Stier of Sonneberg introduced the ball-jointed body to the German doll industry in about 1880. As early as 1881, the doll column in *Harper's Bazar* mentioned the "jointed indestructible [composition] bodies that assume almost any pose are commanded by experienced people as the best outlay of money. These cost from $2.25 upward." In 1884, the same publication commented, "A doll must stand alone, pose in any attitude, and move its head and eyes; all these things are done by the new German jointed dolls, and their bodies are also covered with an enamel in flesh-like tints." Kestner dolls with jointed composition bodies were priced higher than those with kid bodies.

At any rate, Kestner, who may not have been much of an innovator, was certainly always ready to pick up on a good thing. The Kestner ball-jointed composition bodies are of excellent quality. In 1892, Kestner patented his

Excelsior jointed composition body, No. 70685, although it is certain that he was making this type of body long before that, possibly in a somewhat different style. Some Kestner bodies are stamped with this name in red; others are stamped simply with the word "Germany" and a size number.

A Montgomery Ward & Co. catalog of 1903 shows a Kestner doll with yet another body type: a kid body with riveted hip and knee joints and ball-jointed composition arms with moving wrists. Occasionally one of these bodies turns up with a Kestner shoulder head but apparently only a few were made. This model would have been priced higher than the all-kid body but lower than the all-composition body doll. These dolls with sewn wigs, sleep eyes, shoes and socks sold at $3.75 for a 23in (58.4cm) size and $5.75 for a 28in (71.1cm) doll.

The Kestner name was known for fine quality and value. Butler Brothers carried a large line of Kestner dolls, as well as many by other manufacturers, but seldom is any manufacturer other than Kestner mentioned by name. In their 1910 wholesale catalog, Kestner's *Excelsior* bisque-head sleeping dolls are advertised. The bodies were "French" jointed (a term meaning ball-jointed) and described as "hardened layers of paper, next to impossible to break, very light." They were fully jointed even at the wrist with open fingers. The dolls had "bisque turning heads, exposed teeth, lifelike moving eyes, elaborate curly sewed wigs." They were wearing a hemmed chemise with a lace yoke and shirred ribbon trimmings, a bow with streamers, and lace-trimmed sleeves, stockings and real slippers tied with fancy laces. In addition they had the "crown" tag. All of these little details were important and emphasize how well-done the Kestner dolls were. Many dolls did not have real shoes, sewn wigs or hemmed chemises with lace and ribbon trim. With a Kestner doll, the price was higher, but so was the value. A 14½in (36.9cm) doll was 75¢; prices and sizes went up to $8.00 for a 36in (91.4cm) doll. Remember that these are wholesale prices and retail would have been up to twice as much.

After the coming of the character baby head in 1910, Kestner developed his bent-limb composition baby body, a very realistic body with lots of detail in the modeling of the torso, arms, and especially the legs. A variation of this body had chubby straight legs. I have seen only one of these bodies with a Kestner mark, however.

In addition to the baby body, Kestner also used a variety of other types: a fat, chubby toddler one; a slim teen-age type one; a tiny completely jointed one; and a lady one, as well as others which are shown throughout this book.

The character babies and some of the child dolls had a cardboard tag affixed to them in the shape of the Kestner crown with the words "Crown Doll, Kestner, Germany" on it. This tag is shown in *Illustration 12*. Unlike the earlier crown trademark, this one has no streamers. Although this trademark was not registered until 1915, it was used on a doll in a January 1913 ad, so it was probably in use by at least 1912.

Illustration 12. 8½in (21.6cm) J.D.K. 257 baby, all original. *Jimmy and Faye Rodolfos Collection.*

Body Marks

On ball-jointed composition bodies:

Excelsior
Germany 1

On kid bodies:

On composition toddler body:

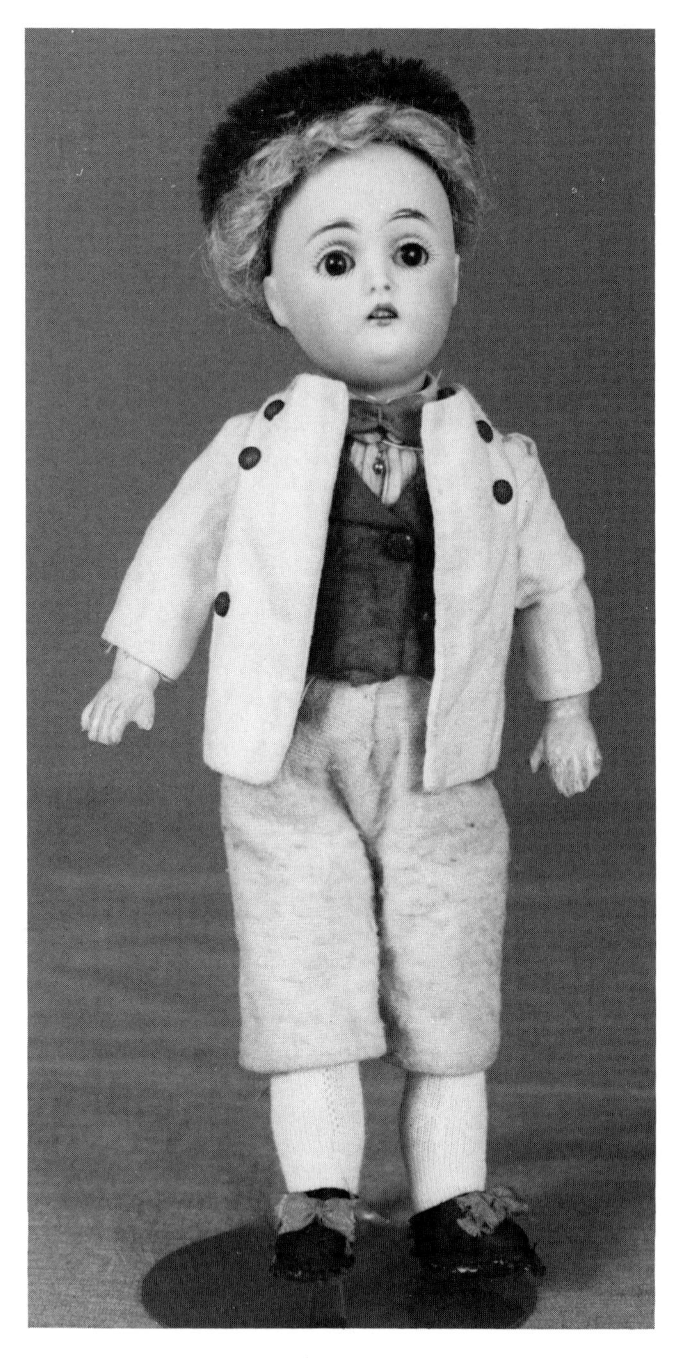

8½in (21.6cm) 155 boy doll, all original. *Jane Alton Collection.*

16in (40.6cm) 179 character child. *Jane Alton Collection.*

Kestner Markings on Doll Heads

The Kestner mold number and size code are incised on the backs of most of the socket-head "dolly" dolls with open mouths and on some of those with closed mouths, as well as most of the character babies. Some of the open- and closed-mouth shoulder heads have these markings also. Not until after 1910 are the initials J.D.K. used on the heads, but only on the 200 series dolls. Even though many of the 100 series heads were made after 1910, they do not have the J.D.K. initials. The 100 series heads must be identified by the mold number and the use of the unique Kestner sizing system, which is given below. It was registered by Kestner in the District Court in 1897, although it was used as early as 1891 when the "country of origin" law went into effect in the United States requiring all foreign products to be marked as to country of manufacture. This complete new marking system giving the new sizing and country of origin was implemented at this time. The complete mark is generally like the sample given, but there are many variations in the type of lettering and the placement of the various components of the mark. The mark of the Kestner & Co. porcelain factory is quite large and elaborate; its use denotes a later doll, probably after World War I.

Sizing	
d	1
c	2
b	3
a	4
A	5
B	6
C	7
D	8
E	9
F	10
G	11
H	12
I, J	13
K	14
L	15
M	16
N	17
O	18
P	19
Q	20

(Sometimes 1/2 is added to both the letter and number to indicate an intermediate size.)

Kestner & Co.

The d 1 size was found on a 7in (17.8cm) child doll. The complete size markings on the 7in (17.8cm) and 8in (20.3cm) dolls are very difficult to read because they are so small. In place of the size letter a symbol is used, such as f, c, or K; the size number is given as 5/0 or 3/0. This is a standard way for a porcelain factory to indicate a mold size smaller than 1.

The largest size Kestner head is Q 20. This was used for a 42in (106.6cm) child doll, a 27in (68.6cm) character toddler and a 25in (135cm) character baby. The latter could be 24-26in (61-66cm) depending upon the method and accuracy of the measurer.

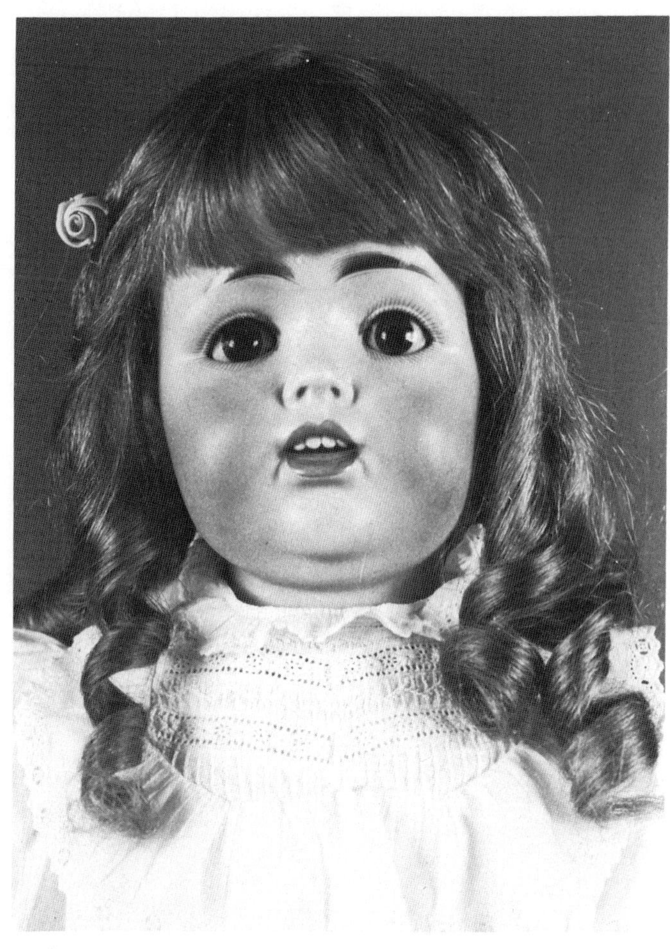

Illustration 13. 28in (71.1cm) J.D.K. 260 character child. *Mary Lou Rubright Collection.*

Typical Mark (100 series)

F made in Germany 10.
149.

made in Germany
E.768.9.

Typical Mark (200 series)

Q. made in Germany. 20.
J.D.K.
220.

The Kestner Catalog of Circa 1930

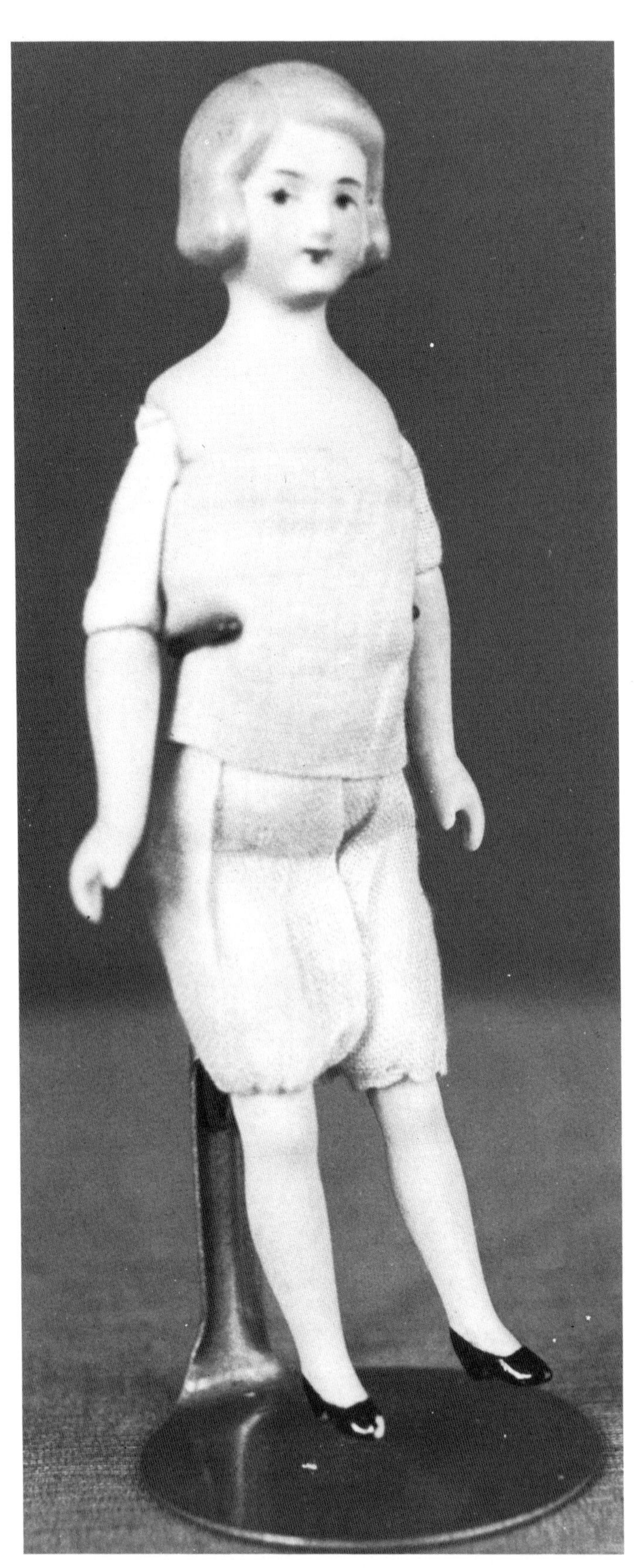

It is always a significant and important find when a manufacturer's catalog is discovered. It is even better when this information can be shared with other collectors as did Georgine Anka and Ursula Gauder when they printed in their book *Die Deutsche Puppenindustrie 1815-1940* a catalog of the Kestner factories, J. D. Kestner, Jr., of Waltershausen and Kestner & Co. of Ohrdruf. They maintain that this is the last catalog of the Kestner firm. The Ciesliks indicate that Kestner was still working in 1934. We have, therefore, called this document the Kestner catalog of circa 1930 and have referred to it many times in this book. The authors have not specifically dated the catalog which could actually be from the late 1920s as many of the dolls are dressed exactly like those from the 1927 and 1928 catalog of Kämmer & Reinhardt.

Table 2a shows ten dressed bent-limb babies and toddlers. One baby has flirty eyes; the heads appear to be possibly mold 257 or 260.

Table 4a shows 14 babies and toddlers with celluloid heads.

Table 5 shows ten bisque-head girls, some of which appear to have character faces; several could be mold number 241. Most of the dolls have the regular ball-jointed child bodies, but several have the slim teen-age type and one has a five-piece toddler body. The dolls are wearing chemises and have very large bows in their bobbed hair, although a few have short curls.

Table 7a shows 12 slim-jointed girls, all beautifully dressed in the latest styles with short hair, large hair bows, and lovely dresses. They appear to be mold 260.

Table 8 shows five girls in chemises, probably mold 260, one dolly-face doll, one black dolly-face doll with short curly black hair, and four googlies of mold 221.

Table 14 shows pairs of dressed doll house dolls. Many are all-bisque children either wigged or with molded hair and varying types of molded footwear, and many are adults with soft bodies. They include both wigged and molded-hair styles as fathers, grandfathers, mothers, grandmothers, chauffeur, and maid.

Table 15 shows heads only, most of which are character faces. Although several are wigless, only one looks as though it could be a solid dome baby with molded hair.

Table 16 shows bisque shoulder heads with or without wigs. It also shows two leather bodies, both with composition arms and one with composition lower legs. Two celluloid shoulder heads, mold 201, are shown.

Table 17 shows parts for jointed composition bodies, pates, rubber cord, and weighted eyes.

Table 18 shows a large variety of all-bisque dolls.

Table 19 shows undressed doll house dolls with cloth bodies.

It is known that Kestner made pincushion dolls at about this time, but none are shown in this catalog.

Illustration 14. 5in (12.7cm) doll house lady with molded and painted features like those shown in the 1930s Kestner catalog. Further study needs to be done on the mold numbers of these dolls to try to determine which ones were made by Kestner.

Closed-Mouth Dolls
Introduction

Although the Germans are credited with inventing the bisque-head doll, what they produced at first were the lovely lady doll shoulder heads with beautifully modeled hairdos, many with ornamentation in their hair and elaborately decorated shoulders. It is certain that Kestner, who owned a porcelain factory in the 1860s when these heads were popular, made these types of heads, but unfortunately none are marked so it is impossible at this point to assign any to him with certainty. The French are credited for developing from these the lovely lady fashion dolls to which were added swivel necks, glass eyes, wigs, and shaped kid bodies. Always in competition, the Germans also produced shoulder head dolls with glass eyes and wigs, some with swivel necks on kid bodies which could also serve as lady dolls, but oddly the German dolls seldom really have lady faces.

Again it was the French who developed the child doll and the jointed compostion body, which the Germans adopted immediately, putting these into production by about 1880. Certainly Kestner, already a leading doll producer by this time, was in the forefront with this new type of doll. Although Kestner heads of this period are not marked with the J.D.K. initials and few have the later Kestner sizing system, there is a group of dolls with characteristics similar to known Kestner dolls which collectors have long attributed to the Kestner firm. Perhaps someday researchers will find proof positive that Kestner really did make these dolls; perhaps they will find proof that another firm such as Kling & Co. made these dolls, but until then we will show them as being attributed to Kestner & Co. All of the dolls shown in this chapter date prior to 1895. When dating Kestner closed-mouth dolls, remember that the letter-number sizing system was in use by 1891. The "Excelsior" body was first used in 1892.

The time period from the 1880s until 1915 was the heyday of the "dolly-faced" doll, one that was just beautiful to look at. The idea of the perfect child was epitomized in the dolly face with her rosy cheeks, sparkling eyes, lips offering a hint of a smile, and flowing blonde curly hair. She was a vision of loveliness. This romanticized concept was typically Victorian; her idealized beauty could not have been created at any other period in history. It was a time when harsh, ugly reality was pushed behind the scenes, and only the glamour and glitter were shown to the world.

B½ Made in 6½
Germany
169.

Illustration 15.

Illustrations 15, 16 and 17. There is no doubt that this 16in (40.6cm) girl is a Kestner. She is incised with what is clearly a Kestner marking. However, the number of the mold is puzzling since all of the other dolls with numbers in the 160s are open-mouth dolls. She has the facial characteristics which make the Kestner dolls so popular. Her glossy thick brown eyebrows are molded, although this is generally thought of as a later characteristic. She has beautiful gray glass sleep eyes; this color often appears on Kestner dolls, seldom on dolls by other makers. Her painted eyelashes are long and well spaced. Her original blonde mohair wig in the curly Rembrandt style covers a plaster dome. Her bisque is very fine with a creamy complexion and rosy tinted cheeks. Her jointed composition body has a yellow finish and carries the Kestner "Excelsior" mark used from 1892, which would place her at that date to just after. She is wearing an old rose silk dress with a white organdy pinafore. *Rosemary Dent Collection.*

Illustrations 16 and 17 are on page 22.

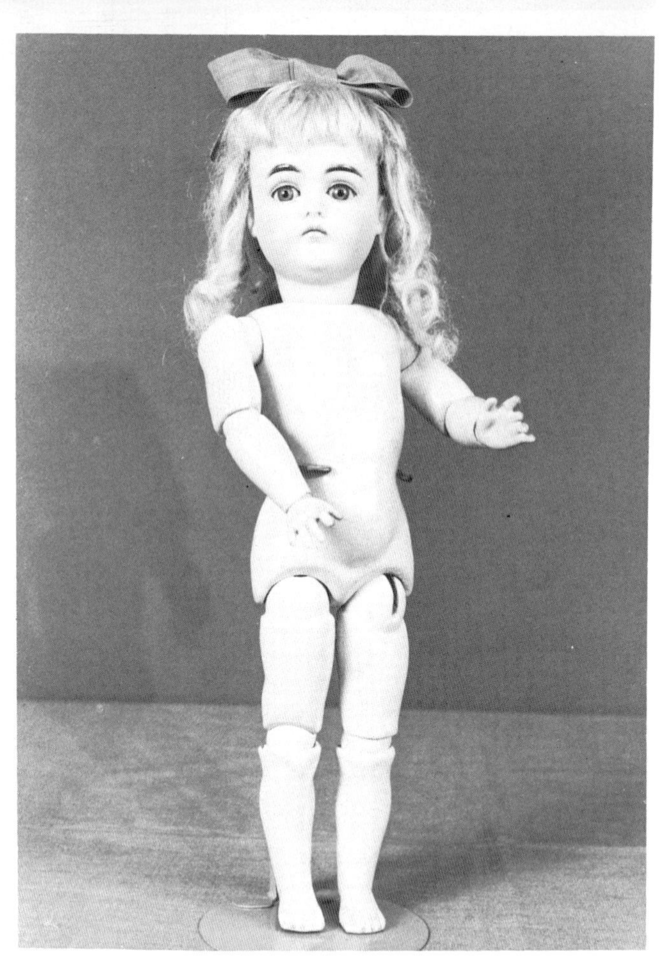

Illustration 16.

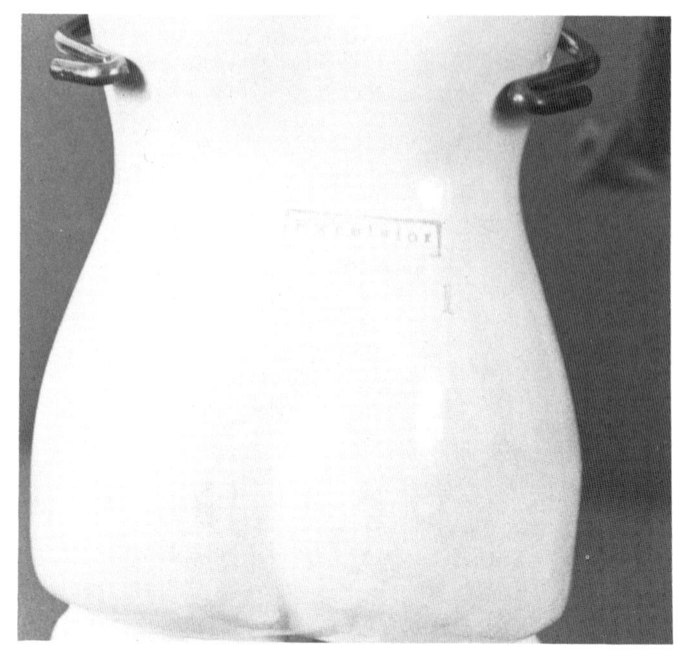

Illustration 17

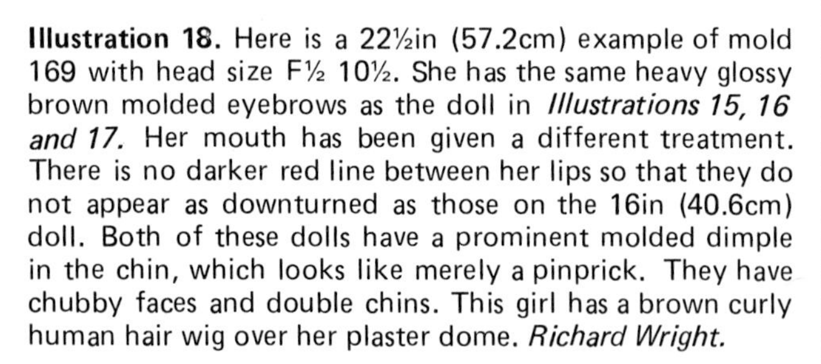

Illustration 18. Here is a 22½in (57.2cm) example of mold 169 with head size F½ 10½. She has the same heavy glossy brown molded eyebrows as the doll in *Illustrations 15, 16 and 17.* Her mouth has been given a different treatment. There is no darker red line between her lips so that they do not appear as downturned as those on the 16in (40.6cm) doll. Both of these dolls have a prominent molded dimple in the chin, which looks like merely a pinprick. They have chubby faces and double chins. This girl has a brown curly human hair wig over her plaster dome. *Richard Wright.*

Illustration 19. This is again without a doubt a Kestner girl as she has the incised marking as *Illustrations 15, 16 and 17* but with size E 9 and mold number 128. She has brown sleep eyes and the full Kestner eyebrows with many tiny brush marks at both the inner and outer ends. The mold is very similar to number 169 except for the mouth which has much longer lips. There is also a darker red line to separate the upper and lower lips. This doll has her original blonde mohair wig in the Rembrandt style with bangs. She is 17in (43.2cm) tall on a jointed composition body with the "Excelsior" mark. *Esther Schwartz Collection.*

Illustration 20. This girl is a sister to the 128 doll in *Illustration 19* and has the same marks. Her painting is a little more vivid on the mouth and cheeks, and her eyebrows are thicker and more arched; also she has a greater number of painted lashes. These are simply stylistic differences since a variety of artists painted the doll heads. She has blue sleep eyes, and a replaced human hair wig. This mold number also came in an open-mouth version which looks just about the same as the mold 129 doll shown in Illustration 111. *Esther Schwartz Collection.*

Illustrations 21 and 22. This large 29in (73.7cm) doll has no mold number, only the size L½ 15½, but she appears to be simply a larger version of mold 128. She is a lovely doll with beautifully painted features and is one of those rarities: a completely original doll bought from her original owner. Her human hair wig is blonde; her sleep eyes are blue. She is also on a jointed composition body with the "Excelsior" mark. *Esther Schwartz Collection.*

Illustration 22.

Illustration 23. This large 28in (71.7cm) doll also is possibly another version of mold 128. She is incised "made in Germany 16" in small letters up at the crown opening in the manner of several other dolls in this study also attributed to Kestner. She has a plaster dome and a very heavy jointed composition body with straight wrists. Her mouth treatment is a little different, especially in the shape of the upper lip. Her eye sockets are cut larger giving her an exceptionally wide-eyed look. She is very full cheeked, and the full front view shows how her cheeks come down to form her prominent double chin. *Mary Lou Rubright Collection.*

Illustration 24. This little girl has no marks on her head, but she is on a marked "Excelsior" body, and her face is definitely related to the known Kestner dolls. She has the Kestner plaster dome and the thick eyebrows with many individual brush strokes. Her closed mouth has a lot of detail in the lip molding with a very protruding upper lip, and her chin sports the "punched-in" dimple. Her cheeks are long and more slender, like those shown on the dolls in *Illustrations 64 - 73.* She has an appropriate replaced white cotton dress with blue embroidered trim in Kate Greenaway style which works very well on little German girls. *H & J Foulke.*

Illustrations 25 and 26. This 13in (33.0cm) girl has incised Kestner markings on the back of her head "a made in 4// Germany." She has the remains of her plaster dome. Her glossy eyebrows consist of several very heavy individual brush strokes; her lashes are lightly painted. Her gray sleep eyes are a color often associated with Kestner. Her face has very interesting detail. The side view illustration shows the profile line of her long and very full lower cheeks, rounding out in a double chin with a "punched-in" dimple. The molding of her mouth emphasizes her upper lip which juts out; it is the open/closed type with parted lips and a white space to indicate teeth, but with no real opening cut into the bisque. She probably dates about 1891 when the Country of Original Law was passed and may have been intended as competition for the open-mouth dolls coming into vogue then. She is on a heavy jointed composition body with straight wrists and cupped hands. *Jane Alton Collection.*

Illustration 26.

Illustration 27. This 15in (38.1cm) girl appears to be a shoulder head version of the doll in *Illustrations 25 and 26.* She is marked on her shoulder "a Germany 4." The back of her head is flat like that of the Kestner 154 mold. Her long human hair curls hide her protruding ears. Her brown eyebrows match her brown eyes; her eyelashes are quite heavy. Of course, decoration will differ from doll to doll made by the same company because the heads were individually painted. The color can vary because of the mix; the heaviness of the lashes would depend upon the thickness of the paint. This head is on a gusseted kid body, which helps account for the fact of the different heights of two dolls with the same size head. *Esther Schwartz Collection.*

A.T.-Type

Illustration 28.

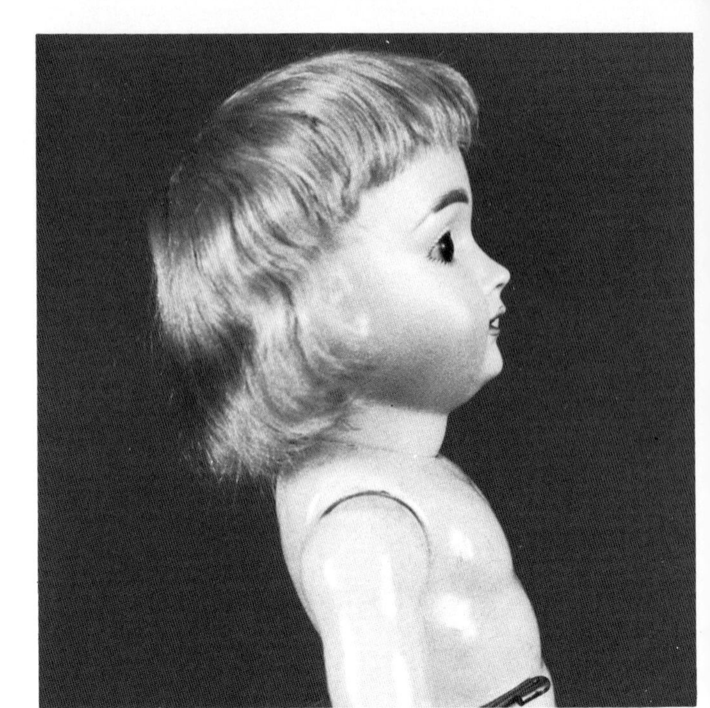

Illustration 29.

Illustrations 28, 29 and 30. This fellow is a rare doll and one which collectors attribute to Kestner. He is 16in (40.6cm) tall and unmarked except for his size number 10. His beautiful long glossy eyebrows painted in the Kestner manner with many individual strokes match his blonde mohair wig. His face is marvelous. The side view shows his sharp nose and chin as well as his stout neck. His mouth is the open/closed style with fairly wide parted lips and quite a deep opening. His body is in excellent condition and, except for the free balls at the shoulders and elbows, is like that shown on the known Kestner doll in *Illustration 218.* Collectors of Kestner dolls enjoy saying that Kestner copied dolls from all of the French manufacturers; this one is referred to as the "Kestner A.T." This mold has also been found with an open mouth and square teeth. *Joanna Ott Collection.*

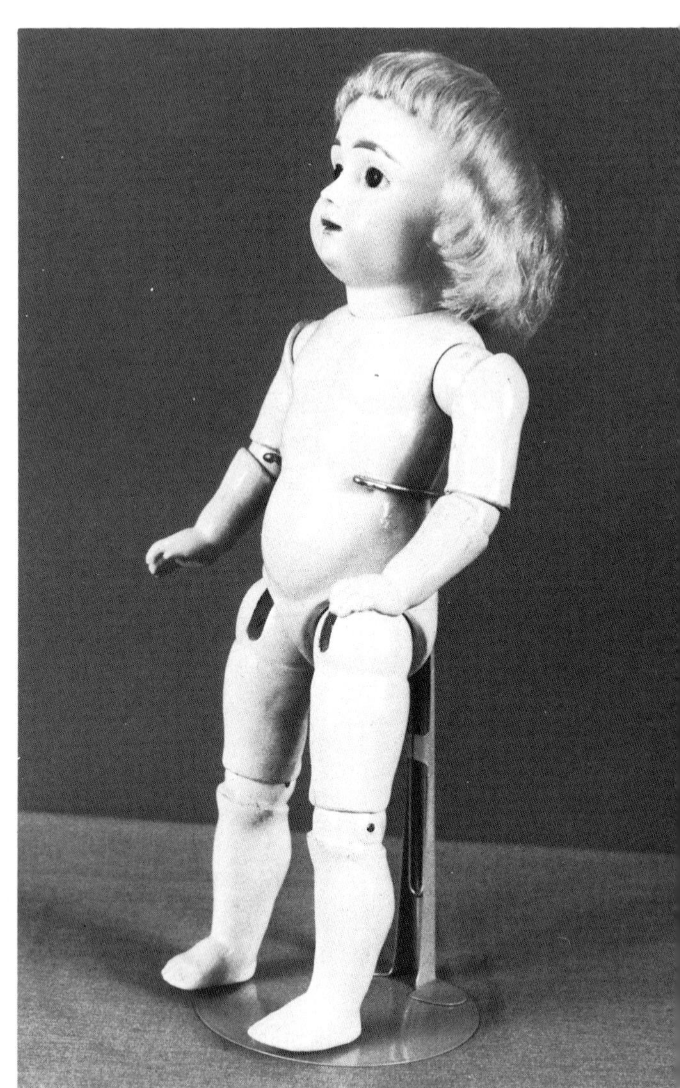

Illustration 30.

X and XI

Again, there is no concrete proof that Kestner made the dolls which bear these marks, but they have traditionally been attributed to Kestner by collectors because of their characteristics. They have a high quality bisque with beautiful painted decoration and much attention given to details. The dolls have the plaster domes favored by Kestner and while it is possible that another company used this method of covering the crown opening, none has yet been identified. The eyebrows are the heavy Kestner type, often glossy. The mouth is in the Kestner style with an upturned upper lip, darker red middle lip line, and two high peaks on the top lip. The eye sockets are small, of the type favored by Kestner. These dolls are probably from the 1880s.

Illustration 32.

Illustration 31.

Illustrations 31 and 32. The XI dolls are real favorites with Kestner collectors. These dolls have quadrupled in price over the past five years. This 16in (40.6cm) XI is typical of the beauty inherent in this face with its soft, serene, and gentle visage. This particular doll has a pale blonde wig, medium-toned eyebrows and striking brown sleep eyes. She has been redressed in old fabrics and lace in the Kate Greenaway style. *Maxine Salaman Collection.*

Illustration 33. Some collectors enjoy having pairs of dolls; this is mentioned is *Harper's Bazar* way back in 1876. This pair of 16in (40.6cm) XI twins both have gray eyes and auburn mohair wigs which emphasize their rosy cheeks. Their replaced dresses are in the popular Kate Greenaway style of pink cotton dotted swiss with lace trim and ribbon insertion. Their fancy bonnets match the dresses. These dolls have excellent quality jointed composition bodies with attached balls in the legs and separate balls in the arms; the wrists are solid with cupped hands. This is typical of the body used on most of these X and XI dolls and is shown in *Illustration 30.* This type of body apparently preceded the "Excelsior" marked one. Because this style is so much like those made by Schmitt & Fils of Paris, France, collectors like to refer to these dolls as the "Schmitt" Kestner. Several of the XI heads have been reported on marked Schmitt bodies, but I have never personally seen one. If this were an original combination, it would mean that Schmitt used some German heads on their own bodies, as there seems little room for doubt that the XI heads are German. Researchers have long believed that quite a few French firms did indeed sometimes use German heads. *Edna Black Collection.*

Illustration 34.

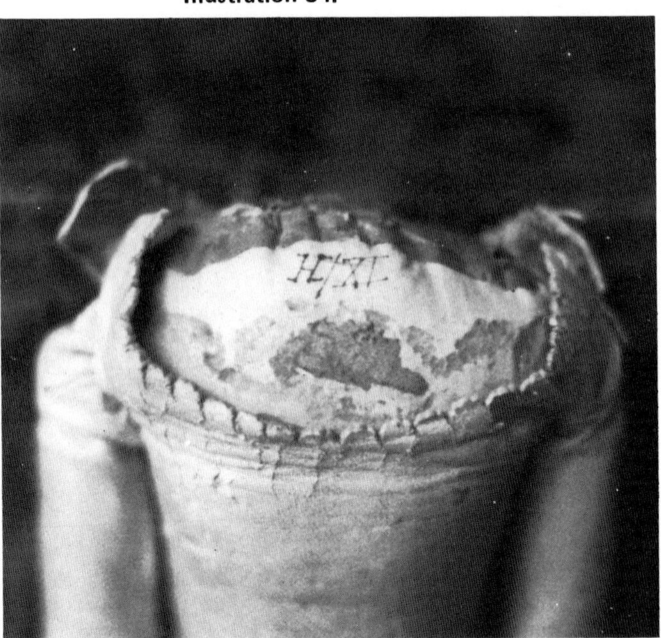

Illustration 35.

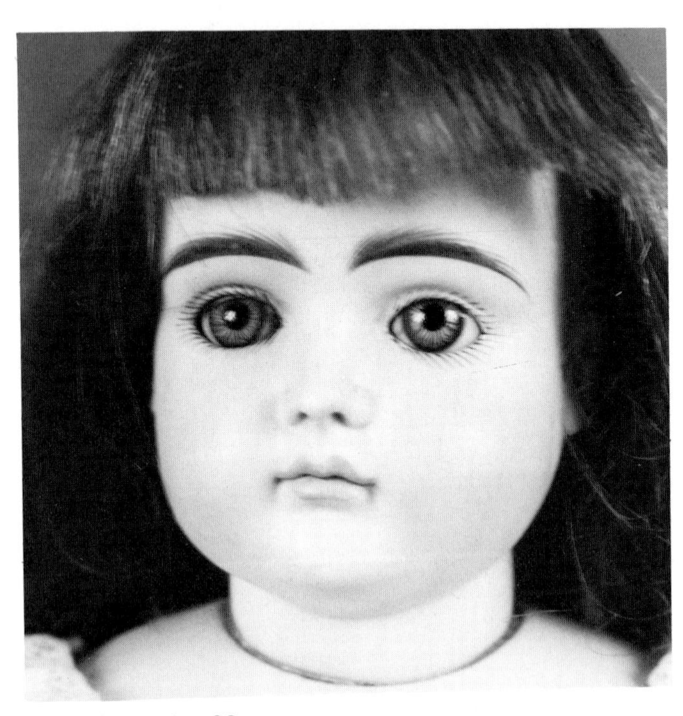

Illustration 36.

Illustrations 34, 35 and 36. This 19in (48.3cm) XI head is mounted on a bisque shoulder plate and kid body with riveted joints and composition lower limbs. There is no doubt that this is her original body as when the shoulder plate was removed, the marking "H/XI" was found written on the inside. Again, she is a lovely doll with beautifully painted eyebrows, chubby neck and cheeks, and a soulful look in her blue sleep eyes. The close view gives a good look at the molding and painting of her lips. She is wearing a very nice old white cotton dress with ruffles and lace trim as well as her original leather shoes. *Joanna Ott Collection.*

Illustration 37. This lovely couple is a 16in (40.6cm) boy with the "XI" mark and a 15in (38.1cm) girl with the "X" mark, which is another desirable mold also attributed to Kestner since it has the same characteristics and qualities as the XI face. This bridal couple was dressed by the ladies of the Home for the Aged in Winchester, Massachusetts, to benefit the Lyceum Street Fair in May 1895. He wears his black wool suit and she is in her beautiful lace bridal gown. *Esther Schwartz Collection.*

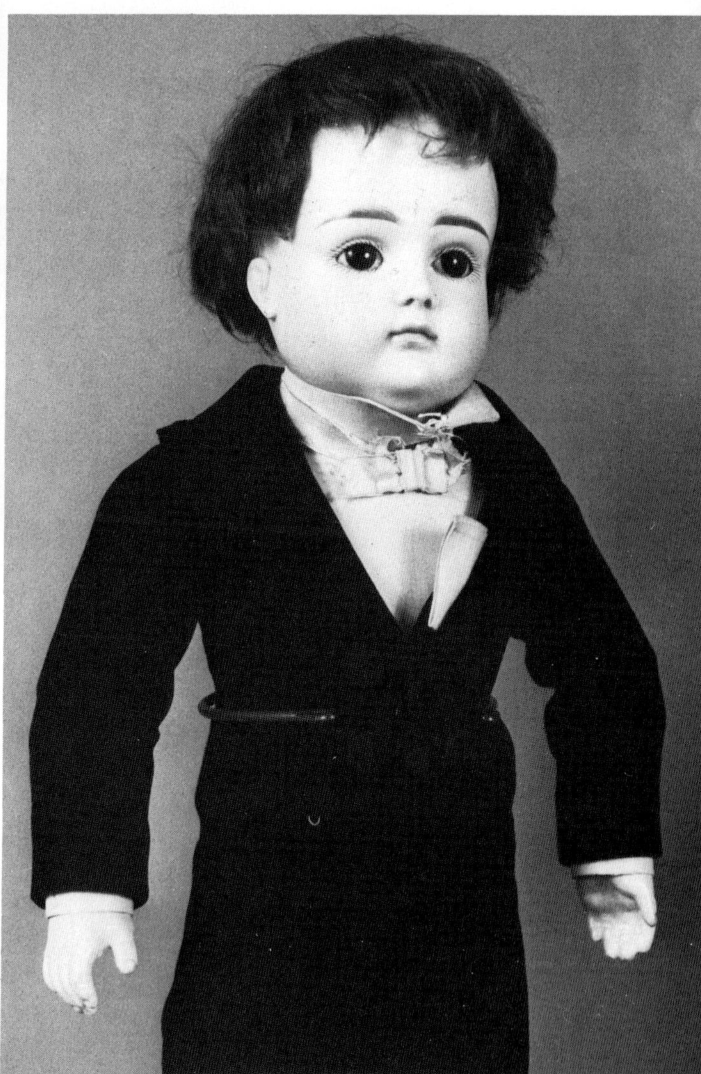

Illustration 38. This close-up of the groom shows the small ear and chubby cheek line characteristic of this mold. His eyebrows are light in color and in quantity; his replaced mohair wig is brown as are his sleep eyes. *Esther Schwartz Collection.*

Illustration 39. This close-up of the bride with the "X" mark shows that this mold has a somewhat thinner face, but fuller lips and a more pronounced double chin. She has her original blonde mohair wig and heavier eyebrows than those of the boy. *Esther Schwartz Collection.*

Illustration 40. This 15in (38.1cm) example of the X mold has the early heavy body with straight wrists and cupped hands typical of this group of dolls. She has brown sleep eyes and very heavy glossy brown eyebrows. Her lips are full, separated by a darker red lip line. Her blonde mohair wig is original as is her white cotton dress with tucked bodice and eyelet skirt. *Sheila Needle Collection. Photograph by Morton Needle.*

It was difficult to decide exactly where to put this doll. She probably should go into the "Swivel Neck" section but she is important here, too, as she is a very rare XI mold with open mouth. Her four upper teeth give her somewhat of a character look, as they protrude quite a bit, making her seem as though she has an overbite problem. Fired in black on her crown rim is XII, possibly denoting a size 12. She has a socket head fitted on a bisque shoulder plate; her kid body has bisque lower arms and hands. She has been appropriately redressed as her original clothing had disintegrated. *Courtesy of Mary Thomasson.*

Very Pouty Faces

Collectors have long assigned this face to Kestner because of its many Kestner characteristics, the same as those mentioned before for attributing the X and XI dolls to him. It does appear that the same maker was responsible for both of these sets of dolls and attributing one necessarily attributes both. These very pouty faces are distinguished by a quite protruding upper lip with a distinct dip in the center which necessitates a dip in painting the red dividing lip line. Collectors refer to this doll as the "Schmidt Kestner" because of both its face and body. It probably dates from the 1880s.

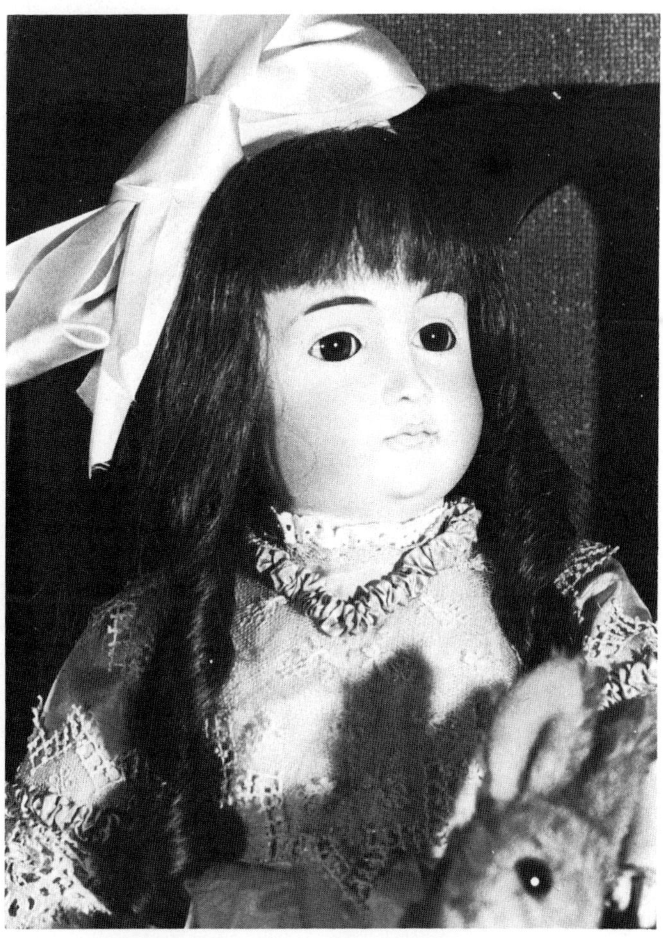

Illustration 41, This 22½in (57.2cm) girl is incised only "15," surely her size number. She has beautiful brown sleep eyes, thin-stroked eyebrows, lightly painted lashes, and a replaced auburn human hair wig. She is wearing an old lilac dress over her early straight wrist body. *Sheila Needle Collection. Photograph by Morton Needle.*

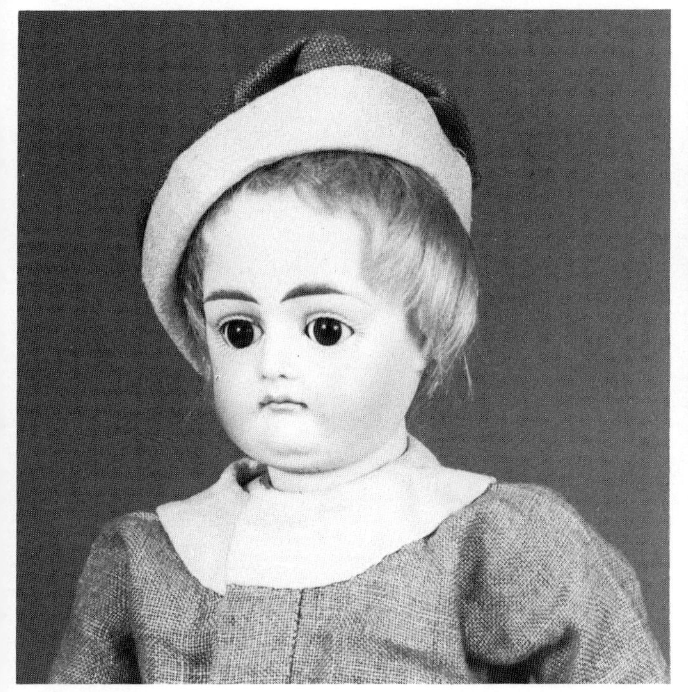

Illustration 42, This 14½in (36.9cm) fellow incised with size number "8", has a "We'll see about that!" expression. He is certainly a very appealing doll with his brown sleep eyes, blonde feathered eyebrows, and pouting upper lip. His first chin is almost lost in the fullness of his second one. His blonde mohair wig covers a plaster dome. He is wearing an appropriate blue linen suit with white trim and matching cap. *Jane Alton Collection.*

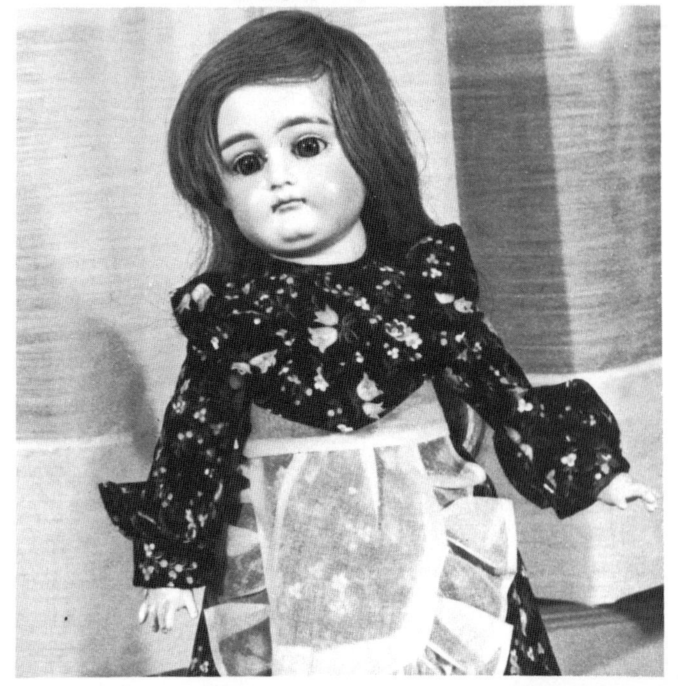

Illustration 43, This 13in (33.0cm) child, marked only "8", is certainly in a real pout. Her protruding upper lip is even more accentuated because her lower lip is molded in. She has one of the most prominent double chins in the group; in fact it has a definite cleft right in the middle. Her eyes are different from those of the others, being a true blue color and of the paperweight type. Her marked body is the *Excelsior* one which is completely ball-jointed, even at the wrists; her fingers are delicate and separated. Her wig and clothes are replacements. *Sue Bear Collection.*

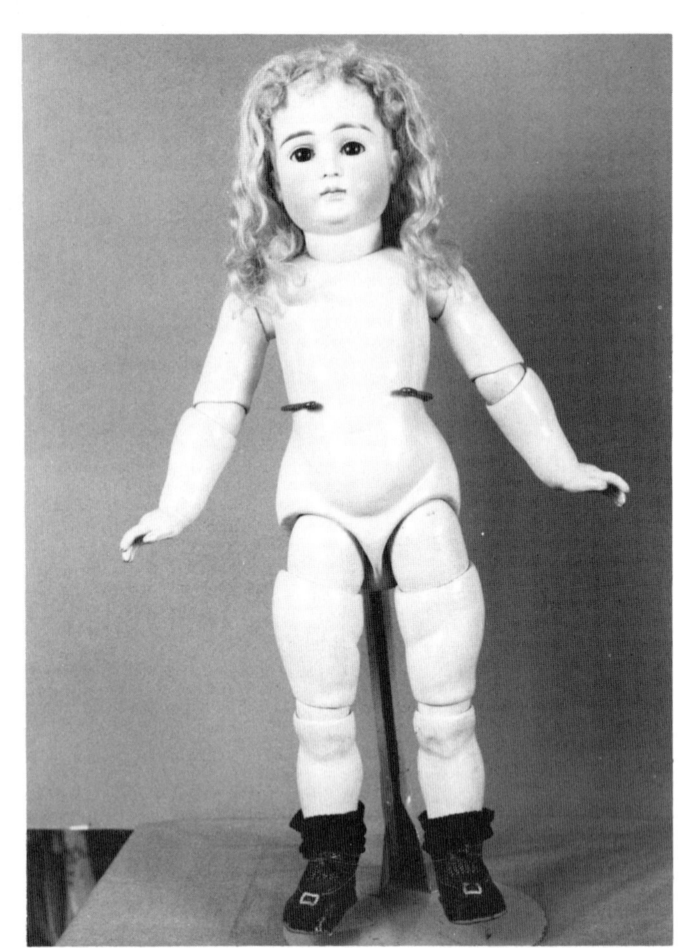

Illustration 44.

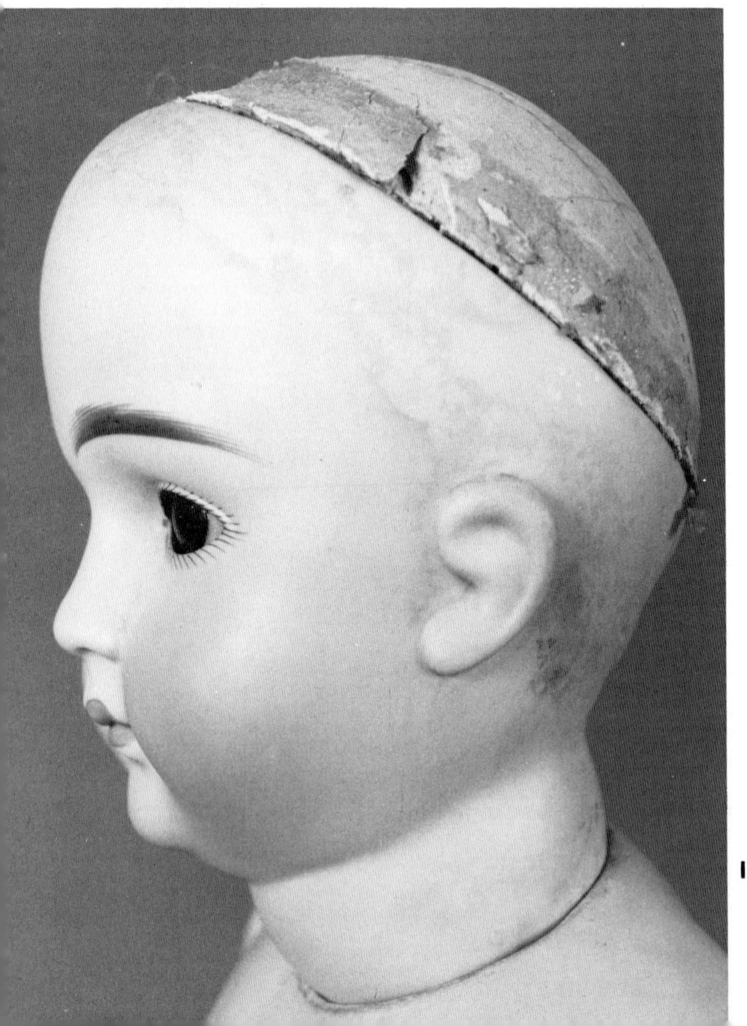

Illustration 46.

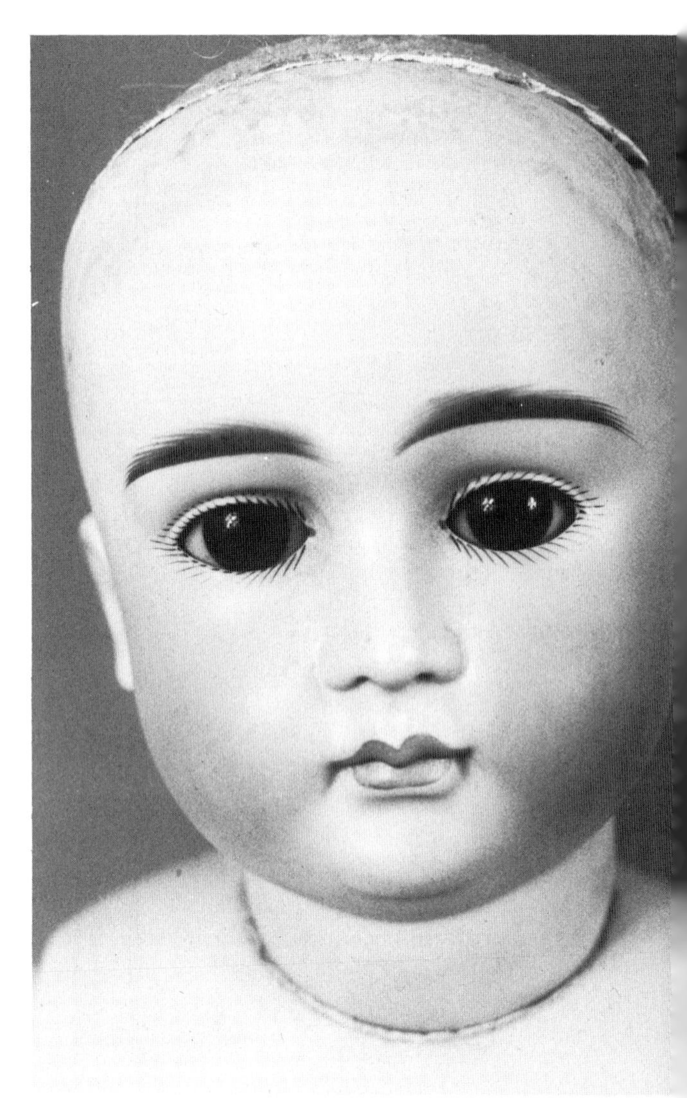

Illustration 45.

Illustrations 44, 45, 46, 47, and 48. There is no doubting the fact that this is a stunning doll. Her crown, cut high in the front and fairly low in the back, is filled in by a plaster pate. Her ears are small, and barely molded; her neck is stout and bulges in the back. Her face is quite full, especially down at her jawline; her lips are also full with a very pouty upper lip, yet it is upturned at the ends in the Kestner manner. Her sleep eyes, in this case brown, are almond-shaped and small in proportion to her face. Her bisque is excellent and pale in color, but with rosy cheeks as is common on most of these pouty dolls. Her eyebrows are flat at the bottom but feathered across the top; they are substantial, but not overpowering. She is wearing her original curly mohair wig. Her body is of extraordinary quality with the large ball joints at the hips and other separate balls at the knees, shoulders, and elbows. She has straight wrists, but with individual fingers. Her flat rear end and large ball joints are like those on the marked body shown in *Illustration 217.* She is 30in (76.2cm) tall, incised only "18." *Esther Schwartz Collection.*

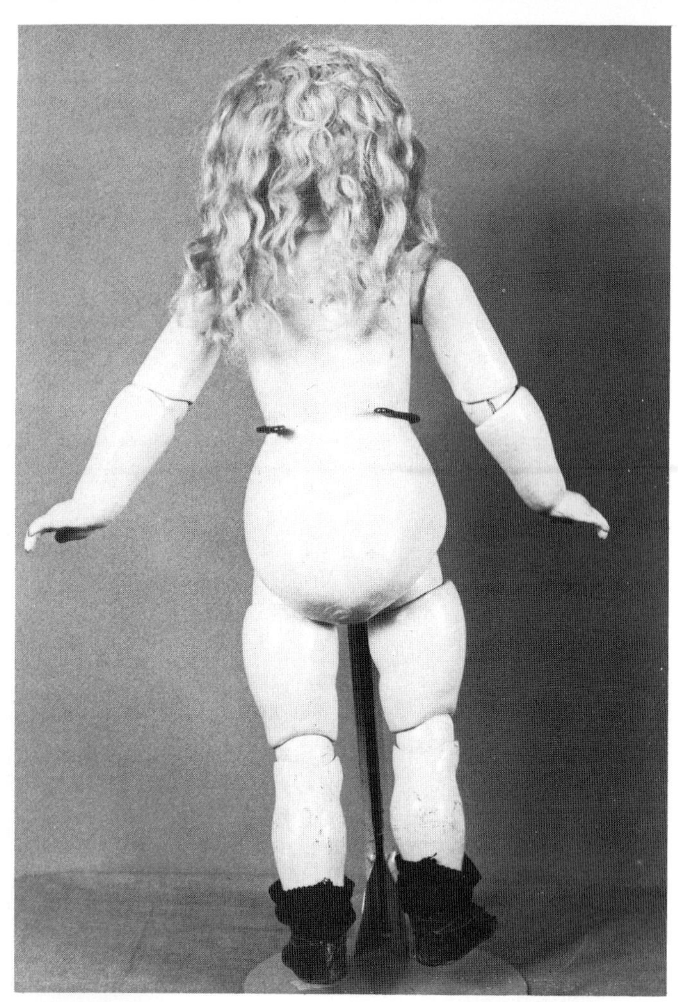

Illustration 47.

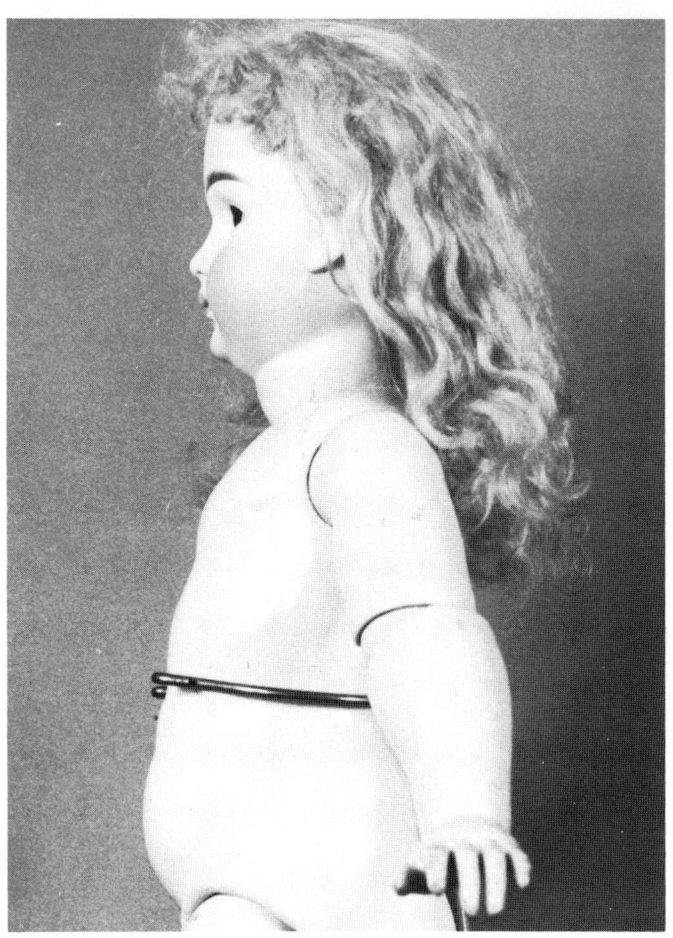

Illustration 48.

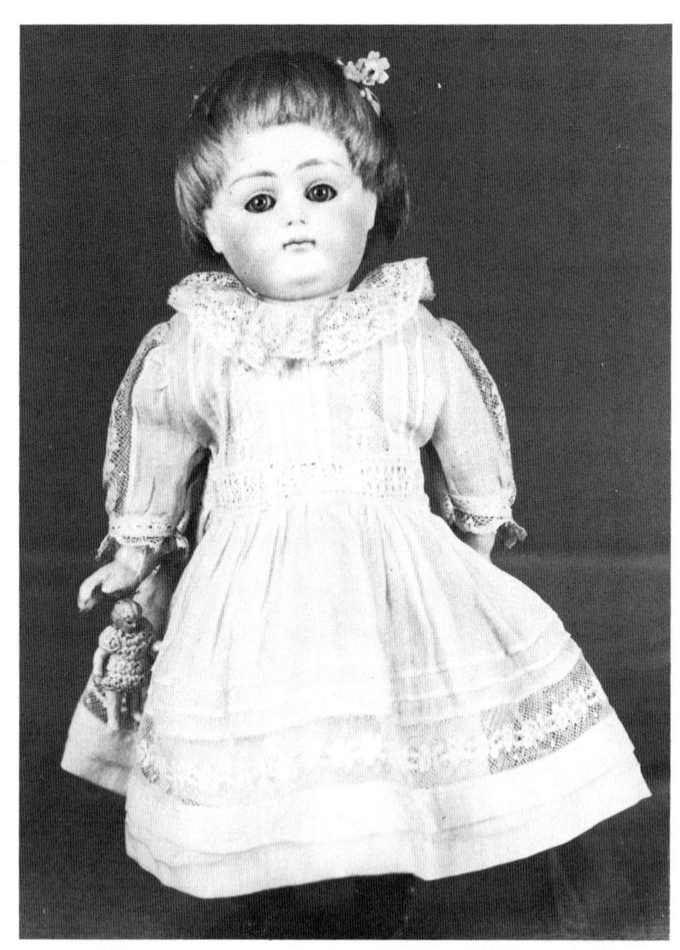

Illustration 49. This 11in (27.9cm) little sister is one of the sweetest. She is very pouty with pale blue eyes, long and thin blonde eyebrows, a blonde mohair wig, puffy cheeks, and a tiny chin, but a substantial double chin. She is completely original in her white cotton dress with lace and tucks and her black leather high-button shoes. What a darling doll! *Maxine Salaman Collection.*

Illustration 50. It certainly is a rarity to find a doll of the 1880s period still in her original store clothes, as the majority of dolls came unclothed or clad only in a chemise. The dressed dolls with bisque heads were truly the luxury dolls. This little 7½in (19.1cm) girl must be the tiniest of these pouty dolls made. Her complexion of pale bisque with rosy cheeks matches her pink dress and hat with fancy ribbon and lace trim. Gray sleep eyes, blonde eyebrows, and a blonde mohair wig complete her features. She is on a composition body with joints at shoulders, hips, and knees as shown in *Illustration 151.* Apparently this was a body style popular with Kestner for many years and is an additional reason for attributing these dolls to him. *H & J Foulke.*

Illustration 51. This 12in (30.5cm) girl has the same face, but treatment of the eyebrows gives her a much more serious and severe look. Instead of the usual curve, her eyebrows are nearly flat across the bottoms, the inner edges are quite angled almost like the side of a pie pan, and the brows are closer to her eyes. They are also feathered with individual strokes all across the top. This specific type of eyebrow decoration appears on quite a number of closed-mouth dolls attributed to Kestner. This doll has brown sleep eyes and a replaced brown human hair wig. Her white cotton dress trimmed with embroidered eyelet seems original. Incised only "7," she is on a jointed body with the early straight wrist and cupped hand. *Ruth Noden Collection.*

Illustration 52 and 53. Going from the smallest to possibly the largest is this 32in (81.3cm) girl incised "18//103." Her very dark human hair wig and eyes are very striking contrasts to her pale bisque. In proportion to the size of her head, her eye sockets are small and her eyebrows are slight, although quite tapered. She has a wooden rod inside her head to which is wired her pate. The view with wig pulled back shows her tiny ear; full cheek line; and crown, high cut in the front, low in the back; and her stout neck. She has the early quite chunky jointed composition body with attached ball joints. *Esther Schwartz Collection.*

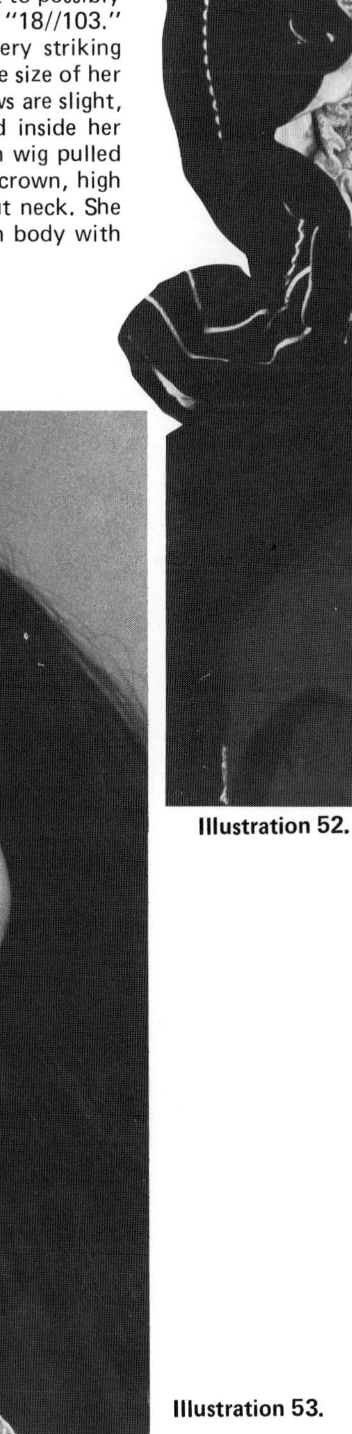

Illustration 52.

Illustration 53.

The Squared-Cheek Look

It has been extremely difficult to sort these closed-mouth dolls into categories for even though they are marked only with sizing numbers, there are definitely different molds represented. It would certainly have been a lot easier if they had been given numbers! This next group also attributed to Kestner has the same general characteristics as the previous ones: plaster pates, sleep eyes, upturned lips, lovely pale bisque, rosy cheeks, and heavy eyebrows. Their bodies are the same type shown in *Illustrations 44, 47 and 48,* some with attached balls and some with free balls; these apparently preceded the *Excelsior* bodies of 1892.

Illustration 54. This 13in (33.0cm) girl is apparently an early 1890s model as she is incised "Made in Germany// 7," so she would probably be after the Country of Origin Law of 1891. She has a lovely serene look on her face, definitely not pouty like the previous group. She has altogether the look of a very pleasant child. Her closed mouth has a slightly protruding upper lip with a darker red lip line. Her gray sleep eyes and dark brown curved eyebrows both suggest a Kestner origin. Her curly blonde mohair wig still retains its lovely fullness. Her dress is replaced but appropriate. *Joanna Ott Collection.*

Illustration 55. This 14in (35.6cm) girl has very different eyebrows from those of the doll in *Illustration 54;* they are quite long and painted with many more strokes to make them look fuller, but they are not startling because they are painted a light color. She has only a slightly pouty mouth with a darker red lip line. Her plaster dome is covered by a replaced brown human hair wig, with long curls pulled to the side and tied with ribbons. This is an excellent variation when the usual long curls become tiresome. Her nice old white dress has red smocking and trim. *Joanna Ott Collection.*

Illustration 56. The straight-on view of this 15in (38.1cm) doll shows off her puffed-out cheeks and double chin. Her tiny deep brown sleep eyes are under heavy brows. Her body is of the same good composition with unjointed wrists and fingers molded together. She has her original plaster pate under her replaced long curl wig. Her brown plaid dress is old and appropriate. *Sue Bear Collection.*

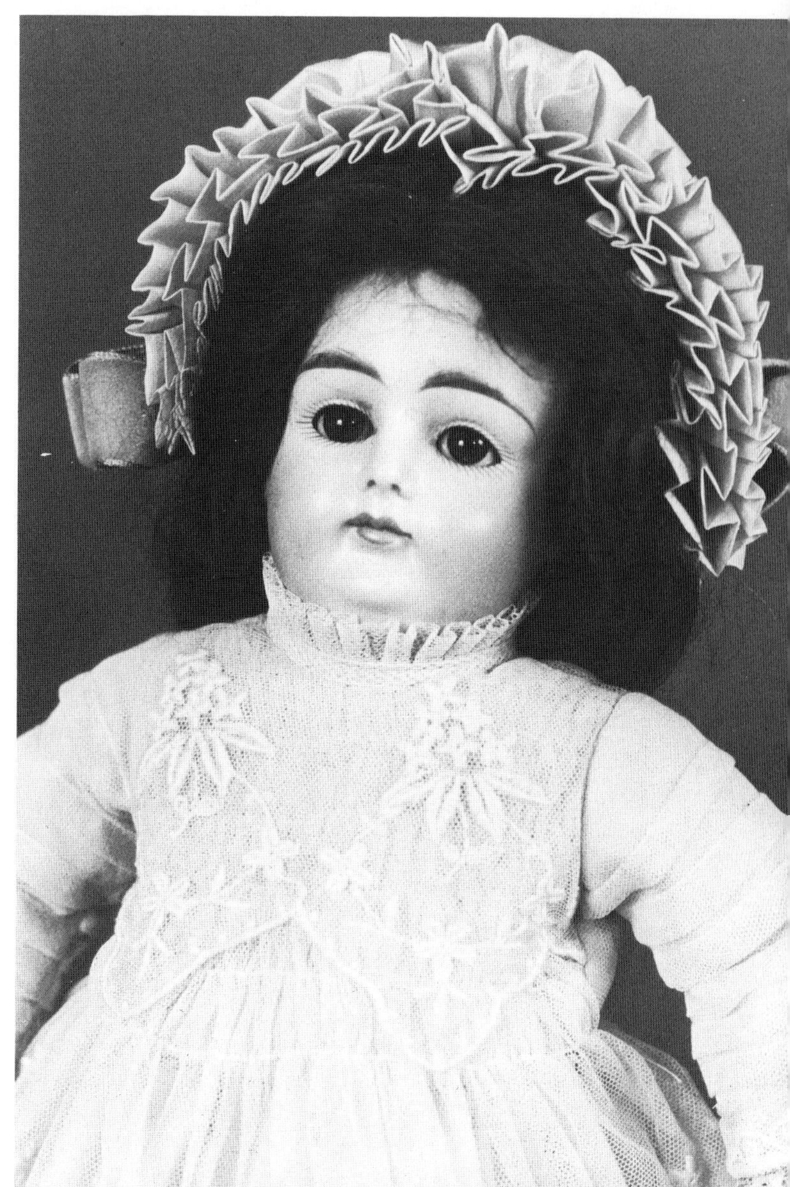

Illustration 57. This 15½in (39.4cm) girl is incised only "10." She has rosy cheeks; her lips, nose dots, and eye dots are light orange, a very natural color. Her crown is low cut and her ears are small, both characteristics of this style doll. Her forehead is quite wide, her neck is chubby, and she has a bulge at the back of her neck; all are typical of this type of early German dolls. Her eyebrows are long and quite tapered, nicely fluffed across the top. Her jointed composition body is of the early type with the pointed, flattened-off buttocks. She does not have original clothes, but an appropriately made dress of ecru lace over peach silk, a favorite style with collectors. *H & J Foulke.*

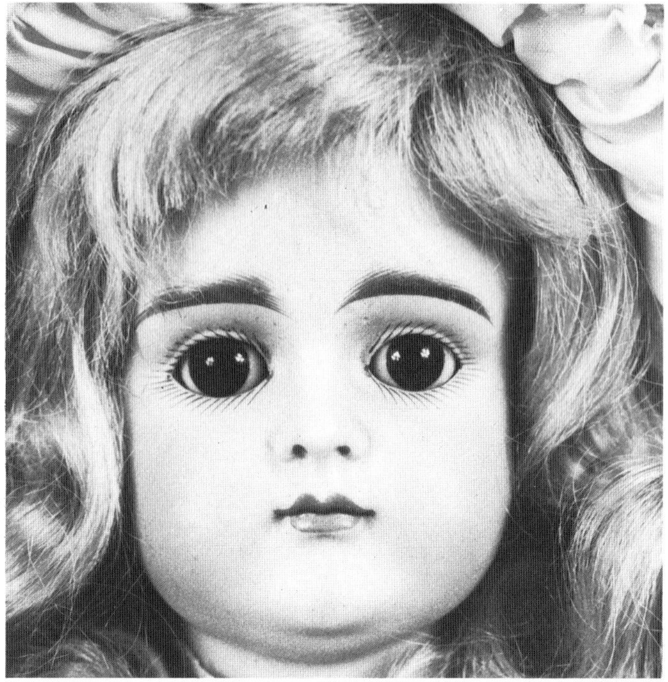

Illustrations 58 and 59. This 16in (40.6cm) girl is also incised only "10." Her beautiful curly blonde human hair wig covers a plaster dome. Although her face shape is the same as that of the preceding dolls, she has a much different expression because her brown sleep eyes are quite a bit larger. She has the same closed mouth with a slightly pouting upper lip and a darker red lip line. Her light brown eyebrows are quite long and tapered, rising in a buildup of strokes at the inner corners. Her pink cotton lace-trimmed dress and matching hat are in the Kate Greenaway style and are quite appropriate for her period. *Edna Black Collection.*

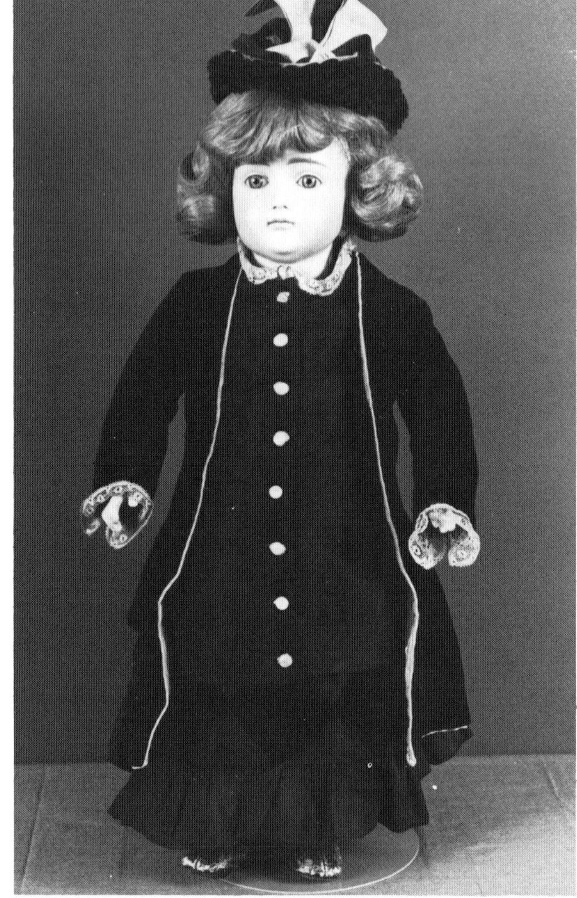

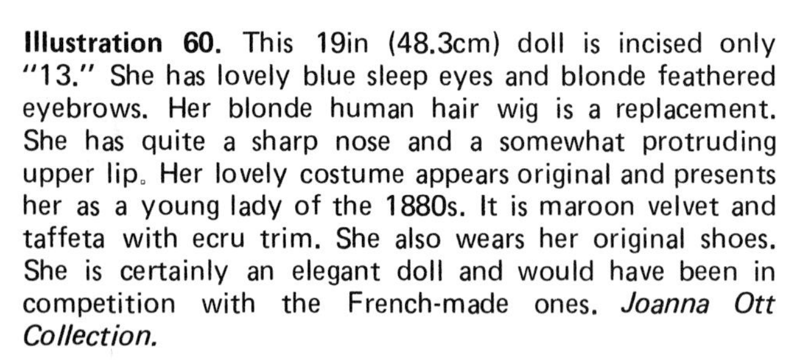

Illustration 60. This 19in (48.3cm) doll is incised only "13." She has lovely blue sleep eyes and blonde feathered eyebrows. Her blonde human hair wig is a replacement. She has quite a sharp nose and a somewhat protruding upper lip. Her lovely costume appears original and presents her as a young lady of the 1880s. It is maroon velvet and taffeta with ecru trim. She also wears her original shoes. She is certainly an elegant doll and would have been in competition with the French-made ones. *Joanna Ott Collection.*

Illustration 61. This doll is truly breathtaking. She is 19in (48.3cm) tall and incised only "13." She is just like the doll in *Illustration 60* except that her eye sockets were cut a little larger which changes her facial expression. Her blue eyes are quite vivid. She has small ears, a protruding upper lip, and light, yet feathered eyebrows. She still retains her original blonde mohair wig in the Rembrandt style. She is elegantly attired in a lovely old white dress trimmed with lace, ruffles, and satin bows with a matching Kate Greenaway style hat. Her light blue coat emphasizes her eyes and has ecru lace trim. *J.C. Collection.*

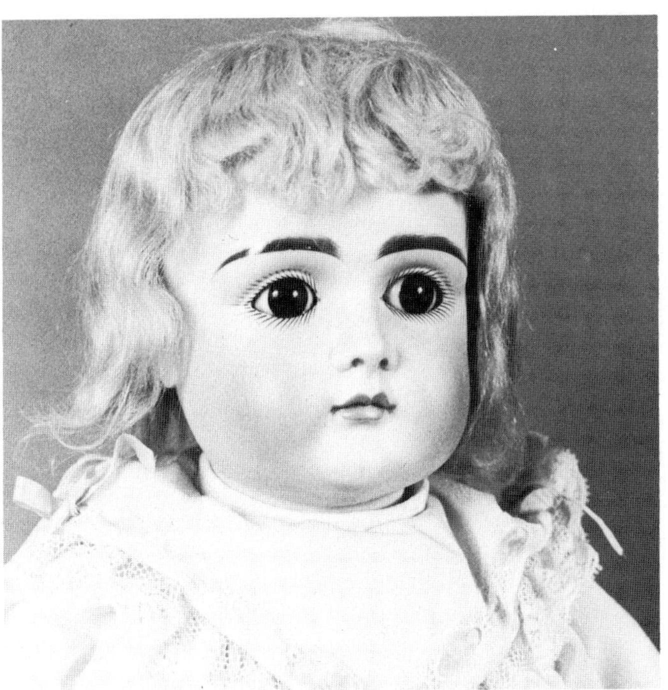

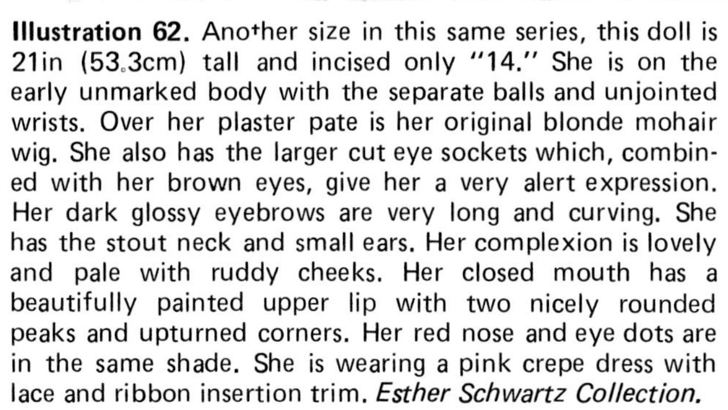

Illustration 62. Another size in this same series, this doll is 21in (53.3cm) tall and incised only "14." She is on the early unmarked body with the separate balls and unjointed wrists. Over her plaster pate is her original blonde mohair wig. She also has the larger cut eye sockets which, combined with her brown eyes, give her a very alert expression. Her dark glossy eyebrows are very long and curving. She has the stout neck and small ears. Her complexion is lovely and pale with ruddy cheeks. Her closed mouth has a beautifully painted upper lip with two nicely rounded peaks and upturned corners. Her red nose and eye dots are in the same shade. She is wearing a pink crepe dress with lace and ribbon insertion trim. *Esther Schwartz Collection.*

Illustration 63. As often happens in doll collecting, the doll shown in *Illustration 62* changed hands, and the new owner decided that she needed some boy dolls in her collection, so she outfitted the doll in a very attractive sailor suit style, popular for boy dolls in the 1880s and 1890s. *Mary Lou Rubright Collection.*

The Long-Faced Look

This face definitely has a longer, thinner cheek line than the others already shown, but it is like the face on *Illustration 24* which we placed in another section because it was on a marked Kestner body. This face is also characterized by quite a prominent "punched-in" dimple in her chin. These dolls share features with the other unmarked dolls included in this chapter which cause them to be attributed to Kestner.

Illustration 64. This 14in (35.6cm) doll is incised with only a "4," her size number. She has her original blonde mohair wig over a plaster dome. Her blonde eyebrows are nearly straight across the bottom, but have feathering all across the top; her dark brown sleep eyes are surrounded by heavy painted lashes. She has somewhat of a pouting upper lip, slightly protruding but painted to look thin. Her body is the type with attached balls at the joints, straight wrists, and cupped hands; her legs are somewhat elongated, a rather common feature on this type of doll. *Edna Black Collection.*

Illustration 65.

Illustrations 65 and 66. This girl is 15in (38.1cm) tall, but has no mark at all. Her face is quite wide across the forehead, and tapers down to her chin; her cheeks are not so chubby as those on the dolls in the previous group. Her upper lip protrudes somewhat, but the painting is full with upturned corners. She has lovely full brown eyebrows, rising high at the inner corners and feathered across the top, which match her dark brown sleep eyes. Her jointed composition body has the attached ball joints and is like the body shown in *Illustration 218.* She is completely original including her full blonde mohair wig, her white dotted swiss dress and her white leather shoes. *Elba Buehler*

Illustration 67. This is another all original girl; she is 17in (43.2cm) tall, incised "8//2." She has her original blonde mohair wig over a plaster pate, beautifully feathered eyebrows, a full upper lip, and a very prominent "punched-in" dimple in her chin. *Esther Schwartz Collection.*

Illustration 68.

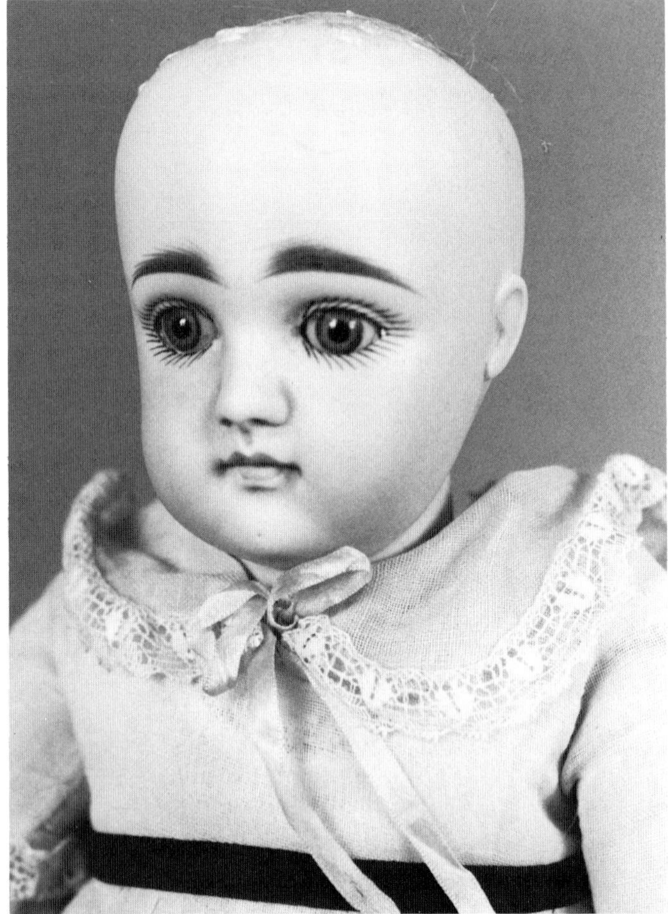

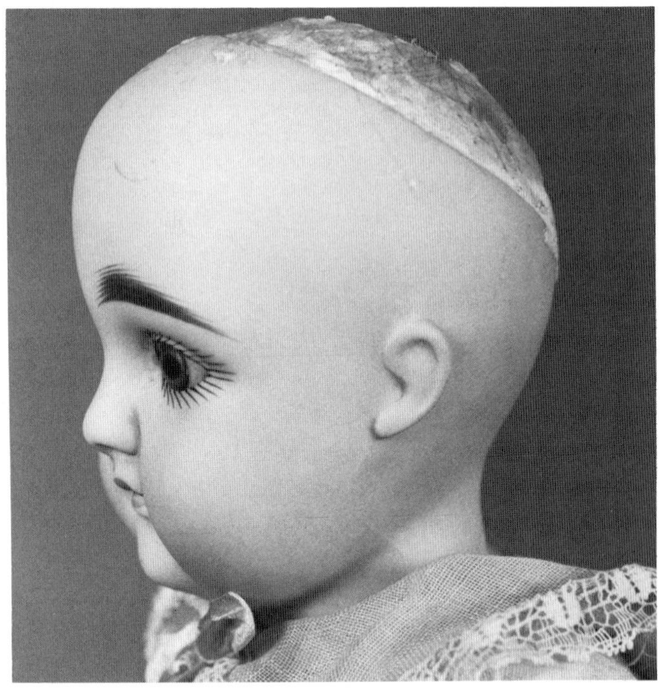

Illustration 69.

Illustrations 68, 69 and 70. This 16in (40.6cm) girl is incised only "7." Her face is a little different from the dolls in *Illustrations 64, 65, 66, and 67,* but she still retains their general characteristics, such as the "punched-in" chin dimple, tiny ears, and beautifully shaped and stroked eyebrows, as well as long painted lashes. These features can be seen quite well in the close-up illustration. Also note the prominent cleft under her nose. She has her original blonde mohair wig over a plaster pate and gray threaded sleep eyes. She is on a fully-jointed but very heavy composition body. *Esther Schwartz Collection.*

Illustrations 71 and 72. This 16in (40.6cm) child has beautiful pale coloring and pink-tinted cheeks. Her threaded gray sleep eyes, a color much favored by Kestner, and her feathered blonde eyebrows combine with her complexion to give her a soft childlike look. Her plaster pate is still intact as well as her original curly blonde mohair wig. Her hat of lovely lace and ribbons appears to be original; her dress, with many tucks and lace insertion and edging as well as satin ribbons, is also old. She is on an excellent quality ball-jointed body with long upper legs. *H & J Foulke.*

Illustration 72.

Illustration 73. This 18in (45.7cm) child has a marvelous pouty expression on her face achieved by a protruding upper lip, tiny brown sleep eyes, and quite heavily stroked brown eyebrows. As characteristic of this group of faces, she has a broad forehead and the "punched-in" dimple in her chin. Her brown human hair wig as well as her clothes are old. She has an excellent early type jointed composition body with straight wrists and cupped hands. *H & J Foulke.*

Illustration 74. This 7in (17.8cm) little girl is probably one of the smallest size socket heads of these pouties. She is certainly a choice doll for those who like tinies. Her eyebrows are very light and glossy with individual strokes; her original blonde mohair wig covers a plaster pate. She has tiny brown sleep eyes with very long painted lashes. Even as small as she is, her closed pouty mouth has a protruding upper lip. Her bisque is pale with lightly tinted cheeks. She is on an excellent quality five-piece composition body with pegged shoulders and hips and bare feet. Her white cotton dress with purple flowers and lace trim is old, but not original. *H & J Foulke.*

The Round-Faced Look

This group of dolls has more of an oval face than the other groups, and their lips are full, but not of the pouty type. As with the other groupings, classification is difficult and there is some overlapping of faces from group to group.

Illustrations 75 and 76. This sweet-faced little girl is 16½in (41.9cm) tall, incised only "7." She has small brown sleep eyes with quite long painted lashes, and red eye dots in the corners. Her eyebrows are blonde with many individual strokes at the inner and outer corners. Her nose is broad with two red nostril dots. Her mouth has full lips with a straight red lip line and a molded cleft. She has tiny ears, but very full cheeks. A plaster dome covers her crown opening. She is on the jointed composition body with attached ball joints. Her dress and matching hat are made of lovely soft old rose silk with embroidery on the front tucks. Her leather shoes are old with rosette trim. *Edna Black Collection.*

Illustration 76.

Illustration 77. This is obviously another child of the same mold, but she has only an undiscipherable number on her head. She is 14in (35.6cm) tall on a jointed composition body with attached ball joints, fat rolls on her thighs and wrists, and cupped hands. She has the same eye treatment as the doll in *Illustrations 75 and 76,* also with very long painted lashes. Her original blonde mohair wig is in the Rembrandt style. She is wearing a lovely old, possibly original, dress with white eyelet and embroidery trim. *Edna Black Collection.*

Illustration 78. This 13in (33.0cm) child's face is just a little different from the others in this group. Her glossy blonde eyebrows are more arched and have many individual strokes rising quite high at the inner corners. Her full lips are lightly colored and have the darker red lip line to divide them, but instead of going straight across like that of most of the dolls, her upper lip is more curved in a downward pout. Her ears are small with little molding, and she has tiny dark brown sleep eyes. Her blonde mohair wig covers her plaster dome. She is on the usual body for this type of doll. Her clothes appear to be original; her dress is printed silk with a feather-stitched hem and her button shoes are brown leather. *Edna Buehler.*

Illustration 79. This lovely child is a large 30in (76.2cm) tall, incised only "18," her size number. Her bisque is beautiful with tinted rosy cheeks. Her eyes are blue with lightly painted lashes (her real lashes are a later addition). She has the typical eyebrows of the group, very long and curved with much feathering. Her closed mouth has the upper lip with two peaks and upturned ends with a white space between her lips which have very good detail in modeling. Her blonde human hair wig with long curls is an old one. She has a very heavy early type jointed composition body with attached ball joints. *Esther Schwartz Collection.*

Illustration 80. This charming 7in (17.8cm) little girl is yet another face which has many of the characteristics attributed to Kestner. She is incised "2½," and has her original mohair wig. Her tiny brown sleep eyes are outlined with gray painted lashes, and each has an inner red eye dot. Her light blonde eyebrows are composed of five or six long individual strokes. Her mouth is like that of the large dolls shown in *Illustrations 41 and 42* with a dip in the red lip line to accommodate her protruding upper lip. Her bisque is very pale and of fine quality with rosy cheeks. She has a jointed composition body with pegged shoulders, no elbow joint, and strung legs with a knee joint, a popular Kestner-type body for small dolls. She has been appropriately and beautifully redressed as a little girl of the 1880s in an exquisite aqua silk and delicate lace outfit of old fabrics; her white leather shoes are old. *Jan Foulke Collection.*

Bru-Type

Illustration 81. Collectors of Kestner dolls enjoy saying that Kestner copied the major French manufacturers. This model is referred to as the Kestner Bru. It is a beautiful doll with pale creamy bisque. The eyebrows have the typical Kestner look but the paint is a little lighter color. The eye has a molded upper lid with painted lashes. There is good molding detail under her eyes; cheeks are very plump; nose is sharp. Of course, her most outstanding feature is her mouth with slightly parted lips showing molded and painted teeth similar to a Circle Dot Bru. The neck on this model is quite stout for a Kestner. This head has been found on a composition jointed body with straight wrists, sometimes with the unusual feature of jointed ankles. This model has also been found mounted on a bisque shoulder plate with gusseted kid body and bisque lower arms. The doll pictured is 24in (61cm) tall, dressed appropriately as a child. *Private Collection.*

Shoulder Heads

The bisque shoulder head dolls were made in Germany over a long period of years from the 1870s until the 1920s. They were less expensive than the socket head dolls, so enjoyed a fairly wide market. Before 1883 the kid bodies had gusseted joints. The *Ne Plus Ultra* hinge joint was patented in 1883 and used extensively thereafter; however, the dolls with the rivet joints were more expensive than the dolls with gusseted joints. Both types were then on the market simultaneously, so it is difficult to date the gusseted dolls. Another type of kid body used with shoulder head dolls had jointed composition arms, rivet-jointed hips and knees and composition or kid lower legs. The closed-mouth shoulder head dolls would date from the 1870s into the early 1890s and probably longer since even after the open-mouth dolls of the early 1890s became popular, there would have been a call for a less expensive doll and the closed-mouth doll certainly would have filled that bill.

Although these German shoulder head dolls for the most part have the proportions of children, some of them are found dressed as ladies, and it is certain that they were in competition with the French lady dolls of the 1870s and 1880s. Of course, these German dolls would have been much less expensive, but would be very attractive when dressed. Their garments would hide the awkward kid bodies, most of which did not have the nipped-in waists and more shapely limbs of the French ladies. Many of these same shoulder head dolls with closed mouths are also found dressed as babies in original clothes as the character baby was not developed until 1910.

The Long-Faced Look in Shoulder Heads

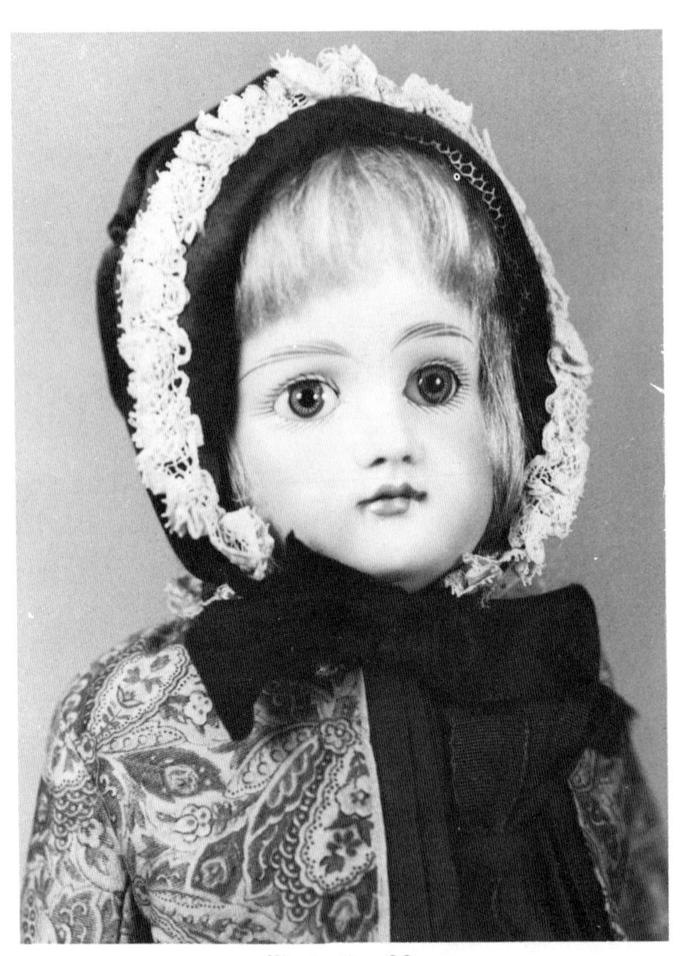

Illustration 82.

Illustration 83.

Illustration 84.

Illustration 85.

Illustrations 82, 83, 84 and 85. This 15in (38.1cm) girl which is incised "5" appears to be a shoulder head version of the doll shown in *Illustration 64.* However, an outstanding difference is in the treatment of the eyebrows which are striated in five strokes. They are blonde to match her original mohair wig which covers a plaster pate. Her gray eyes are highlighted by lightly painted lashes. Her mouth has full painted lips with a protruding upper one and a darker red lip line for accent. As a young girl of the 1880s, she is wearing an original costume of both plain and patterned maroon cotton with a maroon satin bonnet. She has a gusseted kid body with bisque lower arms. *Esther Schwartz Collection.*

Illustration 86. This 20in (50.8cm) girl incised "11," is a larger version of the doll in *Illustrations 82, 83, 84, and 85.* Because of her size, her face seems quite a bit longer. She also has very long and curving striated eyebrows and painted lashes which are darker than those of the doll in *illustrations 82, 83, 84, and 85.* Her upper lip is protruding with upturned outer corners; there is a cleft under her lower lip and a "punched-in" chin dimple. Her full cheeks extend down to a double chin. She is also on a gusseted kid body with bisque hands. *Esther Schwartz Collection.*

Illustration 87. This 22in (55.9cm) girl incised "10" is another example of the same doll as shown in *Illustrations 82, 83, 84, 85, and 86.* Remember that in these shoulder head dolls there can be quite a variance in height with the same size head, as some of these kid bodies have elongated torsos and others longer legs. Manufacturers did this so that they could advertise a larger doll at only a slightly higher cost to them. But sometimes the height varies because the stuffing has dropped down to shorten the total body. This attractive doll also has light striated eyebrows, her original blonde mohair wig, and brown sleep eyes. She has the gusseted kid body with bisque lower arms and is wearing on old pink cotton dress of the Kate Greenaway style popular about 1890. *Joanna Ott Collection.*

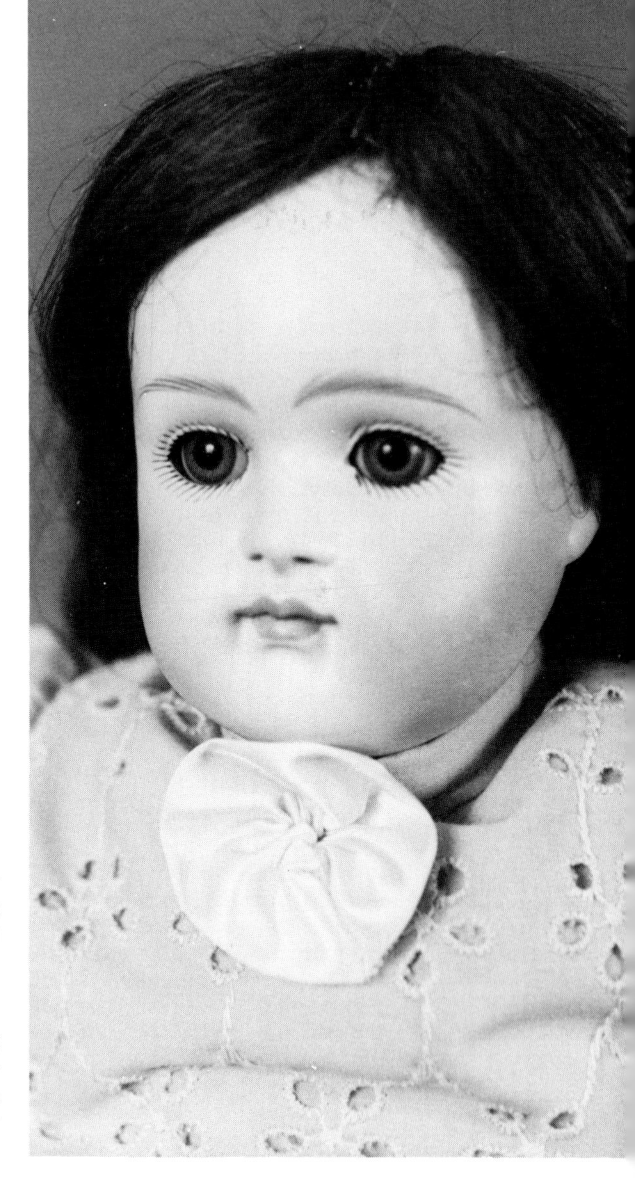

Illustration 88. This 21in (53.3cm) girl is obviously another doll by the same maker, although her face is just a bit different. She is incised "6½" at the crown opening and "8½" on her shoulder, but the reason for the discrepancy in number is a mystery as the head and shoulder are all in one piece. Her face is just a little shorter and fuller than those of the other dolls in this section. Her lips are full but longer and she lacks the chin dimple. She has the characteristic plaster pate and sleep eyes, which are blue. She has a very long body of kid with gusseted joints. *Esther Schwartz Collection.*

Shoulder Heads Marked with Letters

Shoulder head dolls with open mouths made by Kestner using a letter sizing system are shown in *Illustrations 230 to 240.* The letters on these closed-mouth dolls are probably also for sizing, but it appears that several different face molds were used so several series are probably represented here. Collectors attribute these dolls to Kestner because of their characteristics in common with known Kestner dolls.

Illustration 90. This sweet little 14½in (36.9cm) girl is marked only "D." She has the sweet round face like that shown in *Illustrations 75 and 76,* a socket head on a jointed composition body, but this one is a shoulder head doll on a body with a cloth torso and gusseted kid limbs. (This body would have been less expensive than an all-kid one.) She has a plaster dome and threaded gray eyes, both characteristics associated with Kestner. Her cheeks are plump, her nose is wide, and her tiny ears have hardly any molding. Her blonde striated eyebrows are painted close to her eyes which are accented by lightly painted lashes. She has a protruding upper lip and darker red accent line. She still retains her original wig cap with a bit of blonde mohair left. Her dress is a homemade one, possibly original. *Joanna Ott Collection.*

Illustration 91. This little 12in (30.5cm) child has only a size number, but she appears to be a smaller version of the doll in *Illustration 90,* although she could be just a bit more chubby cheeked. She has lovely pale bisque with rosy cheeks. Her sleep eyes are gray; her wide-stroked blonde eyebrows match her replaced blonde wig which covers a plaster dome. Her closed pouty mouth has a darker red line separating the lips. She is wearing her original baby's dress with a complete set of proper underwear, showing that indeed many of these German shoulder head dolls were dressed as babies. Unfortunately, however, many of these found in baby garments have had the original clothes discarded and been redressed as children or ladies. This is certainly one of the tragedies of doll collecting, that original objects are sometimes dismantled because the owners do not like the outfits.

In 1877 *Harper's Bazar* doll report mentions "a fancy for infant dolls dressed in long Christening robe with sash and close lace cap. The face is infantile in expression with blue eyes and blonde hair banged over the forehead. These cost from 75¢ up to $30." The face material is not mentioned, but it is easy to picture a sweet-faced doll, such as this one, from the description. The dolls in baby clothes were still popular in 1887 when "The infant dolls dressed in long robes and caps rivalled in popularity the brides" in all their finery.

Interestingly enough the 1877 report also mentions that "Last year gray-eyed dolls began to rival the blue-eyed blondes," and certainly many of these little shoulder head dolls with closed mouths have gray eyes. *H & J Foulke.*

Illustrations 92 and 93. This 15in (38.1cm) doll is incised with a fancy printed "E." She has a face in common with *Illustration 90* and could have possibly be the next size in a series. Her description matches that of the other doll, except that she has brown eyes. She is wearing her original dress of turquoise wool with large puffed sleeves and wide cuffs. It is trimmed with picot-edged maroon ribbon and is in the style of the 1890s. *Esther Schwartz Collection.*

Illustration 94. This 13in (33.0cm) child is also marked with a fancy letter "E," but she is certainly a different series from the doll in *Illustrations 92 and 93.* Her shoulder head is very turned and tilted to look downward. Her blonde eyebrows are not striated, but painted with a long flat underside and many individual strokes especially at the inside corners. Her closed mouth has full lips, the upper one pouty with upturned corners and a darker red lip line. She has a chin dimple and chubby cheeks. She is on a kid body with the flat over-stitched buttocks and beautiful bisque hands with a free thumb. These closed-mouth shoulder heads are found on a wide variety of bodies which indicates that they were sold separately to producers who supplied the bodies. In fact, some stores sold the dolls in parts also. In 1878 *Harper's Bazar* noted that "Each part of the doll - head, wig, body, and arms - may be selected separately, and made up at home, or ordered to be put together in the store. The plainest doll heads are of bisque without hair and cost from 35 cents up . . . Doll bodies of muslin stuffed with hair are very strong and durable; with kid arms not jointed these cost from 35 cents up. Finer bodies are of kid stuffed with sawdust, and jointed every where; these are $1 and upward. Jointed arms are sold separately for 45 cents a pair." Thus are explained many of these head and body combinations! *Edna Black Collection.*

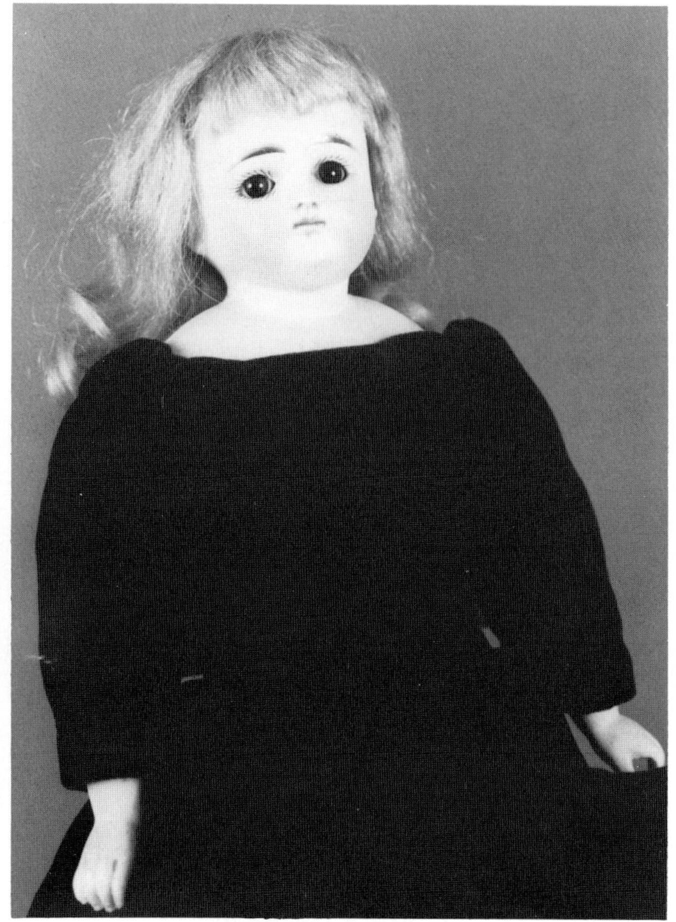

Illustration 95. Here is still a different face, this time incised "Made in Germany G" with a large fancy G. The fact that the country of origin is included probably places this particular mold after 1891 when dolls had to be marked with the originating country. This 16in (40.6cm) girl has tiny, yet very dark, sleep eyes and glossy arched eyebrows with many individual strokes. She has a distinct nose, but only faintly molded small ears. Her head is slightly turned to the right. She has the usual plaster dome and blonde mohair wig. Her gusseted kid body has bisque hands. *Elba Buehler.*

Illustration 96.

Illustrations 96 and 97. This 25in (63.5cm) child, incised only with the letter "O," has a mild, yet piquant face. Her head is turned to the right on a very deep shoulder plate. She has her original brown mohair wig over a plaster dome. Her sleep eyes are a lovely dark brown. As early as 1879 *Harper's Bazar* commented that "dark eyed dolls are a new fancy and those with a piquant cast in the dark eyes are much admired." Her plump cheeks and "punched-in" chin dimple as well as her long and glossy dark eyebrows are characteristics which she has in common with other dolls attributed to Kestner. She has a very heavy gusseted kid body with bisque lower arms and hands. She is dressed in an old deep red satin dress with a lovely lace collar and a newer bonnet. *Edna Black Collection.*

Illustration 97.

Turned Shoulder Heads

Ciesliks note that in 1888 Kestner remarked in reference to bisque doll heads: "They are partly straight positioned on the shoulders, partly slightly sidewards directed." The latter are called turned shoulder heads. Some turned shoulder head dolls with letter marks were shown in the previous section, but those shown here have only size numbers or no marks at all.

Illustration 98. This 17in (43.2cm) child's head is slightly turned to the right with chin tilted and eyes looking up, giving her a rather self-assured look. Her brown sleep eyes are small with delicately painted eyelashes and blonde striated brows. Her full closed lips are separated by a darker red accent line. Her original plaster pate is still intact. The head is mounted on a well-constructed kid body with beautifully detailed bisque lower arms and hands. Her wig and clothes are replacements. Many times collectors dress these German children in imitation of the fashionable French young lady which *Ehrichs' Fashion Quarterly* of 1880 referred to as "The Mignonne" meaning small and delicately pretty. As the German dolls were considerably less expensive than the French ones, especially the dressed ones, a thrifty mother could buy an undressed German doll and with a skillful needle, dress her in beautiful clothes at much less the cost. *H & J Foulke.*

Illustration 99. This 22in (55.9cm) doll has a turned head on a very deep and rounded plate. Her neck is short and stout, a feature usually interpreted to indicate a child doll. As already mentioned, these German dolls were purchased and dressed in imitation of the French dolls, this one as a very fancy lady in velvet and lace. She has a very high and wide forehead, brown sleep eyes, round cheeks, and blonde striated eyebrows. Her mouth molding is quite detailed as with other dolls in this grouping with protruding upper lip, cleft under the nose and under the botton lip. She is on a kid body with her original beautifully modeled hands with dimples, knuckles, joints, and fingernails outlined. *Esther Schwartz Collection.*

Illustration 100. This 17in (43.2cm) girl has a head turned quite far to the right and looking down in contrast to *Illustration 98* which looks up. She has her plaster dome and original blonde mohair wig. Her very dark brown sleep eyes and heavy brown eyebrows are a stark contrast. Again she has a stout neck, a very deep plate, and small ears. She is on a gusseted kid body. *Joanna Ott Collection.*

Swivel Necks

In the 1870s the doll writer for *Harper's Bazar* began to mention the importance of dolls which turned their heads from side to side. Of course, the French started this innovation, but the Germans were fast to copy the turning neck. Actually, very few of their dolls had this feature, since it was more expensive to produce the two-part head and shoulder plate and the Germans were interested in manufacturing inexpensive dolls.

Illustration 102. The same mold as the doll shown in *Illustration 81*, this head is mounted on a bisque shoulder plate. On this doll, the stoutness of the neck and the molding of the lower face with plump cheeks and tiny chin are more outstanding. Also noticeable is her very small ear. Her gusseted kid body has well molded bisque lower arms and hands. She is 21in (53.3cm) tall. *Esther Schwartz Collection.*

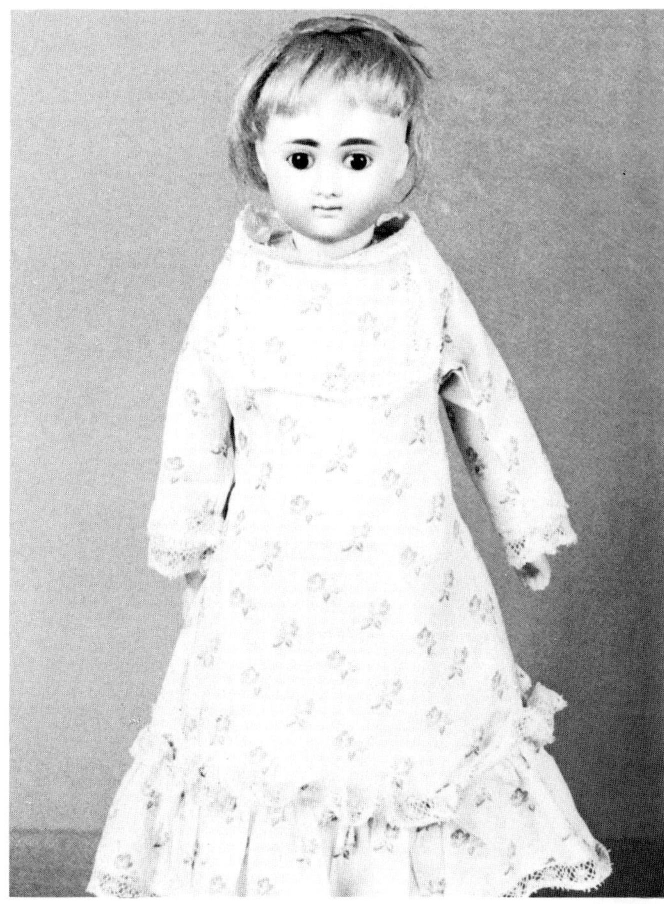

Illustration 104. This sweet girl with round face is more in the Kestner style. She is 14in (35.6cm) tall, incised only "4." She has her original blonde mohair wig, although rather sparse, over her plaster pate. She has lovely dark brown eyes and light brown eyebrows with many individual strokes. Her closed mouth is quite pouty with a protruding upper lip. Her neck swivels in the kid-lined joint of her shoulder plate which is mounted on a body with a cloth torso and gusseted kid limbs with bisque hands. She is completely original and was purchased from the original owner. *Esther Schwartz Collection.*

Illustration 103. Although she has some characteristics of the Kestner dolls, she may not be a Kestner product. A petite 11in (27.9cm), she is incised only "2." She has a plaster dome and a blonde mohair wig. Her blue eyes have lovely threading, and her lashes are dainty. Her blonde eyebrows are like those on some known Kestner dolls with very high individual strokes at the inner corners especially. Her mouth, however, is quite different in style and painting, in that it is an open/closed type with a molded tongue. She has a gusseted kid body with bisque hands and wears a new dress of old pink silk taffeta. *Edna Black Collection.*

Illustration 105.

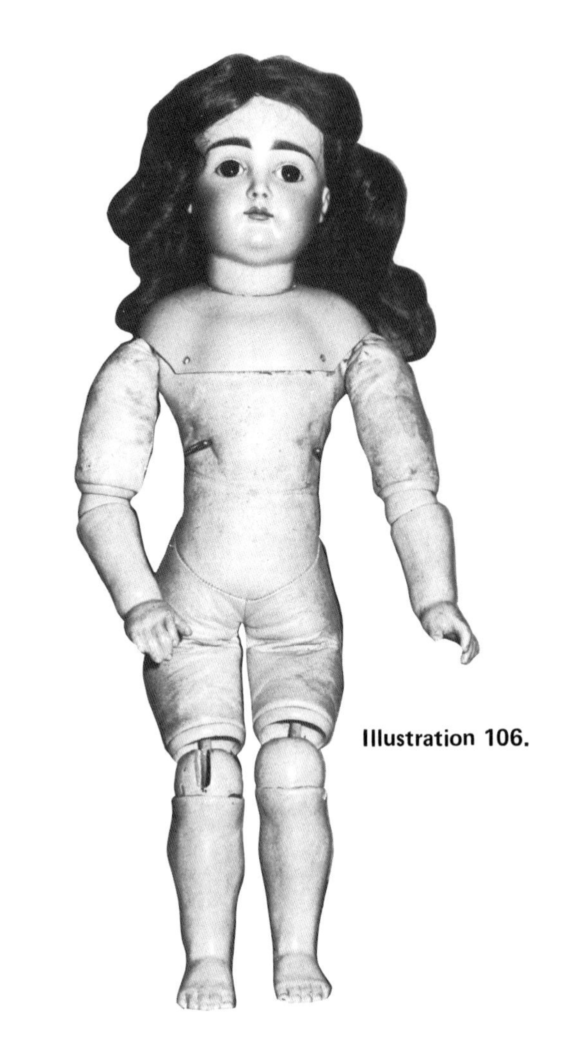

Illustration 106.

Illustrations 105 and 106. This 19in (48.3cm) girl with a swivel neck has more facial characteristics associated with Kestner, leading to an attribution to him. She has the pouty mouth with protruding upper lip painted full with two high peaks and upturned corners; she also has the darker red lip line. Her eyebrows are very full and thick especially at the inside corners. She has a very unusual body with torso, upper arms, and upper legs of kid; her lower limbs are of composition with a ball joint which fits into a circular socket. She is incised "XII" and could also have gone into the section with the X and XI dolls. *Elizabeth McIntyre.*

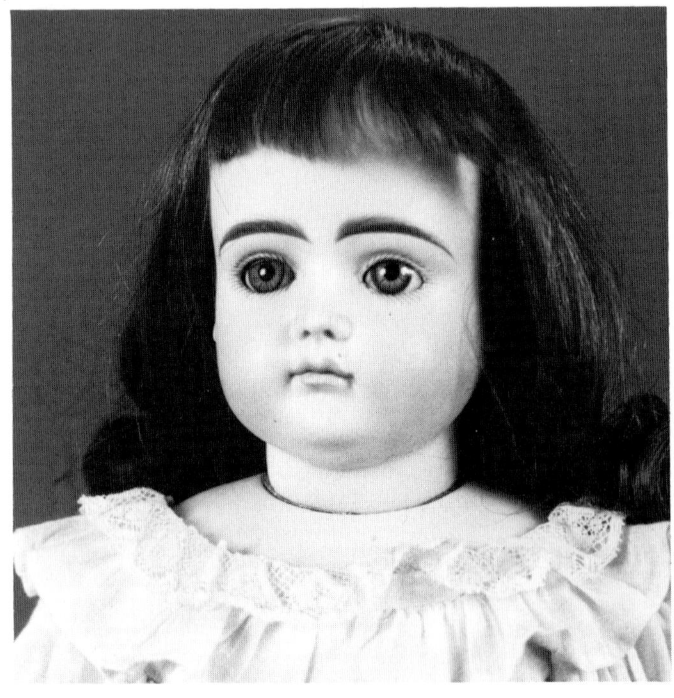

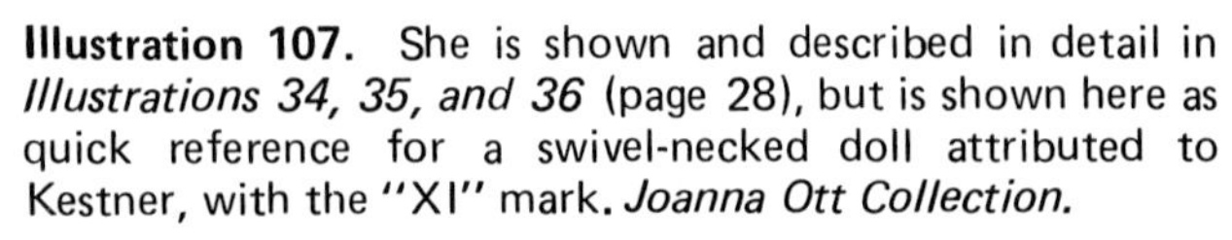

Illustration 107. She is shown and described in detail in *Illustrations 34, 35, and 36* (page 28), but is shown here as quick reference for a swivel-necked doll attributed to Kestner, with the "XI" mark. *Joanna Ott Collection.*

The next two dolls are attributed by some collectors to Kestner, but they are shown with many reservations as to the attribution. They do have plaster domes, but otherwise do not have the presence of Kestner dolls. They both have stationary eyes, with long thin eyebrows and rosy tinting above the eyes. Their mouths are in the open/closed style; they have pierced ears. These are not known Kestner characteristics.

Illustration 108.

Illustration 109.

Illustrations 108 and 109. She is a fine example of how the German shoulder head dolls, especially those with swivel necks could translate into fashion dolls. She makes a beautiful lady with her lovely bisque, blue paperweight eyes, pierced ears, blonde mohair wig, and posable turning head. She has been redressed in the height of fashion and is 15in (38.1cm) tall. *Edna Black Collection.*

Illustration 110. This dainty 15in (38.1cm) lady is a sister to the one shown in *Illustrations 108 and 109.* She is incised "5" on both her head and shoulder. She has rosy tinting on her chin as well as above her eyes. She is fashionably dressed in old, but not original clothes. *Elba Buehler.*

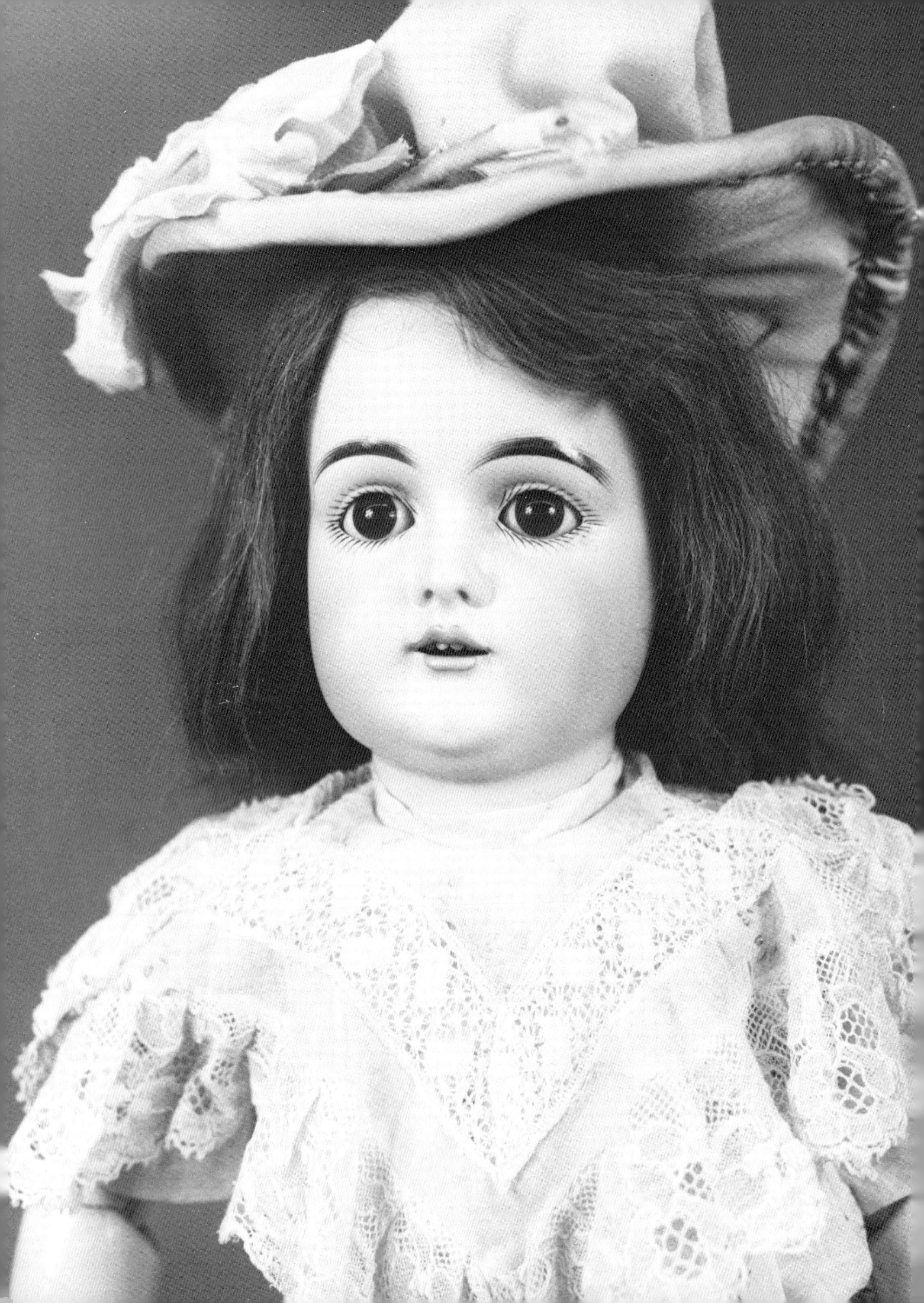

Open-Mouth Dolls

The bisque-head child doll was a staple product of the German doll industry for nearly half a century. Though basically a French invention which was taken over by the Germans, it was the German doll and not the French one which dominated the market. The Germans did, however, due to their own inventiveness, add novelty to their wares by making eyes which slept, cutting the mouths open and adding teeth, and inserting eyelashes and eyebrows.

Bachmann reports that two-thirds of the European-made dolls came from the German factories and workshops so that German dolls dominated the world market at the beginning of the 20th century. The Germans produced a very good doll which would be bought and resold at a price much cheaper than the French charged for their dolls. This was due to the cottage industry aspect of German doll production. Whole families worked together at home to finish dolls, that is making shoes, clothing, wigs, body parts, and assembling the finished product. This method produced cheap labor and low overhead for the German manufacturers and allowed them to sell at a low price, yet still make a profit.

The bisque-head open-mouth doll showing teeth came onto the market in a big way about 1890. There were earlier attempts by dollmakers to insert teeth in bisque and other types of doll heads, but most of these were awkward looking and certainly did not contribute to the beauty of the doll. However, makers finally did hit upon the right technique, cutting an opening in the bisque and either molding teeth as part of the mouth or molding them separately of porcelain and cementing them into the mouth. It is possible that the square-cut upper and lower teeth on *Illustration 226* and the two upper teeth on *Illustrations 224 and 225* are examples of earlier types of porcelain teeth. Collectors also feel that the molded-in and square-cut teeth as shown in *Illustration 171* are earlier also, but these may simply have been less expensive to produce.

An 1888 advertisement from Kestner & Co. porcelain factory notes bisque doll heads with inserted teeth. This is the earliest date for open-mouth dolls yet found. In *Ridley's Fashion Magazine* for the winter of 1888, a 14½in (36.9cm) doll with a composition jointed body, bisque head, and flowing hair sold for 79¢, but a 16½in (41.9cm) doll of the same quality only a little heavier with *mouth open and teeth* sold for $1.25. However, in spite of the fact that the open-mouth dolls were a little more expensive, they became very popular and nearly all of the dolls from 1890 on had this new feature.

The doll with an open mouth was specifically noted in 1888 also in the *Youth's Companion* describing its bisque premium doll for that year: "Its lips are beautifully moulded and slightly parted, showing pearly porcelain teeth which have been naturally inserted." In 1891 the American Merchandise Company was offering superior French bisque dolls with "open mouth showing teeth." Although called "French" bisque, this term is confusing because it often means nothing more than excellent quality bisque, just as the term "French" jointed, used to describe a body, simply means a jointed composition body. In the 1890 Butler Brothers wholesale catalog, "*Excelsior* kid body dolls with open mouth showing teeth, giving a very lifelike appearance" were offered. In 1892 Butler Brothers was offering a "French" bisque doll with mouth open a trifle showing teeth on a kid body with jointed arms and legs.

The same 1891 wholesale catalog just mentioned advertised *Excelsior* bisque jointed dolls. They were 19in (48.3cm) tall and "Made in the latest style, with bisque head, open mouth showing teeth, natural eyes, fine flaxen hair; shoes and stockings; fine silk embroidered chemise; full jointed." Price was $8.50 per dozen wholesale which means the dolls probably retailed for about $1.40 each.

The typical mark on the back of the head of a Kestner open-mouth doll is:

L Made in Germany 15
171

Some variation in placement may be expected. The sizing scale is explained on page 19. The mold numbers are discussed in detail in this chapter.

Dolly Faces with Open Mouths
129

Page 60:

Illustration 111. This 24in (61.0cm) doll is an apt one with which to start this section because she exemplifies all of the charm of the typical Kestner dolly-face. Her bisque, which is very fine, is an excellent material for achieving a natural look because of its translucence and its ability to appear soft and lifelike. Her face shows a cool and ageless beauty. She is the embodiment of the idealized child, a perfect object. Her eyes are alert; she does not have the vacant stare and vapid look of many of her sisters made by other companies. Her lovely glossy eyebrows which are reminiscent of those on the Jumeau dolls are a Kestner characteristic much admired by those who cherish Kestner dolls. Her mouth is also done in the typical Kestner style with the upper lip upturned on each end and the bottom lip of which the molded part is not completely filled in. She is on a Kestner ball-jointed body with the "Excelsior" mark, which was patented in 1892. With this low mold number, if Kestner did follow numerical order in assigning numbers, she would probably have been made about that date. Her head size is J½ 13½. *Emma Wedmore Collection.*

Illustration 112.

Illustrations 112 and 113, This 18in (45.7cm) example of mold 129 is head size G 11. In totally original condition, she illustrates the beauty, charm, and appeal of the German Waltershausen dolls which were noted for their consistently fine quality. Her coloring is soft with a curly blonde mohair wig and blue sleep eyes. Her dark, full and glossy eyebrows add an air of the dramatic to her face. Her upper lip is full and naturally shaped with two high peaks; both lips are accented and shaded with short darker red lines at the top of the upper lip and the bottom of the lower lip. Her original clothes are in the style of a child of the 1890s. Her white lawn dress has a lace and embroidered inset yoke with a lace ruffled edge. Her bodice is gathered at both the top and bottom; her skirt is also gathered with a lace-edged ruffle at the hemline. Pink satin ribbon insertion trims the neckline, waist, and sleeve edges. Her bonnet is trimmed with lace and pink satin ribbons. As for footwear, her stockings are pink, and her shoes are white leather with rosettes. Her jointed composition body has the "Excelsior" stamp as well as the label of G. A. Schwarz, Philadelphia.

The Butler Brothers wholesale catalog for 1899 shows "The World Celebrated 'Kestner Model' Dolls." These were with bisque heads, "French" jointed bodies including wrists and open hands with free fingers. They had fine sewn wigs, elaborate underclothes, costumes and hats. Their sleep eyes were tied with tape which comes underneath the hair at the back of the head. The tape would come through the two little holes in the back of the head and would protect the eyes during the long journey from Germany. A doll the same as the one illustrated here would vary in price according to the elaborateness of her costume. Dolls comparable in size could range in price from $3.75 to $6.50 wholesale; their retail price from about $6.00 to $12.00. *Mary Lou Rubright Collection.*

Illustration 113.

132

This example of mold 132 has beautiful creamy bisque and is the only 132 which we have ever seen. She has a pert face with sparkling brown sleep eyes. Contributing to her overall appeal is her open mouth with tiny teeth. The whole face has somewhat of a character look. Just 9in (22.9cm) tall, she is a wonderful example of a small Kestner doll. She is on a fully-jointed wood and composition body and wears her original costume. *H & J Foulke, Inc.*

133

The only example of this mold we have found is this 7in (17.8cm) child. Even though small, she is immediately recognizable as a Kestner face. She has small brown sleep eyes and original blonde mohair wig in the Rembrandt style so often used by Kestner for small dolls. She is on a five-piece composition body with molded shoes and stockings. *H & J Foulke, Inc.*

133
d made in
germany

134

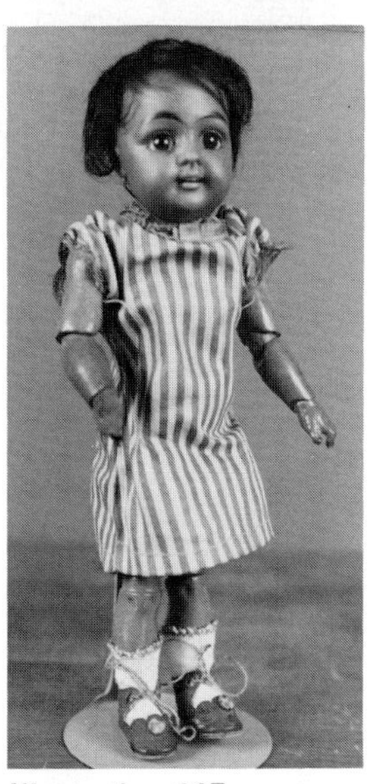

Illustration 115.

Illustration 115. There are not many Kestner black dolls, but we do know that the factory made them as one is shown in the catalog of the 1930s. However, this doll was made quite a bit earlier than that date. She is a particularly appealing little girl, just 9in (22.9cm) tall and incised with mold number 134 and size 2/0. Her color is a rich chocolate brown with an even, smooth finish. Her original short curly black mohair wig covers a plaster dome. Her eyebrows and eyelashes are painted black; her sleep eyes are brown. Her open red lips show four molded-in upper teeth. Her excellent jointed composition body is fully jointed even at the wrists. She is wearing her original red and white striped cotton shift and brown shoes. *Dr. Carole Stoessel Zvonar Collection.*

141

Illustration 116. The 141 is an elusive socket-head mold which does not turn up frequently. It has an expressive Kestner-style face but the mouth is a little longer and thinner. *Courtesy of E. Joanne Knoedler.*

142

Illustrations 117, 118, 119, 120, and 121. This 42in (106.6cm) doll is a marvelous example of mold 142. He is a big beautiful doll, probably the largest child which Kestner made. Every 42in (106.6cm) Kestner doll that I have seen has been this mold number. His head size is Q 20, which is the same size head used on the 27in (68.6cm) toddlers and the 25in (63.5cm) character babies. His eyebrows are one of his outstanding features, a beautiful brown with many individual strokes and very deeply molded ridges. I have never seen eyebrows so deeply molded as these are. He has large brown eyes and very long painted lashes. His open mouth has beautifully shaped full lips with four upper teeth. His ball-jointed composition body is very heavy. His costume is most interesting; it is made of very heavy navy blue wool. The coat has buttons which are inscribed "CITY OF//F.D.// PHILADELPHIA." and its label is from "I.M. Leopold//8th & South Sts.//Philadelphia," perhaps a tailor or a uniform supplier. His hat is stamped "Wm. H. Horstmann Company// Philadelphia" and the insignia on the front is a large "15" with "F.D." and "Phila." The owner speculates that perhaps this was a sample uniform made up for approval. *Dr. Carole Stoessel Zvonar Collection.*

Illustration 117.

Illustration 118.

Illustration 119.

Illustration 120.

Illustration 121.

Illustration 122. This 36in (91.4cm) girl is another doll from mold 142; her size is 0½ 18½. Again she has deeply molded eyebrows and an open mouth with four upper teeth. Her blue sleep eyes have long painted lashes. Her heavy jointed composition body has the Kestner "Excelsior" mark. *Dr. Carole Stoessel Zvonar Collection.*

Illustration 123. Here is a cute tiny 7in (17.8cm) example of mold 142 with appropriate blonde mohair wig and apparently original gauze shift with lace trim. Her five-piece composition body has molded stockings and strap shoes with heels. *Private Collection.*

143

This is a puzzling doll because it is really a character face, but it was in production in the late 1890s long before the character dolls of 1909 and thereafter were produced. It is a fairly plentiful doll, but one which is very popular with collectors because of its special charm and appeal. All of the heads with this mold number are of the socket type.

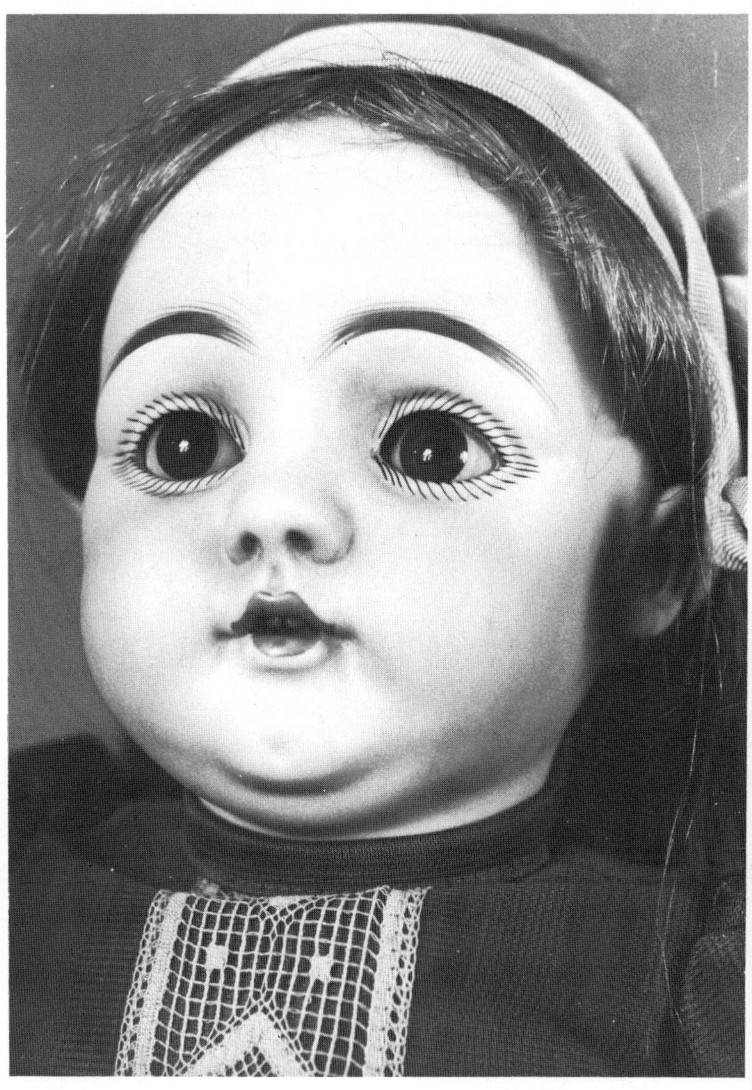

Illustration 124. This 23in (58.4cm) example of mold 143 is a fairly large size for this doll, which for some unknown reason is usually found in smaller sizes. The close-up illustration shows why she is such a sought-after doll. Her large brown eyes are quite striking, and they are set off by her long thick eyelashes and her brown eyebrows, solid in the middle but showing individual brush strokes on both ends. Her whole lower face is chubby: her wide nose, her bowed upper lip, her cheeks, and her chin. She has deep crevices in the modeling all around her mouth, a pronounced philtrum and a prominent chin dimple. She has a socket head size K 14 on a jointed composition body with the "Excelsior" stamp. *Esther Schwartz Collection.*

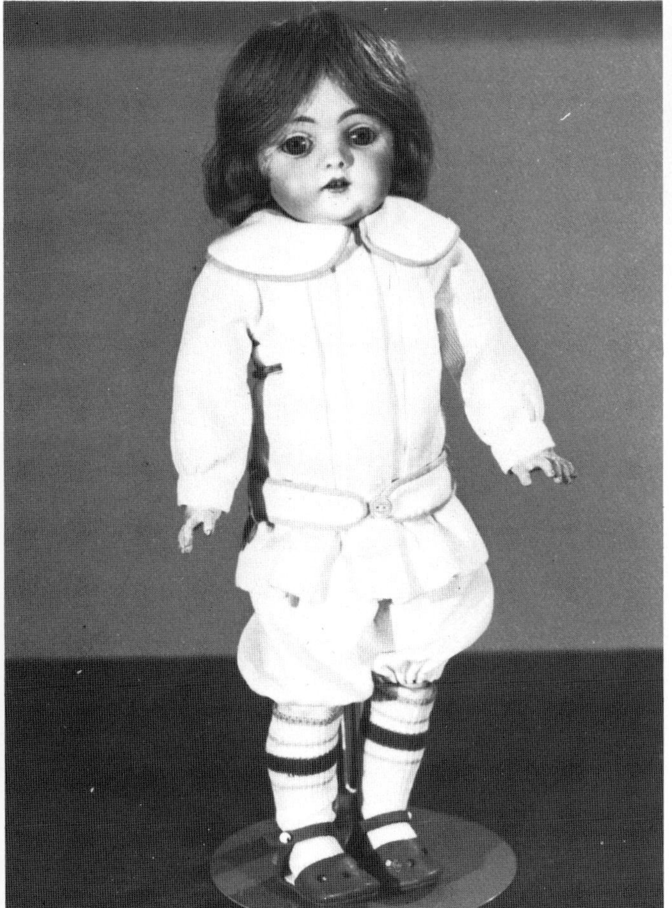

Illustration 125. This is an appealing 12½in (31.8cm) example of the 143 dressed as a little boy in a two-piece suit of white corded cotton with pleats down the front. The large collar and low belt are piped with blue trim. His stockings have stripes in different shades of blue and he is wearing brown leather sandals. He has the mouth characteristic of this mold with bowed upper lip and two upper teeth. He has blue sleep eyes and a replaced human hair wig and is on a fully-jointed composition body. *H & J Foulke.*

Illustration 126. This 11in (27.9cm) 143 is certainly a gem to own because she is completely all original just as she came from some exclusive store. Her christening gown is very elaborate with lace, embroidered insertion and ruffles down the front with tiny satin bows. Her matching hat has several rows of gathered lace to frame her face and her underclothes are just as elaborate as her outer garments. She has her original short blonde mohair wig (as after all she is a baby), heavy brown eyebrows, small blue sleep eyes, and a tiny mouth with two upper teeth. She is on a fully-jointed composition body (except for wrists) and is interesting because she illustrates the point that before the baby body with bent limbs was developed, dolls on regular jointed bodies were used as babies. *H & J Foulke.*

Illustration 127. This 19in (48.3cm) 143 is a good example of how cutting the eye sockets larger can change a doll's expression. Just compare her face to that of the doll in *Illustration 126.* The difference is startling. The doll shown here has blue sleep eyes and a short blonde mohair wig. Her eyebrows are very heavy; her eyelashes are short but dark. She has two tiny upper teeth set like many of the 143s so that she looks gap-toothed. It certainly adds to her charm, and in her sweet dress with gathered ruffles and tiny buttons for trim, she is a pert little girl. *J. C. Collection.*

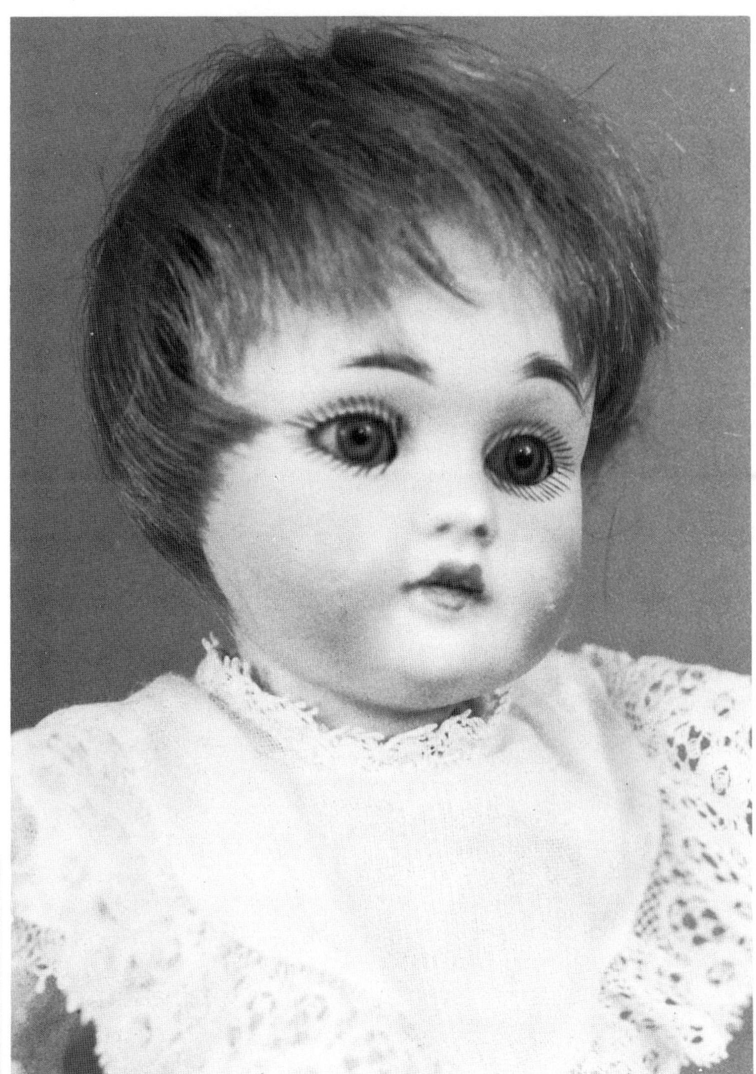

Illustration 128.

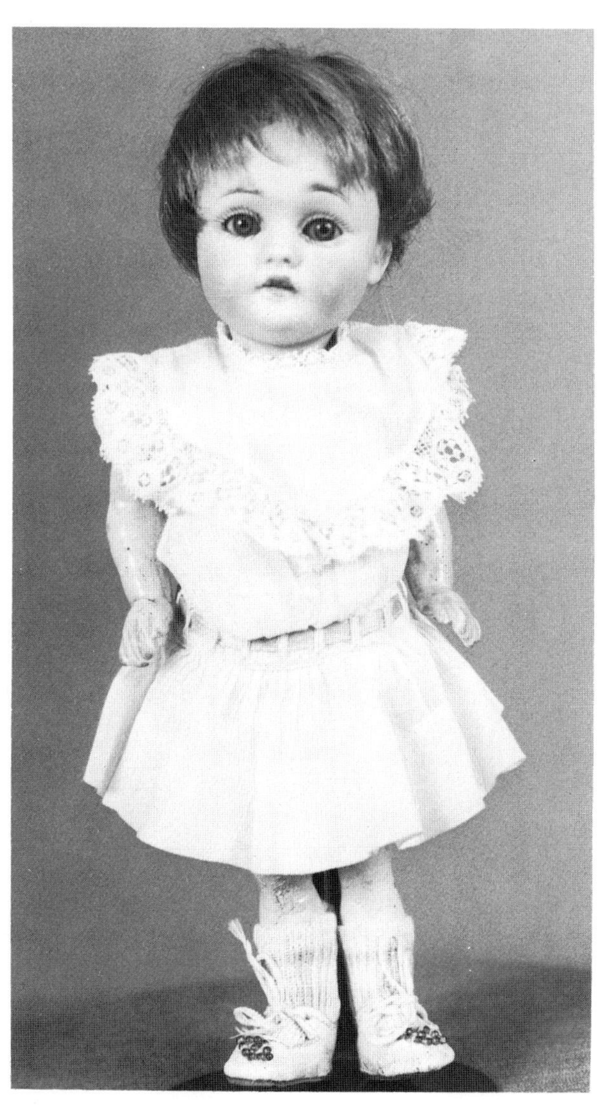

Illustration 129.

Illustrations 128 and 129. This darling little girl is just 8in (20.3cm) tall. Many 143s are found in this small size. These dolls have very good quality composition bodies peg-jointed at the shoulders and ball-jointed at the hips and knees. The elbows and wrists are solid. This is important to note because many collectors unfamiliar with this body type wrongly assume that the dolls have replaced arms, when, in fact, the bodies are correct. This doll is also different because she has four molded-in square-cut upper teeth. This type mouth is often found on the smaller sizes of this mold. Her chubby face has tiny blue eyes, light brown glossy multi-stroked eyebrows, and a pug nose. She has her original light brown mohair wig and is wearing an outfit which appears to be original. Her low waisted white pique dress has an inserted blue ribbon sash and a wide bertha collar with lace trim. Her blue and white socks and tiny white leather shoes also appear original. *H & J Foulke.*

Illustration 130. This 8½in (21.6cm) 143 has the same body as the one in *Illustrations 128 and 129.* Her mark is an illustration of the Kestner system for small heads:

f made in 3/0
Germany
143

Edna Black Collection.

Illustration 131. The body on this 10in (25.4cm) 143 is jointed at the shoulders, hips, knees, and elbows, but has straight wrists. This is also a correct Kestner body for the small size dolls. She has her original blonde mohair wig and the larger style eyes. *Edna Black Collection.*

144

Illustration 132. Although this illustration from our files was taken many years ago, it is used in an attempt to make as complete a record as possible since a doll with this number does not turn up too often. Her face is lovely with excellent bisque. Darked stroked brows emphasize her blue sleep eyes; her open mouth has four upper teeth. She has a long brown human hair wig and replaced clothes. She is 24in (61.0cm) and is on a very good quality composition body; her head size is K 14. *H & J Foulke.*

145

This mold number has been found on a shoulder head doll with typical Kestner dolly face and open mouth. Kestner registered the number in 1897. The body is kid with short lower bisque arms of lesser quality than Kestner's usual production. On this type of body she would have been a model for the economy market. The pictured doll is 15in (38.1cm) tall. *Esther Schwartz Collection.*

146

Illustration 133. There is very little difference between the 144 and 146 molds. They are both very typical Kestner products and both project what collectors refer to as "the Kestner look," which partly includes the heavy eyebrows and the pleasant mouth with upturned corners. This particular doll is 27in (68.6cm) tall on a composition jointed body. Her dark brown hair, matching eyebrows, and dark blue eyes provide a striking contrast to her face and clothes. Her dress is of white embroidery on white cotton with lace and tucks to trim her sleeves. *J. C. Collection.*

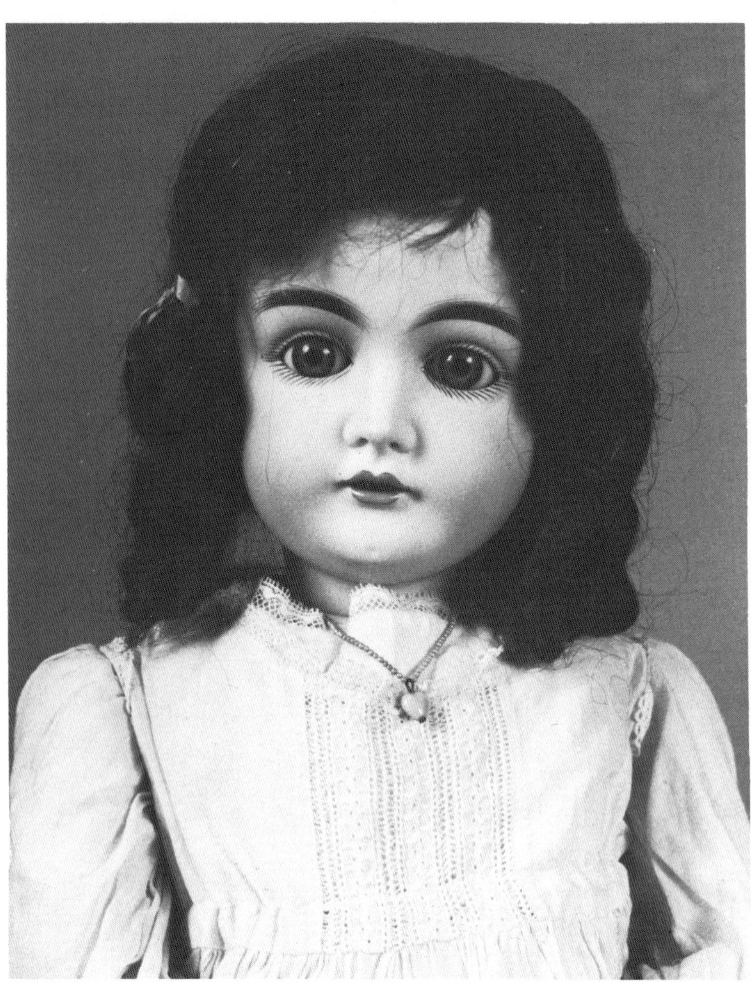

Illustration 134. Although this doll has the exact same face as the one in *Illustration 133,* she has no mold number, only size numbers K½ 14½. Her ball-jointed composition Kestner body has the red stamped "Germany" on it. She is also 27in (68.6cm) tall, although her head is one-half size larger. This variation in size is not unusual and can even be up to 2in (5.1cm). Her bisque is very pale and lovely, and again she is a very striking doll with her brown hair and blue eyes, which are just a trifle larger than those of her sister doll. Her mouth is particularly nice with long parted lips showing four upper teeth. *J. C. Collection.*

147

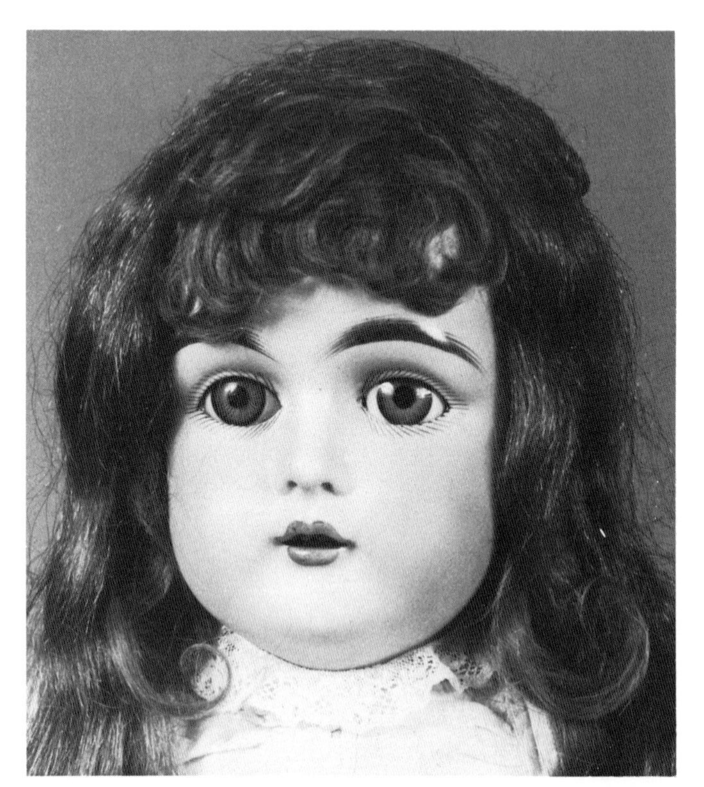

Illustration 135. Just when you think that you have a maker's marking system figured out, a variation comes along like this one. The mark on this shoulder head doll is "15//147//L Made in Germany." The shoulder head dolls with kid bodies were a staple line of the Kestner Company for many years; a doll such as this one could have been made over a period of 20 or so years. The hinge-jointed kid bodies were more expensive than the gusseted ones.

A 1904 wholesale catalog shows Kestner dolls such as this one with gusseted cork-stuffed bodies and "long, real hair ringlet curls falling down to the waist." The dolls were also described as having beautiful faces and closing eyes. The wholesale price for a 22in (55.9cm) doll was $2.50 for a naked doll wearing only shoes and stockings. It appears as though the long curls on the doll illustrated have been combed out, but she does still have the original lovely long brown real human hair as advertised. *J. C. Collection.*

148

Illustration 136. This number is another shoulder head version marked: "6//148." She is on a gusseted kid body with bisque hands of very nice quality with a separate thumb. Over her plaster dome is her original blonde mohair wig styled in long curls. She has long eyelashes, brown sleep eyes, and light brown glossy and very long eyebrows. Her open mouth has four upper teeth. She is very much like the Marvel line of dolls which Kestner made for Butler Brothers and advertised in their 1899 wholesale catalog as the "best kid body dolls made." Their ad goes on to say "Probably no line we ever carried has contributed in such large measure to the greatness of our Doll Department as the 'Marvel' line." A 17½in (44.5cm) doll such as the one illustrated carried a wholesale price of $8.35 per dozen, naked wearing only shoes and stockings. Remember that a doll of this style would have been made over a period of many, many years. The doll is appropriately dressed in a white lawn dress with a dropped waist and much tucking and lace trim. *Elba Buehler.*

Illustration 137.

Illustrations 137 and 138. This 25in (63.5cm) girl is marked "10//148." She is also on a gusseted kid body. With her black hair and blue eyes, she is a lovely doll, a prime example of her type. *J. C. Collection.*

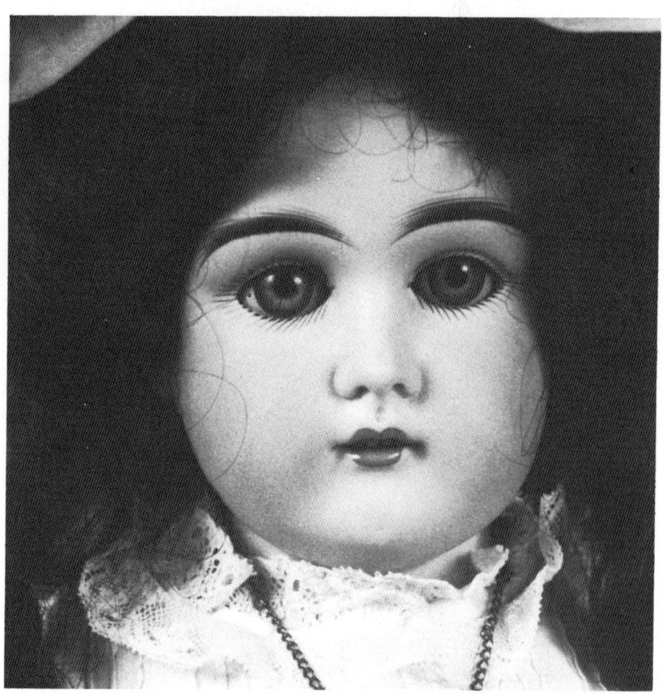

Illustration 138.

149

Illustration 139. This face is similar to the 143 mold through the eyes and cheeks, but it is definitely a 149, size F 10. The bisque is pale with just a hint of red to the cheeks. Many individual strokes contribute to her heavy eyebrows; her sleep eyes are dark brown with very long painted lashes. Her nose is fairly broad, and she has a deep philtrum. Her cheeks are very full, and her chin shows a slight dimple. Her dark blonde center-parted wig is of human hair. She is on a very heavy ball-jointed composition body stamped in red "Germany 3" in the Kestner manner. *Mary Lou Rubright Collection.*

152

Illustration 140.

Illustrations 140 and 141. Certainly a choice doll in any collection is one which is completely original as is this 17½in (44.5cm) example of mold 152. She has her original curly blonde mohair wig. Her eyes are gray-blue in contrast to her very dark eyebrows; her bisque is lovely and pale. She is a socket head on a fine quality ball-jointed body. Her original white lawn dress has a front-tucked bodice with ribbon insertion, a slightly lowered waist also with inserted ribbon, and two rows of lace insertion on her skirt. Her blue satin hair bow matches the ribbons on her dress. She dates from circa 1900. *H & J Foulke.*

Illustration 142. This 24in (61.0cm) 152 is also in her original clothes of about 1900, a white lawn dress with lace yoke trimmed with a wide lace ruffle. Her original blonde mohair wig is center parted and contrasts with her brown eyes and very dark eyebrows; her cheek line is full and her mouth is painted in typical Kestner style. Her head size is H 12. All of the 152 mold dolls have socket heads on excellent jointed composition bodies. *J. C. Collection.*

Illustration 143. This 17½in (44.5cm) example of mold number 152 is head size G½ 7½. Over her plaster dome is a brown real hair wig. Her dark eyebrows have many individual brush strokes at the inner and outer edges. She has the Kestner mouth with the upturned lip corners and four upper teeth. She is wearing a blue wool cape with real mink trim. Her excellent quality jointed composition body has the red stamped "Germany" and size 2. *Edna Black Collection.*

Illustration 144. Judging from the numbers of dolls found today, the shoulder head mold 154 must have been one which was very popular and made over quite a long period of time. Many 154 dolls do not have the word Germany incised on their heads, and collectors have a tendency to assign these dolls to the earlier period before 1891 when the Country of Original Law went into effect. However, they are forgetting that many of these same heads are on the hinged bodies with the *Universal* knee joint which was not patented until 1895 to 1896, and dolls with this type of body were still being advertised in Montgomery Ward & Co.'s 1924 catalog. Assigning dates to dolls is a difficult process indeed! This 28in (71.1cm) girl is marked "9 154 Dep." up around her crown rim. She has molded brown eyebrows, thick and glossy, and brown sleep eyes. Her mouth is open with the typical Kestner decoration and four upper teeth. She is on a very heavy kid body with pinned hips and knees, gusseted elbows and very large and well-molded hands with separate thumbs. She is similar to a Kestner doll of the same size which was offered in a 1904 wholesale catalog. The doll was described as "the largest and finest kid body doll we have ever shown . . . supurb bisque head with beautifully modeled face; long ringlet curls, closing eyes, open-work stockings and patent leather shoes. Price each $5.00." Remember that was a wholesale price. The doll illustrated is in lovely old clothes, but has a replaced blonde human hair wig. *Joanna Ott Collection.*

Illustration 145. This 23in (58.4cm) 154 is dressed as a boy. Some collectors insist that these dolls were always girls, but we know from advertisements many dolly-faced dolls were dressed as boys. In 1899 Butler Brothers advertised Kestner dolls with short and curly boy wigs. These dolls were shoulder heads on kid bodies. The pictured boy is marked "154 G made in Germany" with a very fancy **G**. He has the required short curly blonde mohair wig, large brown sleep eyes, and a very pleasant expression. His kid body with hinged legs and gusseted arms carries the Kestner crown and streamers label of 1895. *J. C. Collection.*

Illustrations 146 and 147. This lovely little 154 girl is marked "Dep. 7½ 154." She has glossy molded eyebrows and brown sleep eyes with real upper lashes instead of painted ones. Both the molded eyebrows and lack of upper painted lashes are usually regarded as later features. Her open mouth has nicely painted lips and four upper teeth. Her gusseted kid body has hinged elbows allowing her lovely bisque hands to be placed in more natural positions. It is also stamped "Gloriosa//TRADE MARK//[Shield symbol] Germany" and has an additional red paper oblong label. She has a new blonde human hair wig and has been redressed in an appropriate style of old fabrics for a girl of 1910. *Edna Black Collection.*

Illustration 147.

Illustration 148. This pretty 19in (48.3cm) 154 has lighter eyebrows than most of the Kestner dolls. She has brown sleep eyes also with real lashes and no painted upper lashes. Her body is the hinge-jointed type. An odd characteristic of many of these 154 heads is that they are nearly flat at the back of the head. She is wearing a dark velvet dress and matching hat with crocheted lace trim, a very attractive costume which does not detract from the beauty of her face. *J. C. Collection.*

155

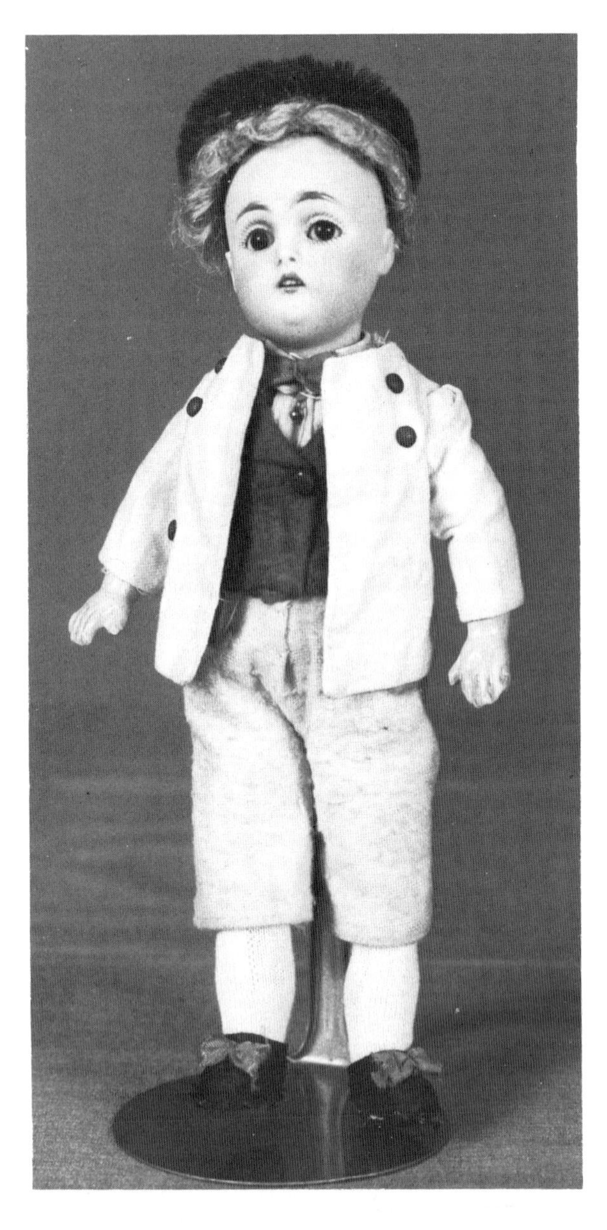

Illustration 149. This mold 155 is quite puzzling as I have only seen it in the small sizes. The boy pictured is 8½in (21.6cm) tall. He has brown sleep eyes with a definite molded upper eyelid. His painted lashes are short and rather far spaced. His most distinctive feature is his small open mouth with four tiny upper teeth. He is on a fully-jointed composition body with straight wrists. His outfit is completely original. *Jane Alton Collection.*

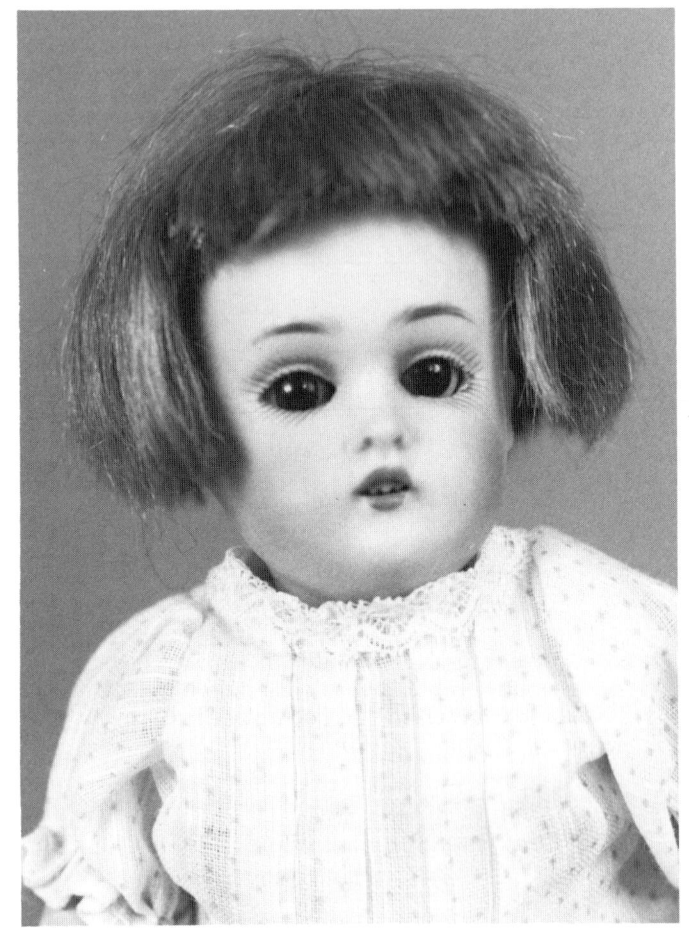

Illustration 150. This close-up of the 155 mold shows that she has fairly soft and light eyebrows for a Kestner doll, but they do have the usual gloss. Her long painted lashes enhance her brown sleep eyes. Her narrow face is also a departure from the chubby Kestner look. Her small mouth has four upper teeth. Over her plaster pate is the original brown mohair wig on a cloth cap. Just 8in (20.3cm) tall, she is on a composition body with pegged bent arms and strung legs with a knee joint. She is wearing a nice old white cotton dress with tiny pink polka dots. Her mark is:

Made in
Germany 5/0
155

H & J Foulke.

Illustration 151. This sweet example of mold 155 is 10in (25.4cm) tall on a chunky five-piece excellent quality composition body with molded black stockings and tan strap shoes with molded bows. Her face is very appealing with creamy smooth bisque; she still retains her plaster pate. Her wig and dress are appropriate replacements. *H & J Foulke, Inc.*

156

This is another elusive face, a very piquant one. There is a lot of detail in the moulding at the eye corners and around the mouth which makes the face look almost smiling. The eyebrows are dark and very heavy in the usual Kestner style. Eye sockets are quite deep with a molded upper lid and both real and painted eyelashes. She is 28in (71.1cm) tall on a jointed composition body. Her head is marked "made in Germany L 1/2 Dep 15 1/2 156." *Nancy Forester Collection. Photograph by James Brownell.*

159

Illustration 152. This doll is marked "4½ 159 Dep. 8½" on her shoulder head. Her gusseted kid body carries the Kestner crown and streamers label. For a Kestner she certainly is a poorly decorated doll. Her eyebrows are only one stroke, possibly decals, and her eyelashes are carelessly painted. She has the typical Kestner mouth decoration and a short blonde mohair wig. Another doll 15in (36.1cm) tall with this same number is marked "159 dep 5 Made in Germany." *Photograph by Thelma Bateman.*

160

Illustration 153.

Illustration 154.

Illustrations 153 and 154. Mold 160 actually has less of the Kestner look about it and clearly is related to the 155 number. This 17in (43.2cm) girl has beautiful bisque and a lovely complexion. She has light eyebrows with separate brush strokes, blue sleep eyes, and long painted lashes. Her eye sockets appear smaller than on many Kestner dolls. Her mouth is also shaped differently, not being so wide; her upper lip is not so peaked, but does upturn at the ends. Her lips are barely open enough to show her four upper teeth. She has lovely long blonde mohair curls. Her socket head is on a good quality jointed composition body. *Jane Alton Collection.*

161

Illustrations 155 and 156. Red Riding Hood costumes have always been a popular motif in which to dress dolls. This 15in (38.1cm) example of mold 161 is all original. She is wearing a white cotton dress which has a smocked bodice, lace insertion in the sleeves and skirt as well as two rows of ruffled lace. Her blonde mohair wig is center parted with small ringleted curls. Her red hooded cape is trimmed with satin bows. She is wearing white lacy stockings and white leather shoes. She is really a charming little girl. *Maxine Salaman Collection.*

Illustration 156.

Illustration 155.

Illustration 157. This 18in (45.7cm) example of mold 161 shows a face that is a little longer in the cheek line than some of the other Kestner molds. However, she does have most of the other Kestner dolly-face characteristics: the thick glossy eyebrows, the long painted lashes, and the upturned upper lip. Her mouth is open showing four inset upper teeth. She has brown sleep eyes and a brown mohair wig. The 161 mold is a socket head used on a jointed composition body. *Emma Wedmore Collection.*

Illustration 158.

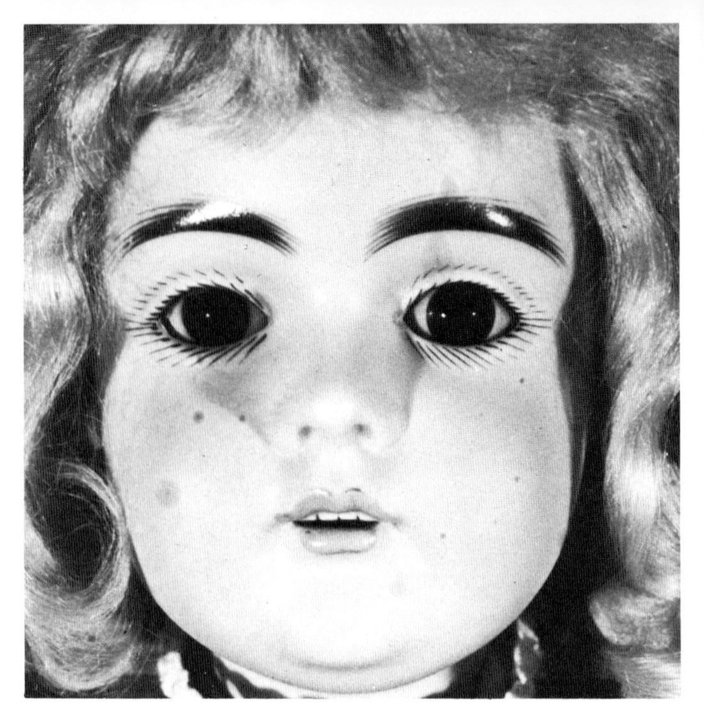

Illustration 158 and 159. Although these illustrations are from our files and taken many years back, we thought they were important because they show this same 161 face with upper molded teeth which is a characteristic of some Kestner dolls. The molded-in square-cut teeth are often regarded as earlier than the separately inset porcelain teeth. This particular 20in (50.8cm) doll has a lovely pale bisque face with very glossy and thick eyebrows and long eyelashes. Her blonde mohair wig is original. Her blue and white checked dress is a replacement made from old fabric in an appropriate style. *H & J Foulke.*

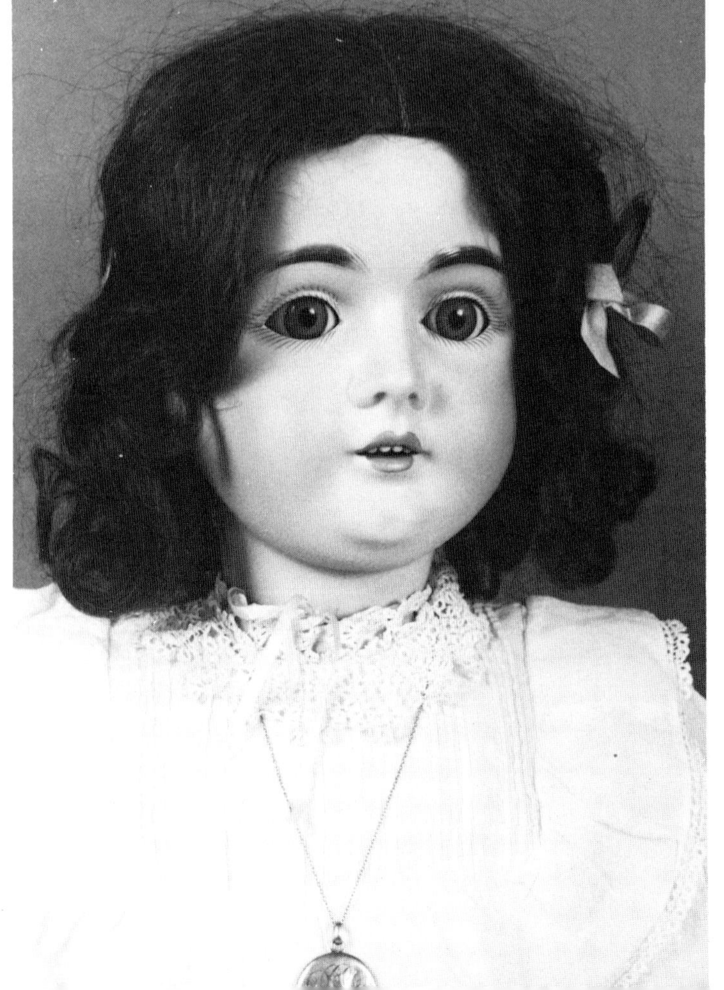

164

Illustration 160. Mold 164 is one of the more commonly found Kestner faces, so it must have been a very popular style. She is a fairly full-cheeked child with a very pleasant face. Her mouth, which is broad and barely open, has four tiny upper teeth. Her molded eyebrows are thickly painted and show individual brush strokes. Her long painted lashes set off her blue sleep eyes. She has lovely brown human hair, center-parted. This mold is a socket head used on a jointed composition body; the doll is 29in (73.7cm) tall; head size M½ 16½. *J. C. Collection.*

Illustration 161. This 16in (40.6cm) girl from mold 164 is size C 7. Her original plaster dome is covered by a blonde mohair wig; her dark blonde eyebrows are quite heavy. She has been redressed in a white cotton dress with a wide collar and lace trim, an attractive style for a little girl. When originally purchased most dolls were nude or wore only a chemise and shoes and stockings. These dolls were quite a bit less expensive than the dressed dolls. Mothers and grandmothers took on the job of dressing the dolls for Christmas, birthdays, and like occasions. *Joanna Ott Collection.*

Illustration 162. There were, however, many dolls shown in the Buttler Brothers 1899 wholesale catalog which had elaborate costumes and very fancy hats. The prices for these dolls were two or three times as much as for the undressed dolls. Many collectors like to dress their dolls in these stylish outfits, reminiscent of the luxury dolls that many of them were. This 26in (66.0cm) 164 has been redressed in just such an outfit. *Sheila Needle Collection. Photograph by Morton Needle.*

Illustration 163. This large 30in (76.2cm) girl is a marvelous, if perhaps a later, example of mold 164. This is based not only on the molded eyebrows but also on the very long real eyelashes and lack of the upper painted lashes, producing a variation in the treatment of the face. This doll is unusual also because she has only two upper teeth. With her curly dark human hair wig and her brown sleep eyes in addition to her large size, she is a doll to be remembered. She is also beautifully dressed in a blue satin frock and a dark coat with matching hat. *J. C. Collection.*

166

Illustration 164. This 19in (48.3cm) girl is incised "8// 166." She is a pretty shoulder head model on a gusseted kid body. This mold does not turn up too often, but has the usual Kestner characteristics including the glossy brown molded eyebrows. Her open mouth has a peaked upper lip and four inset teeth. Dolls such as this one were offered in a 1904 wholesale catalog. Kestner dolls were referred to by name as the "most famous line of kid body dolls made in Germany." They were noted for their beautiful faces "drawn from life." The kid bodies were cork-stuffed with gussets at the hips and knees; the larger dolls had gussets at the elbows also; the short lower arms were bisque. The dolls had what was described as fine curly hair, probably real mohair. All had stockings and patent leather shoes except the very cheapest model, which has imitation stockings, perhaps cloth lower legs. The dolls also had fine modeled bisque heads and sleep eyes. Wholesale prices ranged as follows. Retail prices would be 50 to 100 percent higher.

11in (27.9cm)	$ .19 each
13in (33.0cm)	.37
15in (38.1cm)	.50
17in (43.2cm)	.70
19in (48.3cm)	1.00
22in (55.9cm)	1.50
24in (61.0cm)	2.00
25½in (64.8cm)	3.00

It is interesting to note how rapidly the price increased with the size. The doll shown is named "Roberta" after her original owner and she is still just as she was when she was new except for her shoes. *Joanna Ott Collection.*

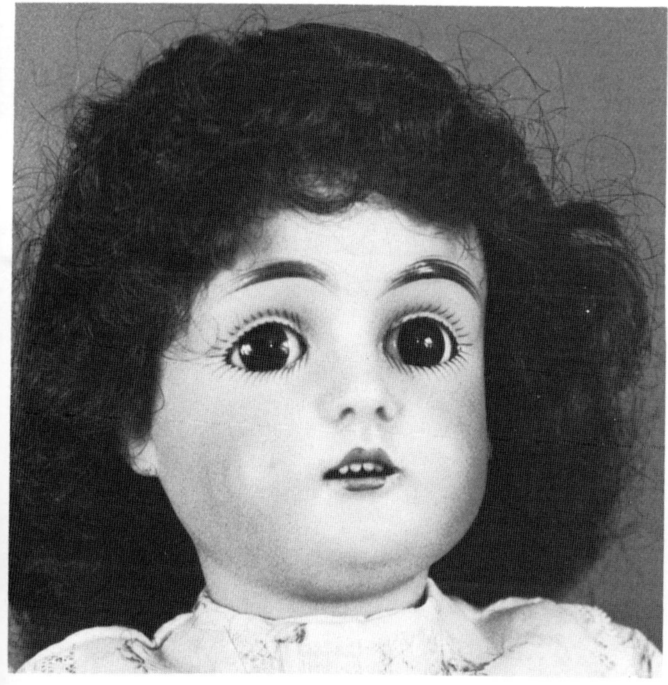

Illustrations 165 and 166. Mold 167 is really a favorite one with Kestner collectors who love the pertness which shows in this face. In fact, dolls from this mold seem so alive and real that I have heard them referred to as "character" dolls, when, of course, they really are not, but it is easy to see why there could be confusion. This 18½in (47.0cm) girl has brown sleep eyes, long painted lashes, and lovely glossy brown eyebrows. Her lips are parted quite a bit and she has four upper inset teeth. Her bisque is lovely with nicely tinted cheeks. Mold 167, a socket head used on a jointed composition body, is considered by many to be the sweetest of the Kestners faces. Her white lawn dress has sprigs of blue flowers for decoration and although new is made in an appropriate style. *H & J Foulke.*

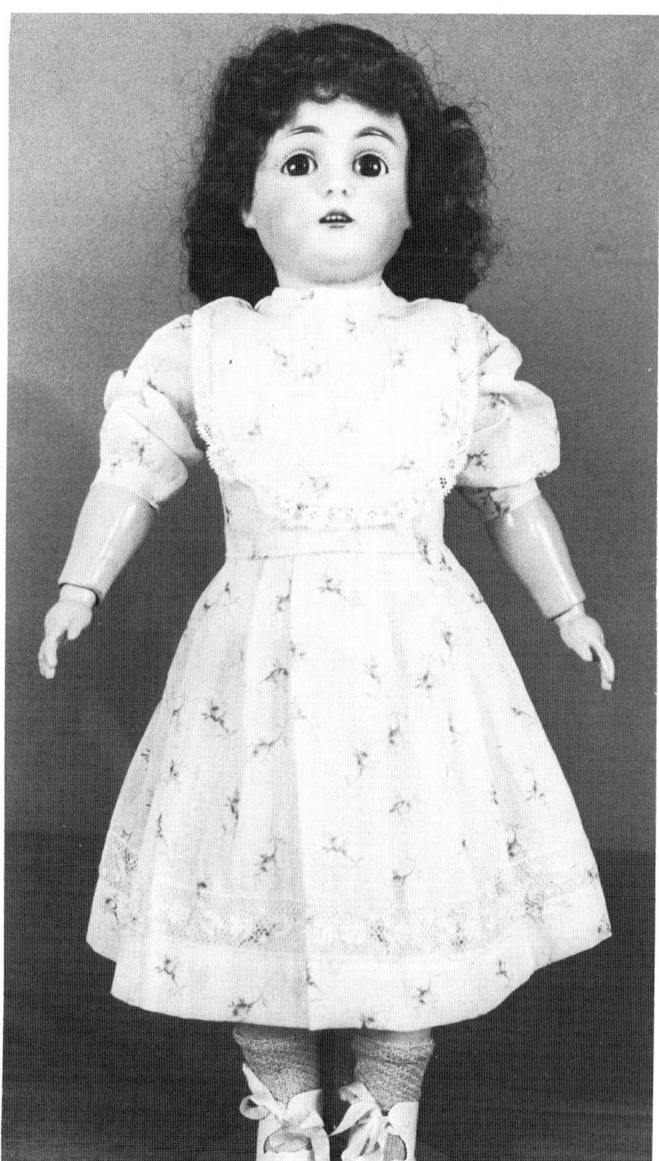

Illustration 166.

Illustration 167. Another pert little 167, this girl is just 12in (30.5cm) tall. She is wearing what appears to be an original white cotton dress with lace, ruffles, and inserted satin ribbon for trim. She also has dark eyes and hair. *H & J Foulke.*

Illustration 168. This sweet little girl is 13in (33.0cm) tall incised: "a½//made in 4½//Germany//167." She has an open mouth with four molded-in square-cut upper teeth. Her eye sockets seem deep and her eyebrows have been painted lower than usual. Her blue eyes go nicely with her new blonde human hair wig. She is on a jointed composition body with the red stamped "Germany 2/0." Her blue cotton dress is newly made of old fabric. *Edna Black Collection.*

Illustration 169. This little girl is just 8½in (21.6cm) and is the smallest size I have seen in this mold. She also has an open mouth with four molded upper teeth. Her eyebrows are hidden under her bangs, but they are blonde, quite light for a Kestner doll, and match her blonde braids. *Emma Wedmore Collection.*

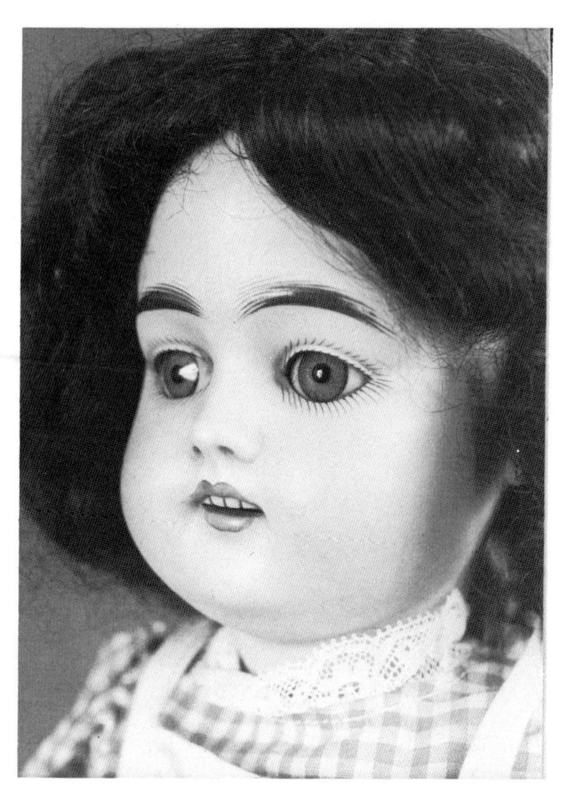

Illustrations 170 and 171. This is another sweet Kestner face which does not turn up too often. This particular 20in (50.8cm) doll has beautiful bisque as do most Kestner dolls of this period. Her blue sleep eyes are surrounded by long painted eyelashes, and topped by glossy brown molded eyebrows. Her open mouth has four molded-in square-cut upper teeth and an upper lip with turned up corners. The close-up view gives a good look at all of these features. She is a socket head size E 9 on a jointed composition body stamped in red "Germany 2½." *H & J Foulke.*

Illustration 171.

Illustration 172. This 19in (48.3cm) doll is another lovely example of a 168 mold. She also has blue sleep eyes and very thick eyebrows, but with many individual brush strokes rising high at the inner corners. Her open mouth has four inset rather than molded-in teeth. Her blonde mohair wig is original and she is wearing a lovely old dress with satin rosettes for trim. Her head size is D½ 8½. *J. C. Collection.*

170

Illustrations 173 and 174. This little 6¾in (17.2cm) doll was certainly a surprise as I had never run across a Kestner like her before on this unusual body; her head turns as her legs walk. She has pegged shoulders, molded black stockings and tan two-strap shoes. Her head is incised "made in// Germany//170 2½ o." Her plaster dome is covered by her original blonde mohair wig styled with braids. Her eyebrows are lightly stroked; she has an upturned upper lip and four inset teeth. Her eyes are brown. What a gem she is! According to the Ciesliks, this body was registered in 1903 as DRGM 212 878. *Joanna Ott Collection.*

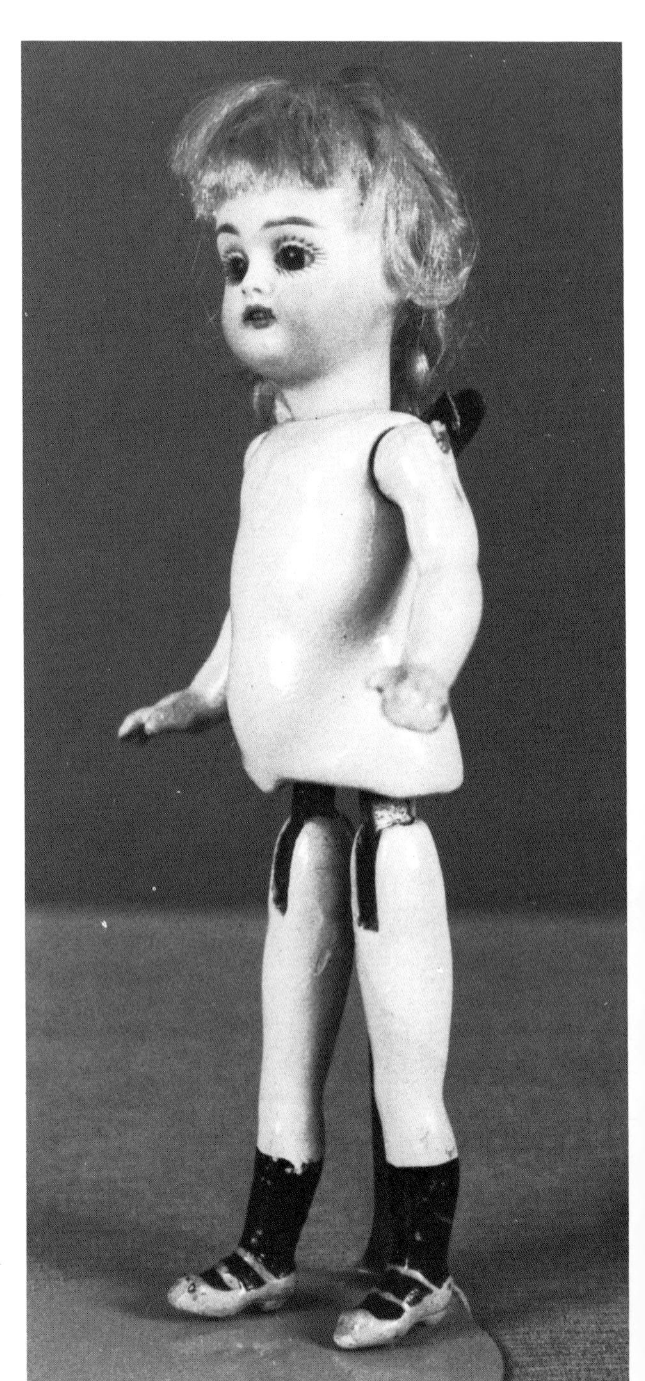

Illustration 174.

Illustration 173.

171

Illustration 175. Judging from the number of mold 171 dolls found today, it was a very popular face made over a long period of years. Molds 171 of the socket heads and 154 of the shoulder heads are the most usually found Kestner numbers. This is a very attractive mold with a longer, thinner face than that of other Kestner dolls. Her other distinguishing feature is her more open mouth with curved upper lip; usually the four inset upper teeth are curved in also. Her eyebrows are usually molded, and there is a dimple in her chin.

This is also the mold that was used in the 18in (45.7cm) size for *Daisy* the doll of the Lettie Lane Paper doll printed in the *Ladies' Home Journal* and offered as a premium in 1911. The pages pertaining to *Daisy* are shown in Illustrations 186 through 190. (Dolls from other companies were also used.) Even though only the 18in (45.7cm) size was used for *Daisy*, collectors have the tendency to refer to all dolls of this mold as *Daisy. H & J Foulke.*

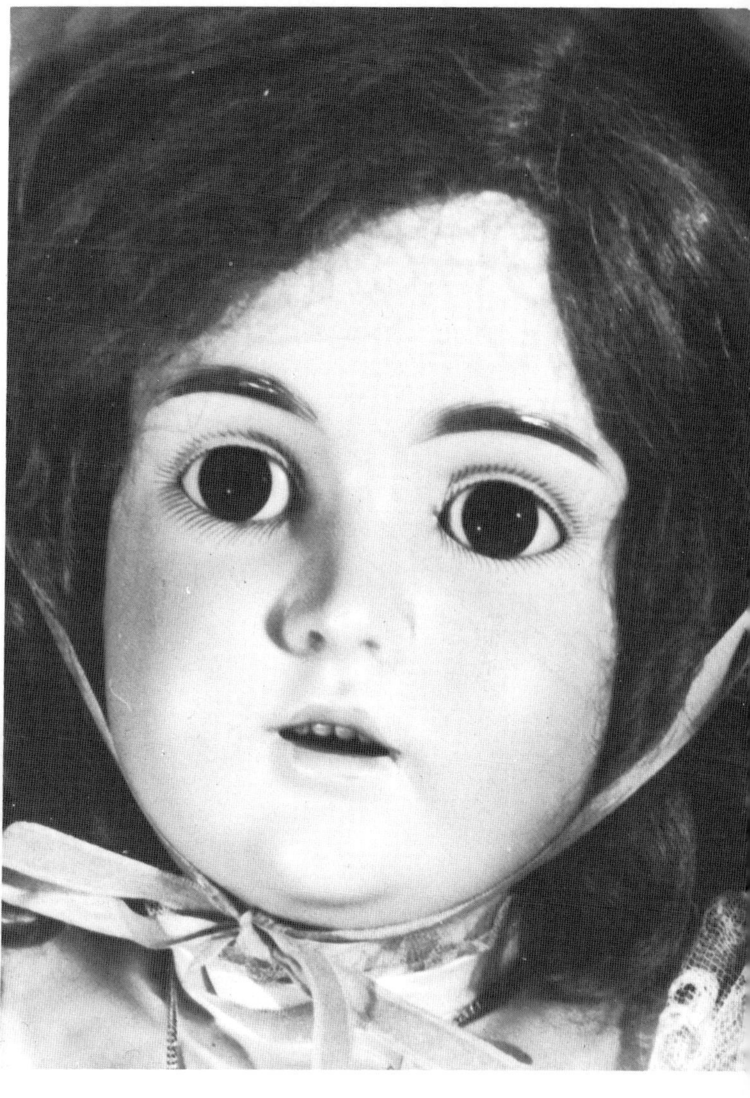

Illustration 176. This 24in (61.0cm) 171 is all original. Her curly blonde mohair wig still has the original blue satin ribbon. Her blue sleep eyes have real upper lashes, painted lower ones; the eyebrows are molded, but the painting is not as heavy as that of the earlier dolls. She has beautiful bisque and lovely coloring. Her blue dotted swiss dress has lace insertion and shirring in the waist, lots of bodice; puffed elbow length sleeves with lace edges; and a full skirt with lace insertion and five rows of tucks. A blue satin sash and a pearl necklace complete her ensemble. She is wearing black lace stockings and shoes. Her head size is H 12. *H & J Foulke.*

Illustration 177. This 21in (53.3cm) 171 is size F½ 10½. Her blue sleep eyes are somewhat smaller than expected, and she has real eyelashes as well as painted ones. Her blonde human hair braids are very attractive. Many dolls came with long braided hair, but collectors today tend to curl the hair or comb it instead of leaving it in its original style. She is a very dressed up little girl in her hat and coat with a wide lace collar. Her jointed composition body is stamped in red "Germany 3." *J. C. Collection.*

Illustration 178. This 27in (68.6cm) girl appears to be one of the later Kestner dolls. Her smooth bisque has an overall ruddy complexion with rosy cheeks. This high coloring is usually associated with the later dolls. To substantiate this she also has molded glossy eyebrows of a warm brown color, and real, rather than painted, upper eyelashes. Her long brown mohair curls help to accentuate her long, thin face. Her parted lips show four curved upper teeth. She is on an excellent jointed composition body stamped in red "Germany 5." Her head size is K 14. *H & J Foulke.*

Illustration 179.

Illustration 180.

Illustrations 179 and 180. This is another lovely example of the 171 mold. She has brown molded eyebrows and a matching brown human hair wig, center-parted with lovely long curls. Her mouth is nicely shaped with two distinct peaks forming her upper lip. She is 24in (61.0cm) tall and was obtained from her original owner in completely original condition. Her white lawn dress is lovely with lots of lace and tucks for trim. *Esther Schwartz Collection.*

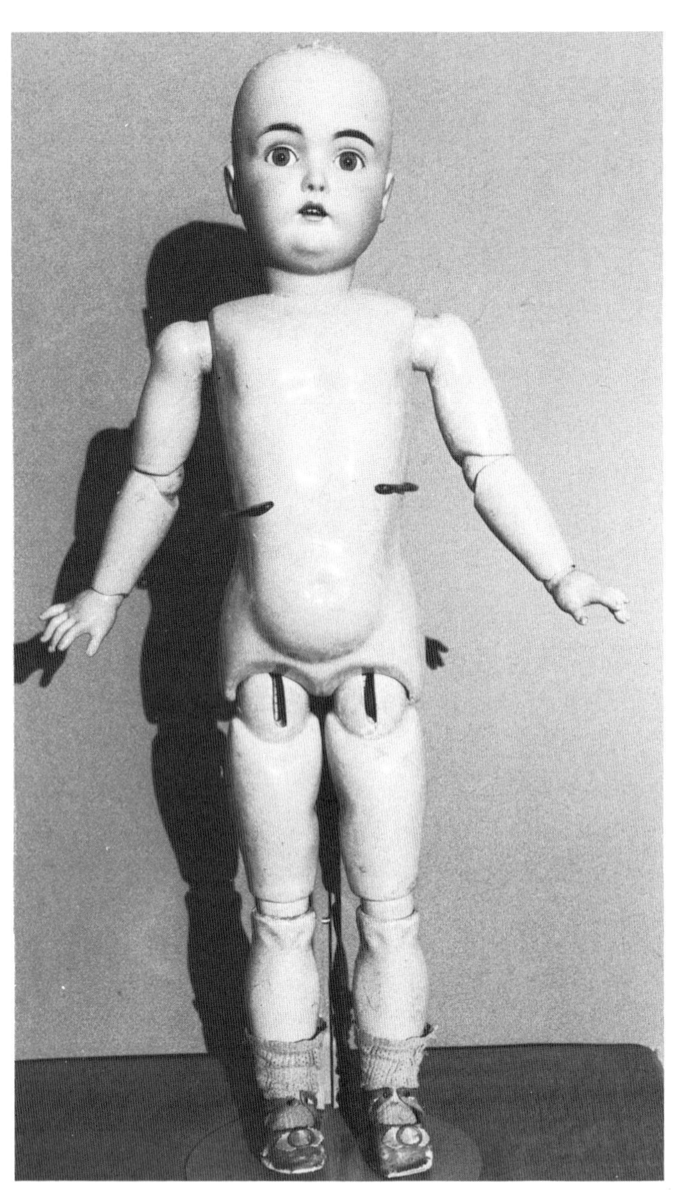

Illustration 181.

Illustrations 181, 182, 183, 184, and 185. This 29in (73.7cm) girl apparently represents a little earlier period in making. Although she has the molded eyebrows, she has very pale bisque and light cheek coloring, an attribute usually associated with the earlier dolls. She has both upper and lower painted lashes. Although we know that the real lashes are a later characteristic, we should not forget that the painted lash dolls continued to be made at the same time as they were less expensive models. Here we are also showing the molding of the ear, which is really hardly anything considering the large size of the doll, as well as the plaster pate and the incised marking. The two holes at the nape of her neck were for tying her sleep eyes in place so that they did not bang around on the long trip from Germany to the United States. She is also a very good study of the typical Kestner child body. Her shoulders are made very broad by the large attached ball joint at her shoulders. She has separate balls at her elbows and knees. Her hands and fingers are nicely detailed with red outlining her nails and knuckles. Her torso has detail modeling on the rib cage, stomach, and buttocks which is straight and flat across the bottom and on which is stamped in red "GERMANY" in the red box with the size number beneath used by Kestner. As typical of Kestner, the legs always have better molding than the arms (the lower halves of the arms are turned wood). She has fat rolls on her thighs and cute dimpled knees. *H & J Foulke.*

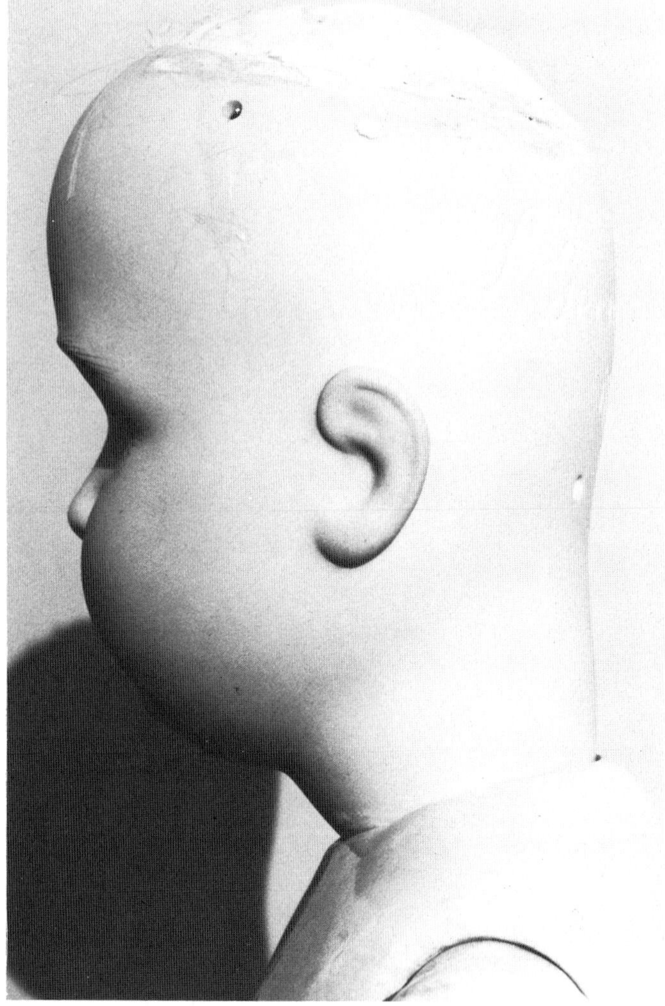

Illustration 182.

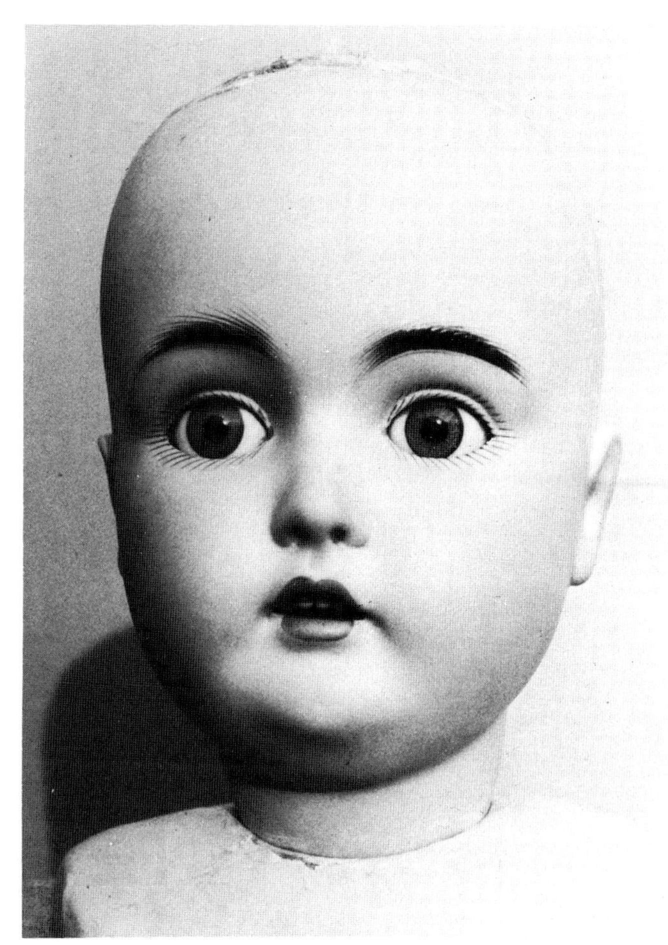

Illustration 183.

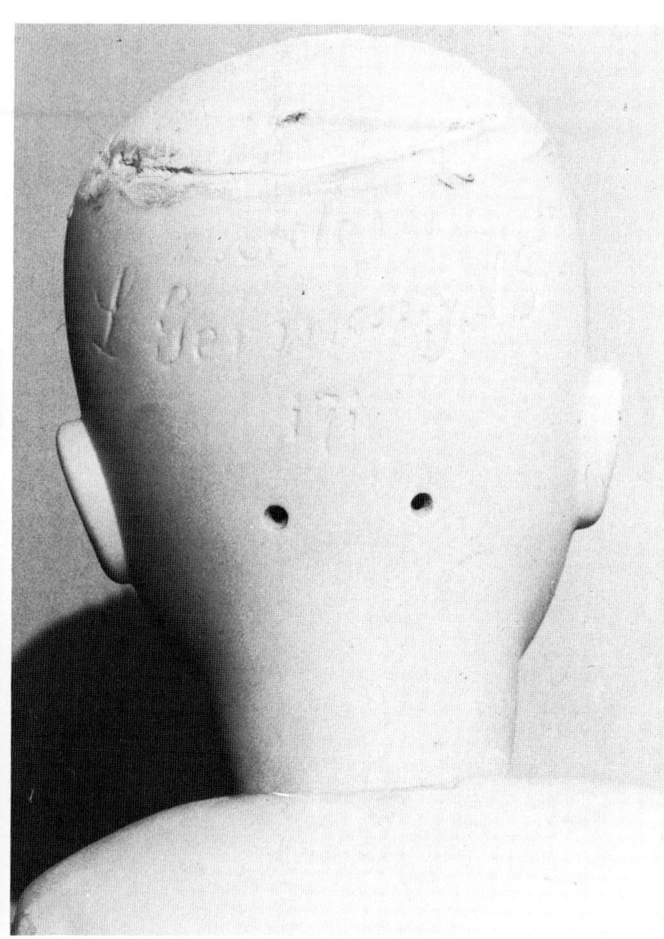

Illustration 184.

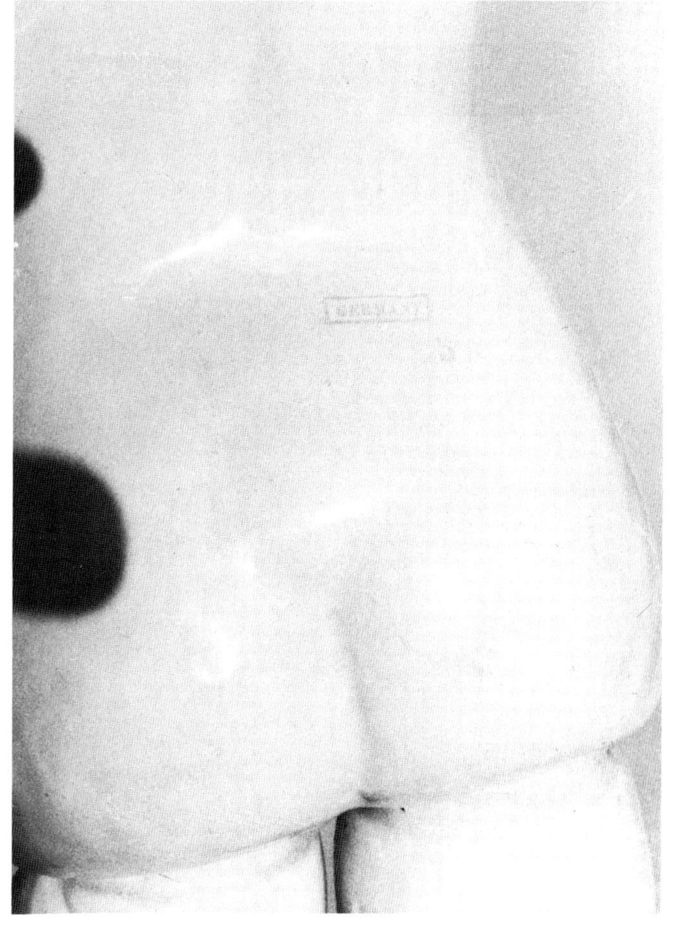

Illustration 185.

Daisy, the Doll That Came to Life

The *Ladies' Home Journal* had been running Sheila Young's paper doll pages of Lettie Lane and her family since October 1908. Lettie Lane had a doll named *Daisy*. Then someone at the *Journal* had a marvelous promotion idea: they would offer a real *Daisy* doll as a premium to little girls who sold three subscriptions to the *Ladies' Home Journal. Daisy* was offered in the March 1911 *Journal;* the caption read: "Lettie Lane Presents Her Most Beautiful Doll to Every Little Journal Girl." The little girls would send in three names and $4.50 and they would receive *Daisy* free. She was described as a fine bisque doll with lovely golden curls and real hair eyelashes; she had pearly white enameled teeth, and cheeks as pink as roses. Just 18in (45.7cm) tall, she was a fully ball-jointed doll. Her blue eyes would open and close. *Daisy* came dressed in a white muslin shift with blue trim on her bodice and belt; she wore white lacy stockings and white slippers. Patterns for the clothes shown in the March 1911 *Journal* and for underwear were sent free with each doll.

In the June 1911 *Journal,* an announcement was made that owing to the popularity of the doll, stock had run out, but "orders placed in Germany, where Daisy is made in the quaint villages, will bring a fresh supply soon." They expected dolls by September 1, which indicated that they were running three months behind on orders. By October the editors were announcing: "Thousands of Dolls, Fresh From the German Villages Where They are Made, are Now in THE JOURNAL'S Offices Ready for Immediate Shipment." This great demand for the doll doubtless accounts for the fact that dolls by several makers including H. Handwerck and Kestner were used. The Kestner doll offered as *Daisy* was mold 171. Only the 18in (45.7cm) size is correctly referred to as *Daisy*.

The Doll That Has Come to Life

Lettie Lane Presents Her Most Beautiful Doll to Every Little Journal Girl

By Sheila Young

Illustration 186. *Daisy*, the real bisque-headed doll, was offered in the March 1911 *Ladies' Home Journal*. She was shown in full color with four outfits designed by Sheila Young. At the top row she had a green gingham morning dress with a striped lawn guimpe. For afternoon wear there was a light blue lawn dress with small front pleats and white collar and cuffs with hand done embroidery of eyelets and scallops, which were described as simple enough for nearly all little girls to do! On the botton row was a tan belted side-buttoned coat with white collar and cuffs and contrast buttons, bow, and belt. A very best dress was shown of pink silk with lace trim, satin bows, and front pleats.

Lettie Lane's Most Beautiful Doll as a Bride

The Doll That Has Come to Life for Every Little Journal Girl

By Sheila Young

Daisy and Her Clothes

How You Can Have Daisy

Illustration 187. In the April 1911 *Journal, Daisy* was presented as a bride wearing a white satin bridal gown with a lace underslip and chiffon veil. On the top row was a gown for evening wear made of white swiss over pink silk. For going-away *Daisy* would wear a gown of tan cloth with brown silk trim. On the bottom row was a morning dress of striped challis or gingham with solid contrasting trim. Her best Sunday dress was of blue and white silk with a wide sash and three rows of ruffles at the hem.

Lettie Lane's Doll in Her Vacation Clothes

Illustration 188. From the June 1911 *Journal* came the page showing *Daisy* in her vacation clothes. She was wearing her afternoon dress of white lawn with blue satin ribbon inserted on the shoulders and a white guimpe. On the top row was a party dress in a yellow flowered dimity with satin ribbon trim and a large side bow. For afternoon wear she had a pink plaid gingham dress trimmed with bands of embroidery. On the bottom row was a blue serge bathing suit trimmed with red, white, and blue braid. She also had a cover-all apron of blue and white checked cotton with white poplin collar, cuffs, and pockets.

Lettie Lane's Doll in Her School Clothes

Any Little Girl May Now Have the Real Daisy for Her Very Own

Illustration 189. The October 1911 *Journal* presented *Daisy* in a blue serge sailor dress with a long collar and red silk knotted ribbon. On the top row was a double breasted brown wool belted coat and a one-piece apron in a cotton print to wear over her frocks to keep them clean. On the bottom row was a Kate Greenaway frock made of gay flowered or striped lawn with a white lawn guimpe and a ribbon sash. She also had a brown gingham dress with pleated bodice and skirt and wide belt.

A MERRY CHRISTMAS FOR THE CHILDREN

Lettie Lane's Most Beautiful Doll

In Her Party Clothes

Illustration 190. The *Journal* for December 1911 presented *Daisy* in Christmas party clothes designed by the fashion editors, but drawn by Sheila Young. *Daisy* was shown in her dinner dress of pale green challis with lace collar and cuffs. On the top row was a plaid woolen coat with red collar and cuffs. She also had a "Dolly Varden" dress made of pink flowered lawn with ruffles and a pink satin sash. On the bottom row were an Indian costume and a Red Riding Hood outfit.

173

made in
a. germany 4.
173.

Two dolls have been located with mold number 173. They are both 13in (33cm) tall. The mold is a sweet dolly face much like #174 with a punched-in chin dimple. It is a socket head on an excellent jointed composition body with the Kestner Germany mark. *Courtesy of Joan Williets.*

174

Illustration 191. This is a mold which is very hard indeed to find. A head with this number is shown in *Illustration 280* as a part of the character doll set. It would make up into an 11in (27.9cm) doll, and is incised "b made in 3// Germany//174." She has a plaster dome, blue sleep eyes, brown eyebrows, an open mouth with four upper teeth, and a dimple in her chin. Her light brown mohair wig is styled with braids. The 174 mold doll shown here is 14in (35.6cm) tall. She is a cute doll with brown eyes and a blonde mohair wig. Her mouth is long and narrow with four upper teeth. *Mary Lou Rubright Collection.*

192

Illustration 192. This is a puzzling mold. The Ciesliks attribute it to Kämmer & Reinhardt and date it 1892. The date seems appropriate for the quality and style of the doll. However, K & R did not have a porcelain factory and did not make their own heads. It is very possible that Kestner made these heads for Kämmer & Reinhardt. It certainly does have the general look of a Kestner doll and is included here because its origin is still not clear. The heads are incised only "192" with no Kestner initials or letter/number sizing. Although some dolls have plaster domes, not all do. None are on marked Kestner bodies but, of course, they would not be if made for K & R. However, if the doll is a Kestner, the mold number is fairly high for a doll of this period. The pictured doll is a rare variation of the 192 mold as she has a closed mouth with slightly parted lips. She is 19in (48.3cm) tall. *Karen Potthoff Collection. Photograph by James Brownell.*

Illustration 193. This 22in (55.9cm) girl has a very high cut crown with a plaster dome. Her lovely blonde mohair wig is the same type found on dozens of Kestners. She has deep brown eyes and dark eyebrows with individual strokes at the inner corners and all across the top. They are also nicely tapered from thick at the inner corners to very thin at the outer corners. Her open mouth has four upper teeth and the upturned Kestner corners. She is wearing her original gold silk dress. Her body is unmarked. *J. C. Collection.*

Illustration 194. This is a sweet little 7in (17.8cm) all-original 192 girl. She has dark one-stroke eyebrows, very long painted lashes, and brown sleep eyes. Her open mouth has three upper teeth which do not show in the illustration. Her head has side holes for stringing. She has a five-piece composition body in excellent condition with unusual painted dark blue ribbed stockings and tan and brown low shoes with heels and painted laces. She is in a completely original outfit. Her white lawn dress has three-quarter length sleeves and a satin ribbon sash. Her matching hat has a satin ribbon and a paper brim. Her olive green felt cape has a lace collar. She is wearing a necklace and bracelets. Her blonde mohair wig is over a cardboard pate. *H & J Foulke.*

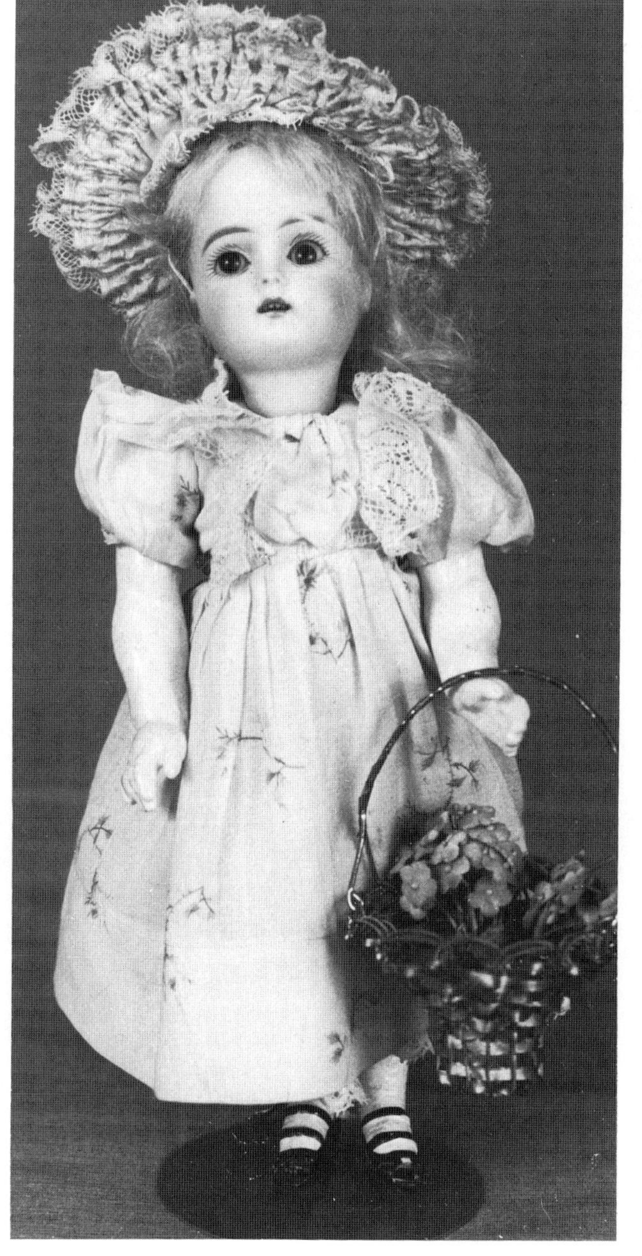

Illustration 195. This 9½in (24.2cm) 192 is also on a well-shaped five-piece composition body with molded stockings and straps but it is different from the body on the doll in Illustration 194. This girl has brown sleep eyes and blonde stroked eyebrows which match her blonde mohair wig. Her open mouth shows four upper teeth and has upturned corners; she has pierced ears also. She is wearing a lovely old cotton print dress and matching hat. *Jane Alton Collection.*

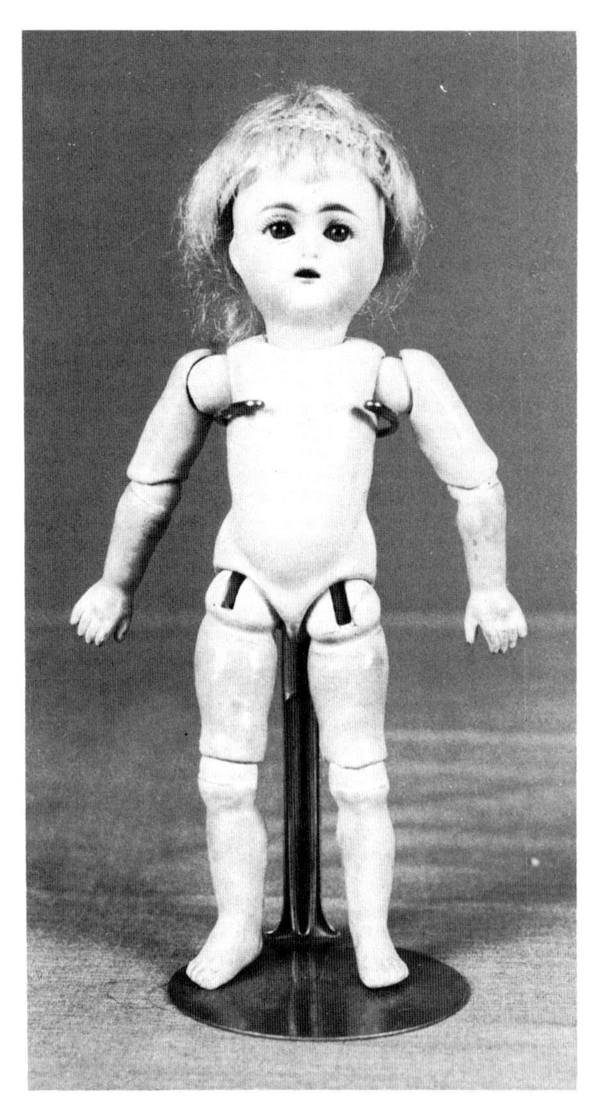

Illustration 196. The cut of the features and the decoration cause this 8½in (21.6cm) 192 to have a different look. The open mouth is smaller with the upper teeth set quite far back. The eyebrows are painted lower, but are shaped like those on the large 192. The jointed composition body is excellent quality with separate balls at the shoulders and attached balls at the knees. *Emma Wedmore Collection.*

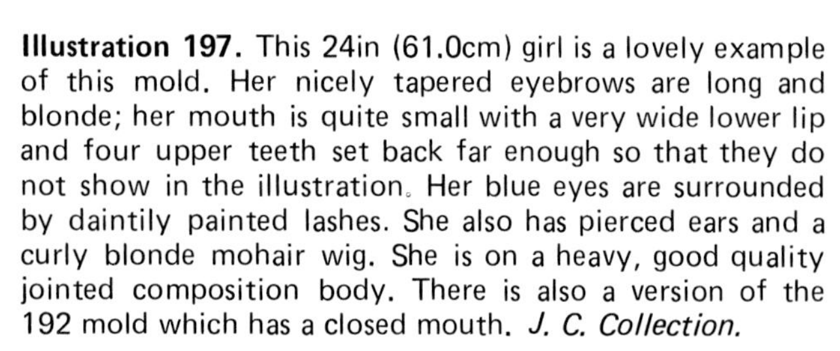

Illustration 197. This 24in (61.0cm) girl is a lovely example of this mold. Her nicely tapered eyebrows are long and blonde; her mouth is quite small with a very wide lower lip and four upper teeth set back far enough so that they do not show in the illustration. Her blue eyes are surrounded by daintily painted lashes. She also has pierced ears and a curly blonde mohair wig. She is on a heavy, good quality jointed composition body. There is also a version of the 192 mold which has a closed mouth. *J. C. Collection.*

195 and 196

These two molds are exactly alike except for the fact that 195 is a shoulder head model to be used on a kid body and 196 is a socket head model to be used on a jointed composition body. The mold has curved oblong slits on the lower forehead; hair placed behind them would go through the openings to make natural eyebrows. According to the Colemans, Louis Steiner patented this idea in the United States and Germany in 1910, but the patent was assigned to George Borgfeldt who represented Kestner. Armand Marseille also made dolls with hair or fur eyebrows. The lack of upper painted lashes indicates that these dolls also should have real hair upper lashes. These two features added some cost to the doll, making her a luxury item. In a 1910 wholesale catalog Kestner's "Excelsior" bisque head on a "French" jointed body (another term for a ball-jointed composition body) with a sewn wig, sleep eyes, shoes, stockings, and a fine hemmed chemise was priced at:

$1.80 for a 21½in (54.6cm) doll
2.25 for a 24in (61.0cm) doll
3.00 for a 27in (68.6cm) doll
3.75 for a 31in (78.7cm) doll

With the addition of real eyebrows and eyelashes, the price was:

$2.40 for a 22½in (57.2cm) doll
3.35 for a 25in (63.5cm) doll
4.00 for a 29in (73.7cm) doll

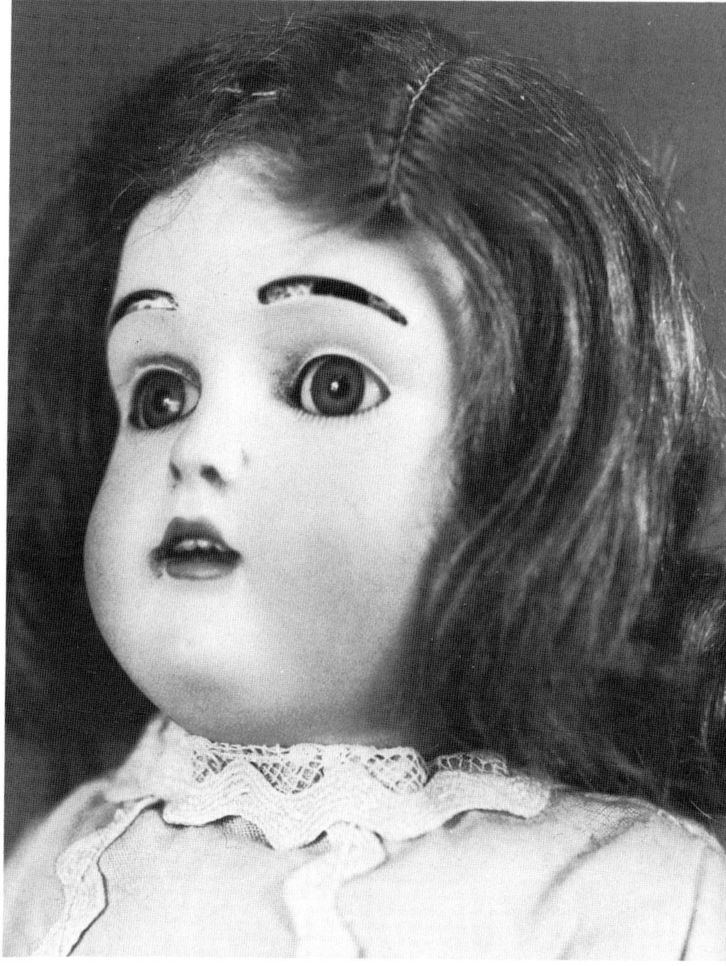

Illustration 198. This 13in (33.0cm) example of mold 196 is size A 5. She is on a jointed composition body with "Germany" in an oblong box, all stamped in red. Unfortunately as with most dolls of this type, the eyebrows and eyelashes have disintegrated, but the idea was an interesting one. As not too many of these slit eyebrows dolls are found today, perhaps the idea was not too popular. Many collectors today feel that these dolls look quite awkward. This doll has a light brown mohair wig over her plaster dome, and an attractive open mouth with four upper teeth. *Emma Wedmore Collection.*

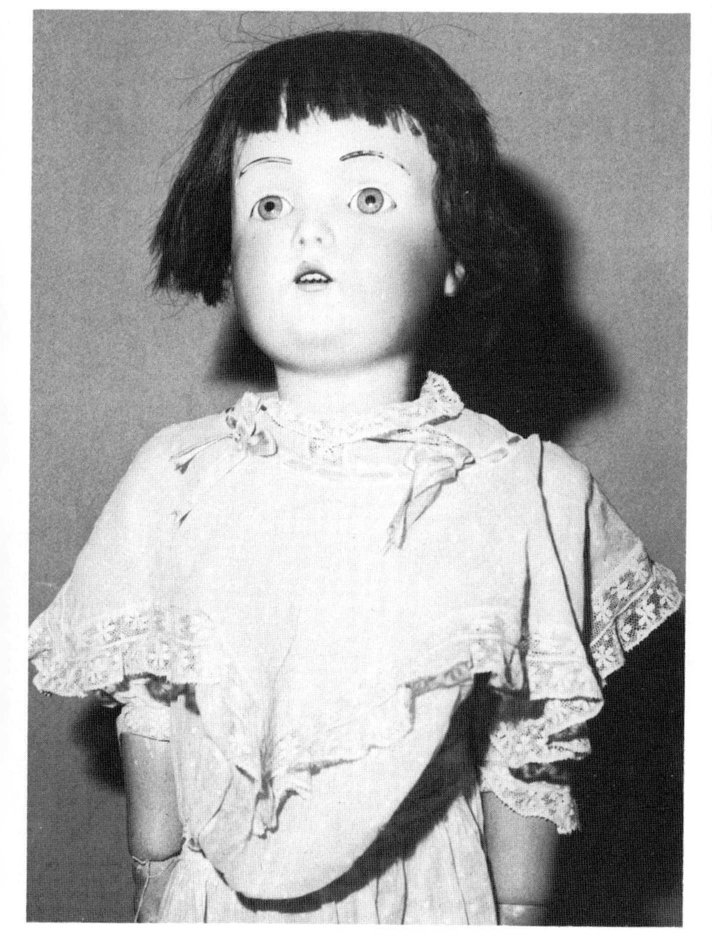

Illustration 199. This large 28in (71.1cm) example of mold 196 has lovely bisque with rosy tinted cheeks. She also has the open slits for insertion of eyebrows and should have real upper lashes. Her sleep eyes are the gray-blue shade often found in Kestner dolls. The brown mohair wig appears to have been bobbed long ago in the past by little fingers trying to make dolly look modern! Her lovely white dotted swiss dress has a very wide bertha with double rows of lace trim and satin ribbon insertion. *H & J Foulke.*

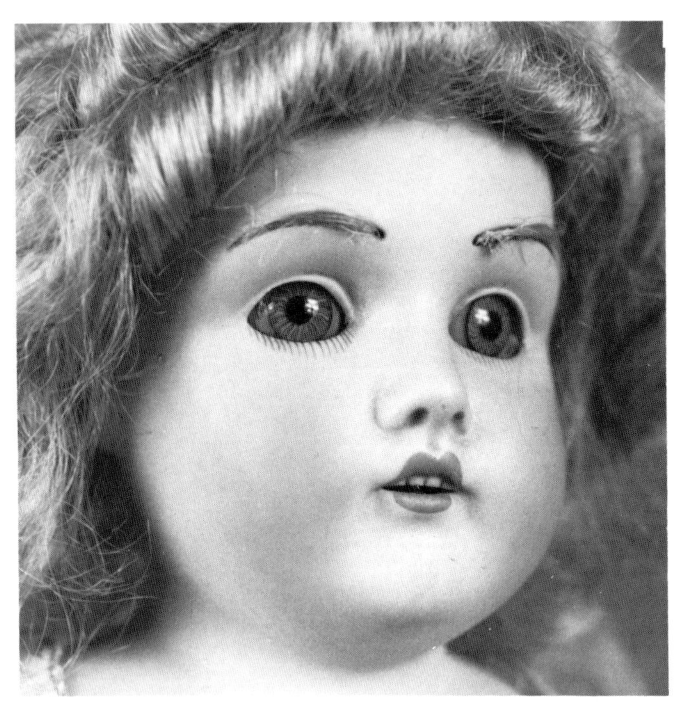

Illustrations 200, 201, and 202. This 17in (43.2cm) girl is an example of mold 195, the shoulder head model. She is incised high up around the crown opening "5 ¾ 195. Dep." and "Made in Germany" under the kid on her shoulder plate. She has the usual Kestner plaster dome. She has the open slits for inserting eyebrows and no painted upper lashes, since it was intended that she have real eyelashes as well. Her open mouth has four molded upper teeth. She is on the rivet-jointed Kestner kid body which still retains it original crown and streamers label. *Dr. Carole Stoessel Zvonar Collection.*

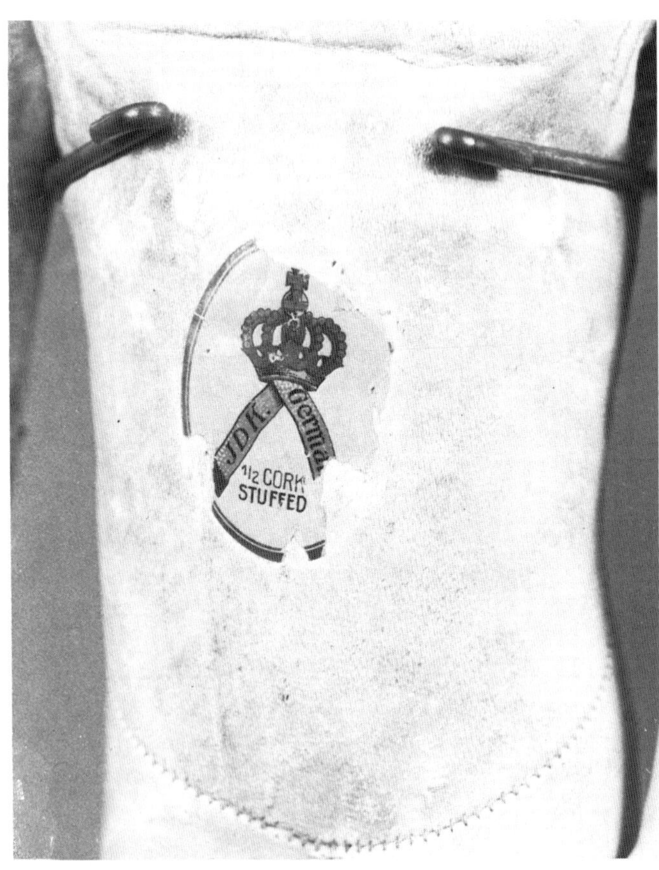

Illustration 202.

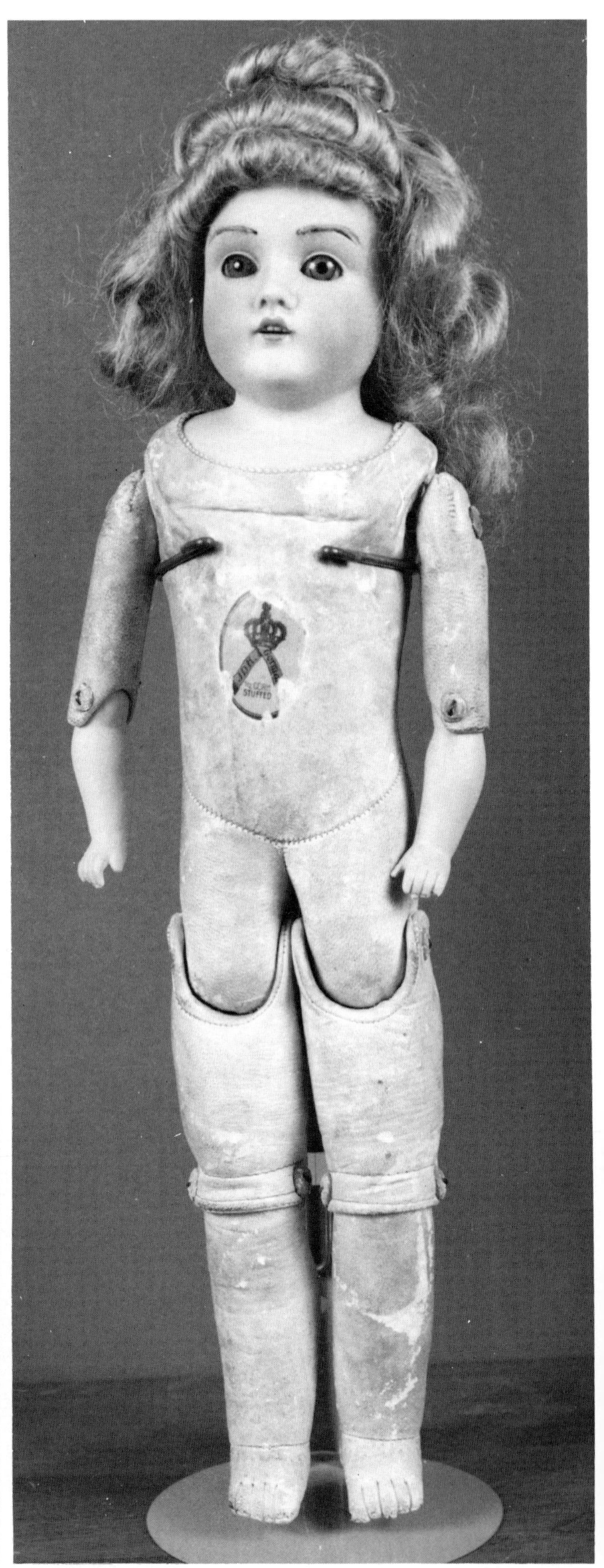

Illustration 201.

Playthings for January 1910 carried an advertisement by George Borgfeldt for dolls with natural hair eyebrows as "The greatest novelty of the century in the doll line." The ad shows a drawing of a little girl holding a doll and pointing to its eyebrows saying: "My dolly has eyebrows just like mine." These dolls sold for $1.00 up.

214 and 215

These are an interesting pair of socket-head molds as they bring up the question, are these character faces, or are they dolly faces? The easier thing to do is to beg the question and say that some dolly faces have more character than others! And these dolls certainly do not seem to fit in with the 100 series of conventional faces. If Kestner did follow numerical order in his mold making, then they were already well into producing character dolls when they made these two molds and one wonders why they would need two new dolly faces when they already had a substantial number of them in stock. Perhaps these faces were merely an attempt to modernize the look of the old-fashioned doll face.

Illustration 203. This nice large example of mold number 214 has quite a pleasant face with a chubby healthy look. Her large brown sleep eyes give her an alert expression; she has both upper and lower painted eyelashes; her mouth is cut fairly wide open to show her four inset teeth and tongue. She has glossy molded eyebrows and a blonde mohair wig. She is wearing a lovely old nightdress and cap with lace and ribbon rosettes for trim. *Sheila Needle Collection. Photograph by Morton Needle.*

Illustration 204. This doll presents an excellent example of how two dolls of the same mold can look so different. She definitely has smaller eye sockets which give her a more wistful look. Her blue sleep eyes have traces left of real hair lashes on the top, and her lower lid has painted lashes. This eye treatment is also in contrast to that of the doll in *Illustration 203.* Another difference is in the size of her mouth which is smaller and somewhat pouty. She has molded eyebrows, but the painting on this series is not as heavy as those on the 100 series dolls. Her blonde human hair wig is original. She is 31in (78.7cm) tall and her jointed composition body has the red "Germany" stamp. The doll is incised "M made in 16//Germany//J.D.K.//214." *Rosemary Hanline Collection. Photograph courtesy of the owner.*

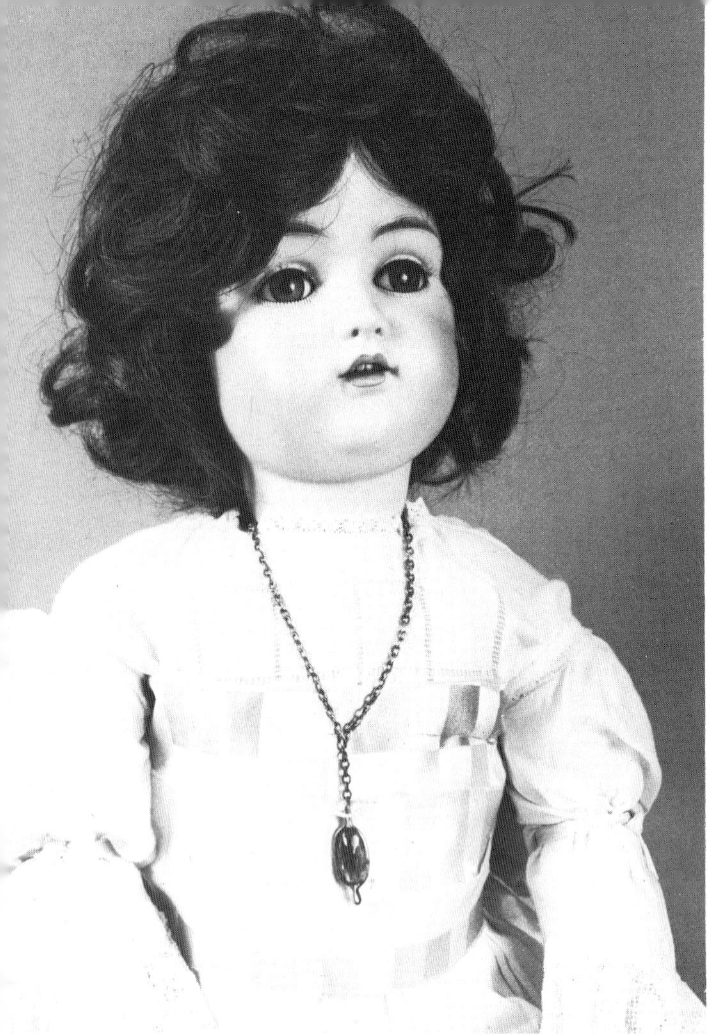

Illustration 205. Another example of mold 214, this girl is 25in (63.5cm) tall. She has light brown molded eyebrows made with many brush strokes, real upper eyelashes, and a brown human hair wig. Her lips are pursed as though she is just about ready to talk. Her lovely white cotton dress has much lace and openwork. *J. C. Collection.*

Illustration 206. While the other 214 mold dolls have the same style of incised mark, this 24in (61.0cm) one has a different type, including the mark of the Kestner & Co. porcelain factory. This side view shows the large but not intricately molded ear on this series and the nicely rounded full lower cheek. She has a brown human hair wig with braids which appears original, and brown sleep eyes with painted lashes as well as real upper lashes. As typical of most of the dolls of this series, she has quite red cheek color. *Emma Wedmore Collection.*

214
13

Illustration 208.

Illustrations 207 and 208. This number seems to be much harder to find than the 214 one. She is particularly interesting because she has cutouts in the bisque for inserting the fur eyebrows as well as real hair lashes. Kestner was offering a doll with real eyebrows in 1910, but this particular one probably dates about two years later if Kestner was making molds in numerical order. The 215 mold has a very appealing face and makes an attractive doll. This 25in (63.5cm) girl has an old human hair wig, and is appropriately dressed for a little girl of her time. Her incised mark is "K Made in 14//Germany//J.D.K.//215." *May Wenzel Collection. Photographs courtesy of the owner.*

Illustration 209. This 26in (66.0cm) example of mold 215 is really a gorgeous doll. She has light brown fur eyebrows, painted lower lashes, and real upper lashes to accent her blue sleep eyes. Her light brown mohair wig has lovely long curls. The modeling of her face is exceptionally good with indentations at the temples. She is completely original in her white lawn dress with lace trim. Her body is stamped in red in the Kestner manner with "Germany 6." *Dr. Carole Stoessel Zvonar Collection.*

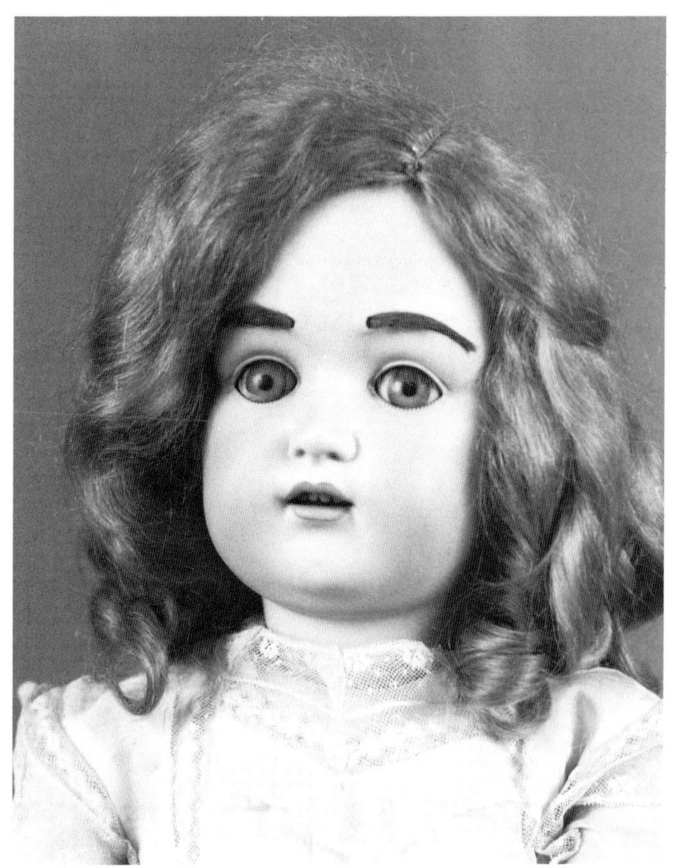

264

This socket-head child doll was made for Catterfelder Puppenfabrik by Kestner. She has sleep eyes with both real and painted eyelashes, an open mouth with four upper teeth, and pierced ears (not a usual Kestner feature). She is 17in (43.2cm) tall and has been appropriately redressed. *Sheila Needle Collection. Photograph by Morton Needle.*

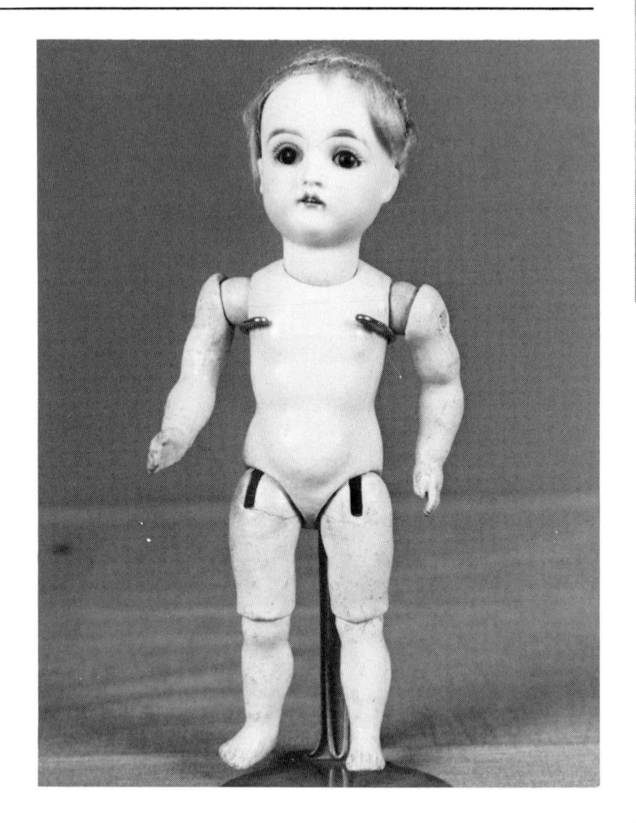

250

This socket head child was made for Kley & Hahn, a doll factory in Ohrdruf, the same town where Kestner & Co.'s porcelain factory was located. It carries the K&H *Walküre* trademark, which was registered in 1903 and predates this mold by at least 12 years. The face is somewhat of a departure from the typical pert Kestner look and may have been Kley & Hahn's own design. It is much more of an ordinary German dolly face than those of Kestner's own molds, although the bisque is of excellent quality. The dolls do not have plaster domes. *Dolly Valk Collection.*

No Mold Numbers

Lest we think that the Kestner firm made classification easy for doll collectors by using mold numbers on all of their open-mouth dolls, here are some without any mold numbers.

Illustration 210. This 8½in (21.6cm) child, although unmarked, has a face like that of mold 129. It has the same pert expression through the eyes, the same curved glossy eyebrows, and the same dainty mouth with peaked upper lip and four upper teeth. What is left of her blonde mohair wig covers a plaster dome. Her composition body is a variation of the one shown in Illustration 151. *Emma Wedmore Collection.*

Illustration 211. This gorgeous 18in (45.7cm) child is incised with size G 11, but no mold number. She has an attractive bow-shaped mouth with upturned upper lip and four inset upper teeth. Her eyes are small and almond-shaped, but her painted eyelashes are very long. Her arched eyebrows are also long with many brush strokes at the inside corners. With her light brown curled human hair wig, she is a very striking doll. Her jointed composition body is stamped in red "Excelsior//PRP No. 70685//Germany." *Richard Wright.*

Illustration 212 and 213. Completely original dolls are hard to find nowadays, so this one is a treasure indeed. Her head size is J 13 making her 21in (53.3cm) tall on a composition ball-jointed body stamped "Excelsior//Germany 3" in red. Her blue sleep eyes are accented by wide-spaced, but long and heavy eyelashes. Her original wig matches her glossy, long and thick eyebrows which nearly come together above her nose. Her white lawn dress has a dropped waist and is decorated with many tucks and ruffles as well as much lace. She is also wearing her original high-button shoes. *Joanna Ott Collection.*

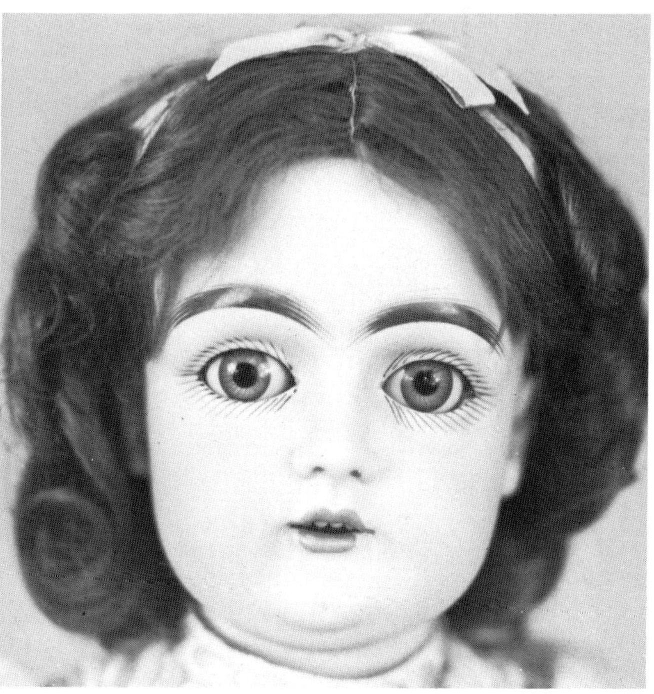

Illustration 213.

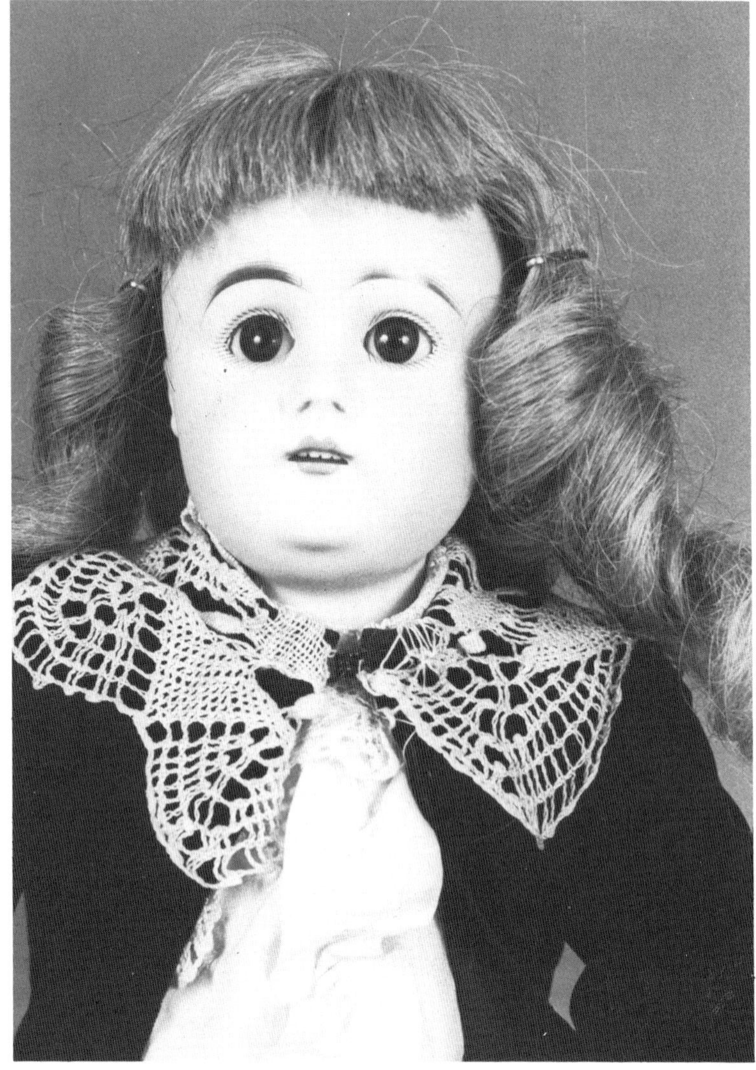

Illustration 214. This 18in (45.7cm) boy has been attributed to Kestner because of his plaster dome, incised marking, and quality of bisque, but he is somewhat of a departure from the usual Kestner. His head is wide at the top and definitely narrows to his cheeks. His molded ears are quite tiny, like those on the closed-mouth dolls. His eyebrows are lighter, more arched, and not so heavy as the usual Kestner ones. Yet he has the Kestner mouth with upturned upper lip and four inset upper teeth. *Jane Alton Collection.*

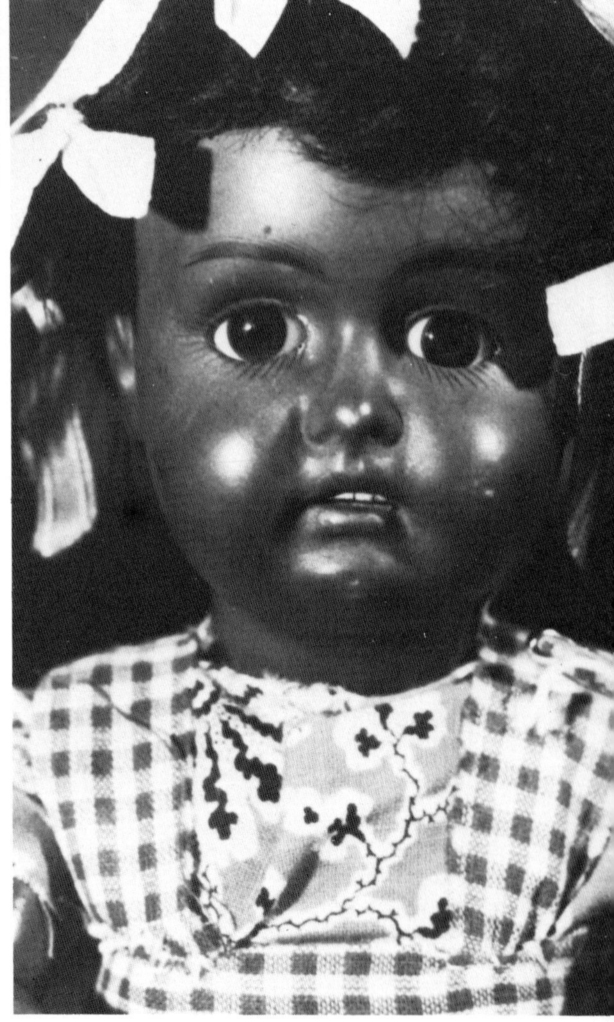

Illustration 215. This little girl is 12in (30.5cm) tall and has head size B 6. Of course she is rare because she has black tinted bisque. Kestner apparently did not make many black dolls, since very few appear in collections or on the market. She is a very attractive and pert little girl with her brown sleep eyes, very long black painted eyelashes, and brush-stroked eyebrows. Her open mouth shows upper teeth. She is wearing her original black mohair wig and is on a black jointed composition body. *H & J Foulke.*

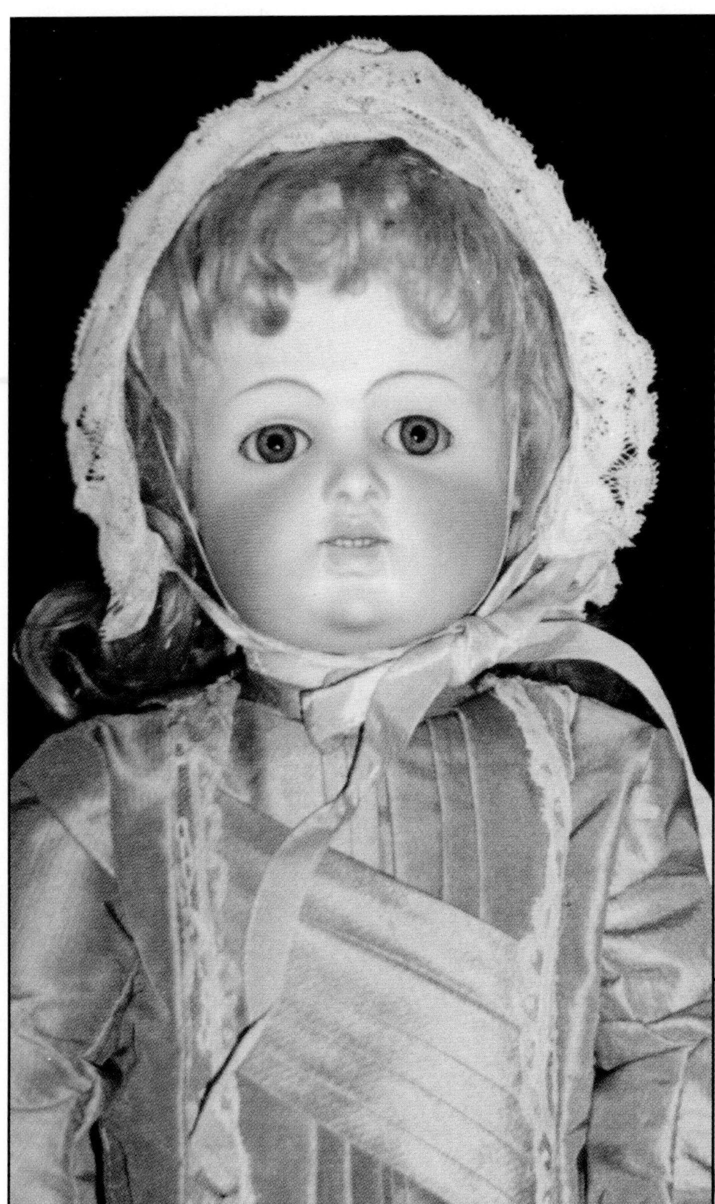

17in (43cm) Bru-type Kestner, open/closed mouth and painted teeth. *Courtesy My Doll Room.*

20in (51cm) closed mouth Kestner marked "13." *Courtesy Hearts Desire.*

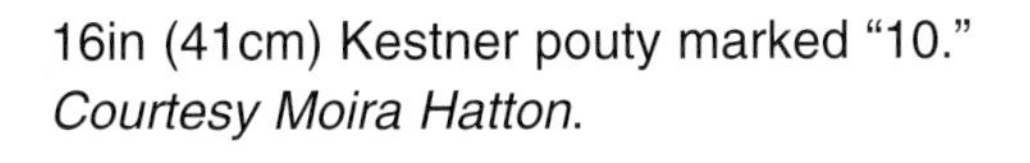

16in (41cm) Kestner pouty marked "10." *Courtesy Moira Hatton.*

25½in (65cm) closed mouth Kestner. *Courtesy Kate and Richard Smalley.*

25in (64cm) closed mouth Kestner. *Courtesy Dorothy Cohen Antiques.*

Closed mouth Kestner. *Courtesy Gigi's Dolls and Sherry's Bears.*

15in (38cm) closed mouth Kestner marked "7." *Courtesy Regina Steele.*

14in (36cm) closed mouth Kestner marked "13." *Courtesy Nancy Smith.*

27in (69cm) all original Kestner. *Courtesy El Nen Antiques.*

23in (58cm) Kestner closed mouth mold 128. *Courtesy Shelia Needle.*

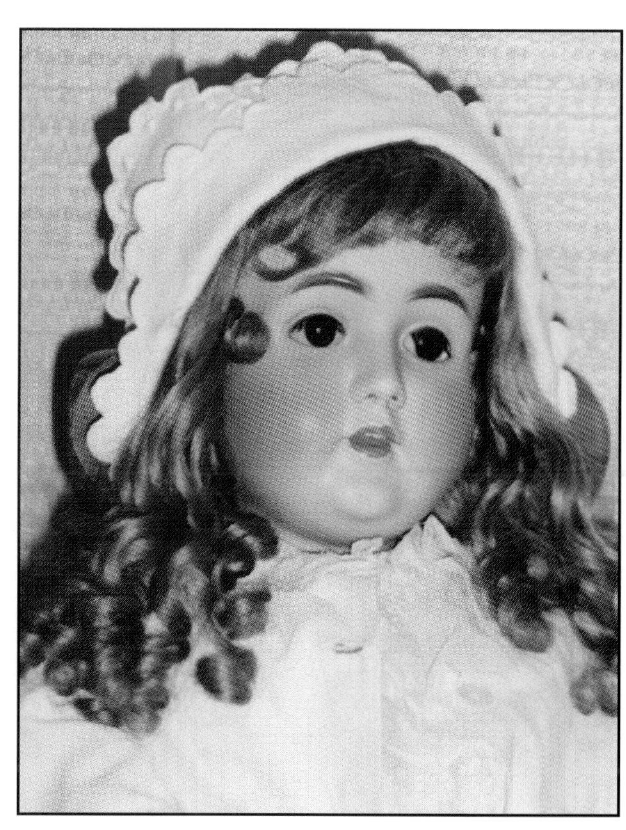

38in (96cm) Kestner mold 142 with hair eyelashes. *Courtesy My Doll Room.*

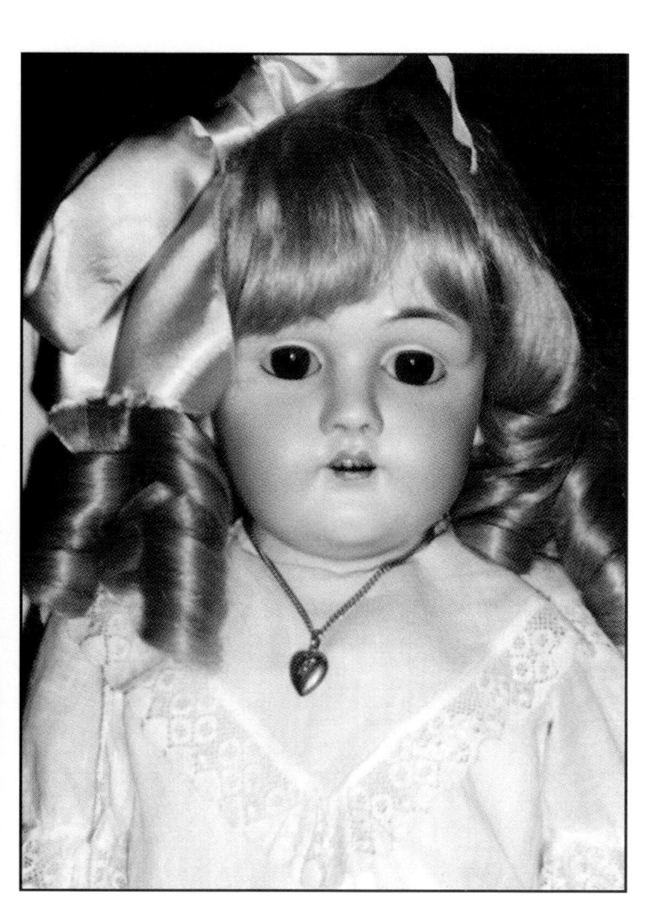

27in (69cm) Kestner mold 154. *Courtesy le Cheval de Bois.*

26in (66cm) JDK mold 167. *Courtesy Ann Condron.*

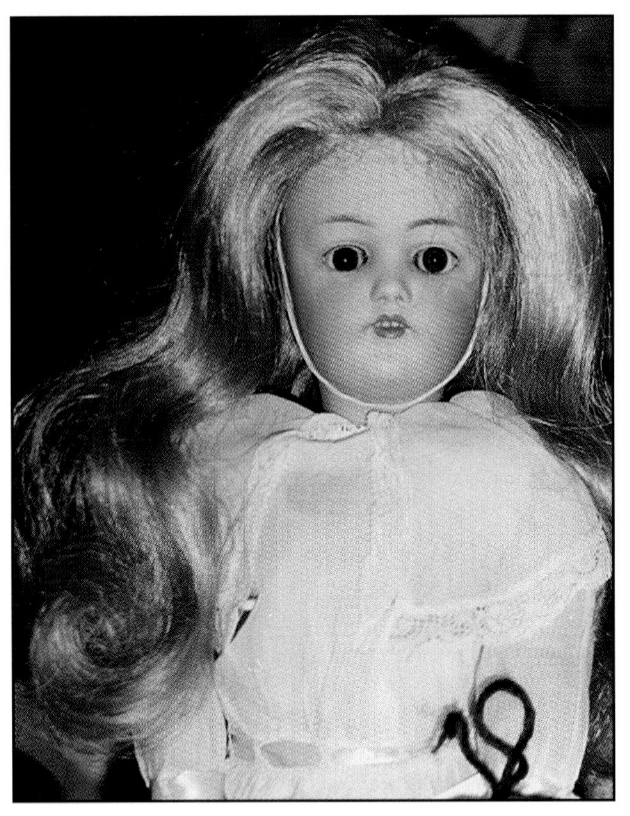

Kestner mold 168. *Courtesy My Doll Room.*

"Daisy" mold 171. *Courtesy Ann Condron.*

25in (64cm) Kestner mold 171. *Courtesy Kay & Wayne Jensen.*

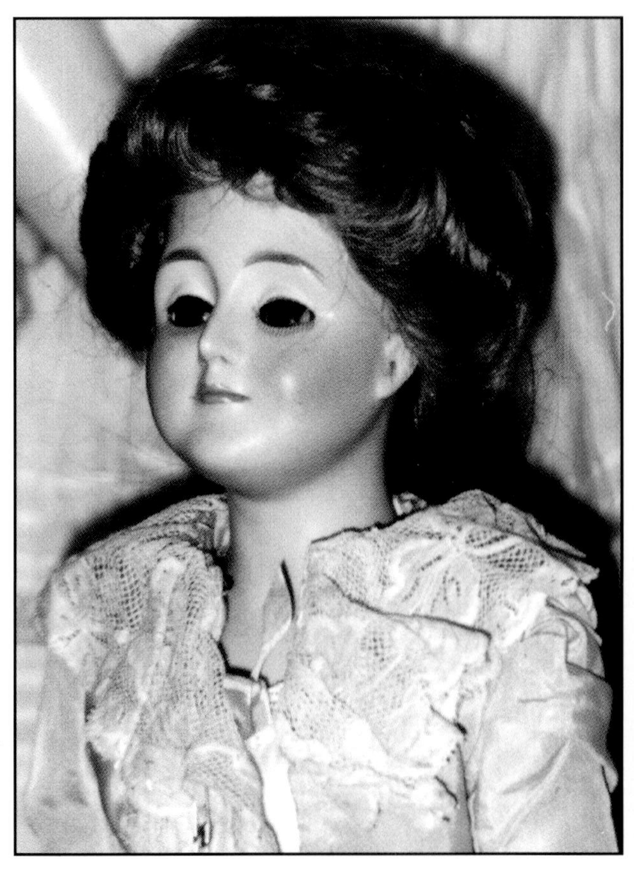

20in (51cm) Kestner "Gibson Girl." *Courtesy Becky Lowe.*

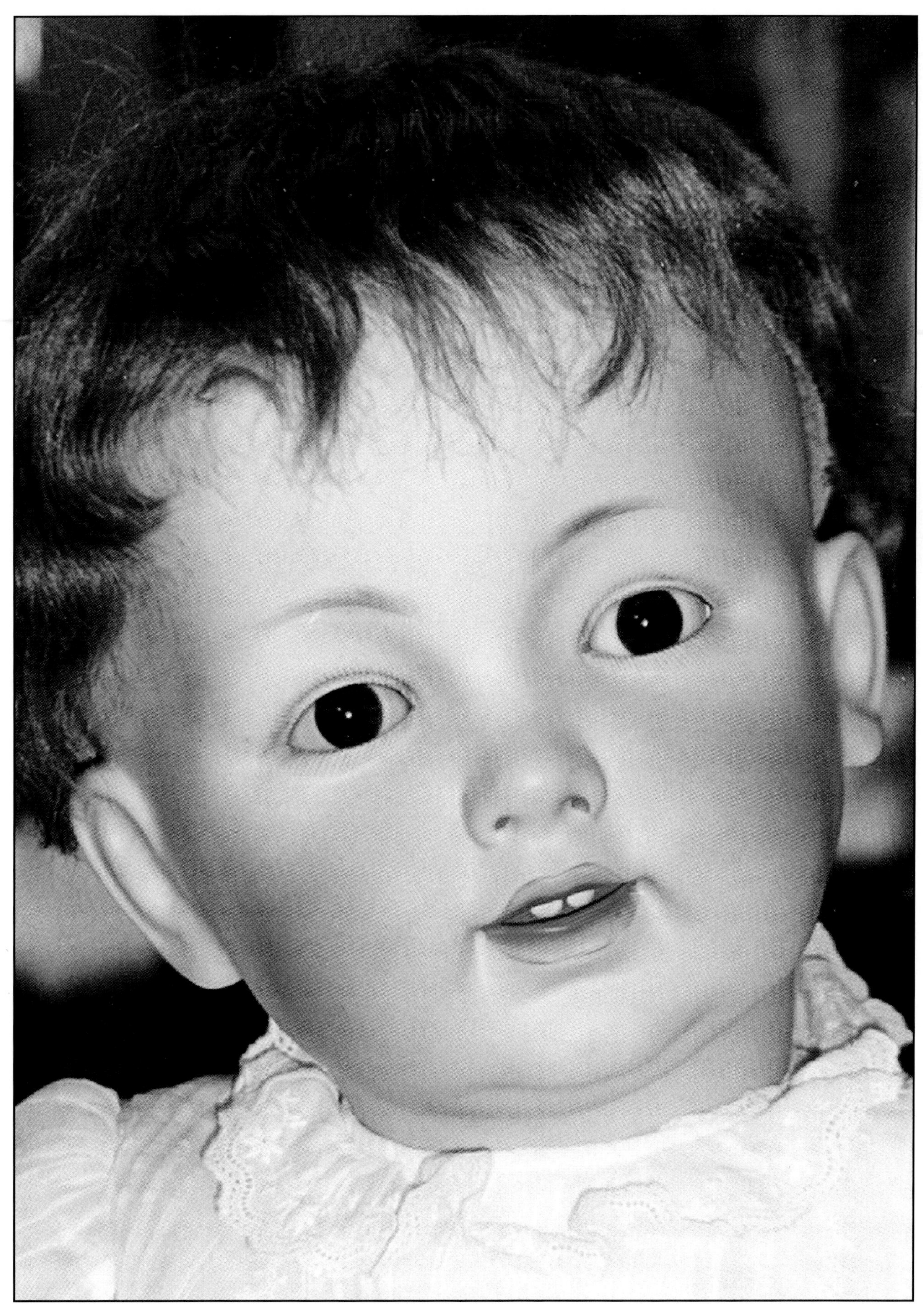

28in (71cm) Kestner mold 220. *Courtesy Appalachian Doll Company.*

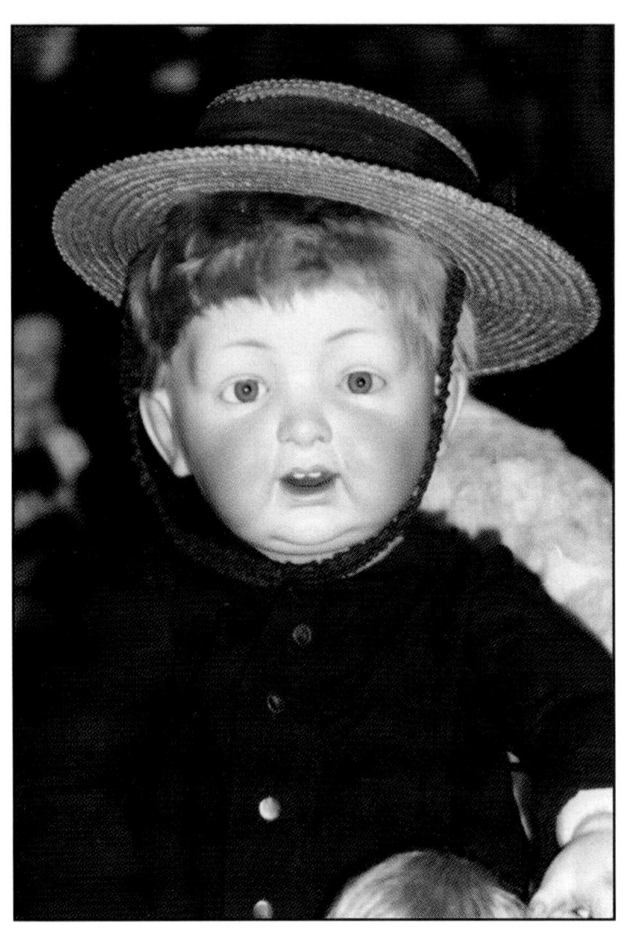

27in (69cm) Kestner 220 boy. *Courtesy Val Star.*

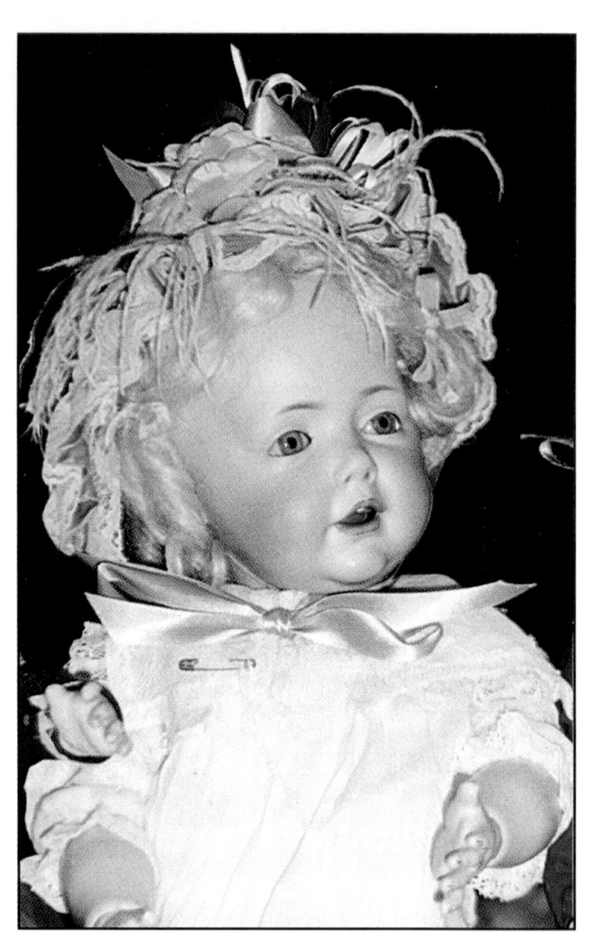

19in (48cm) "Hilda" JDK 237. *Courtesy Helen Lee.*

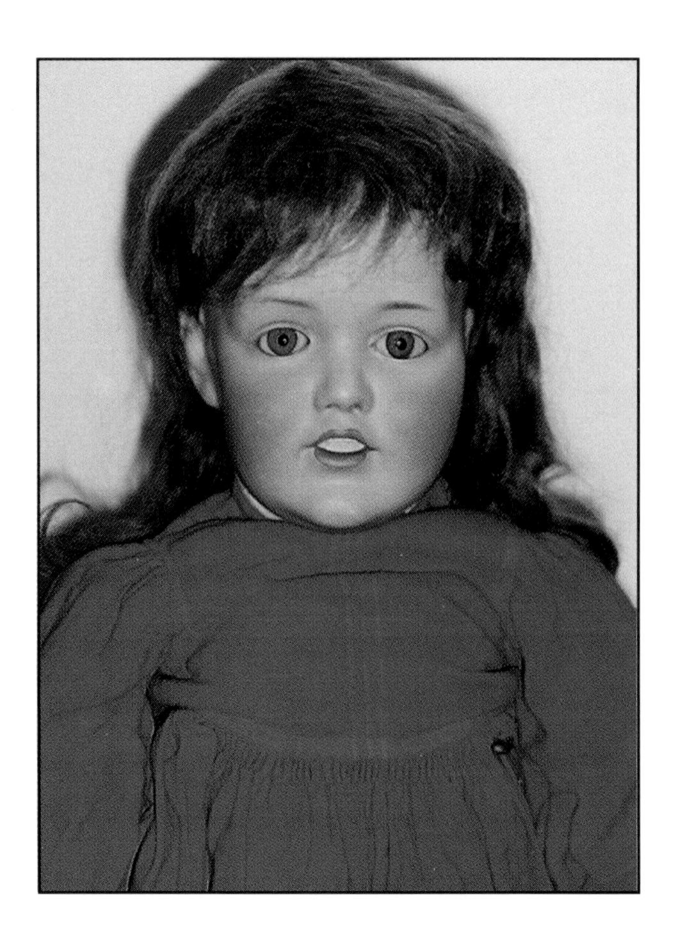

20in (51cm) Kestner 241. *Courtesy Linda's Antiques.*

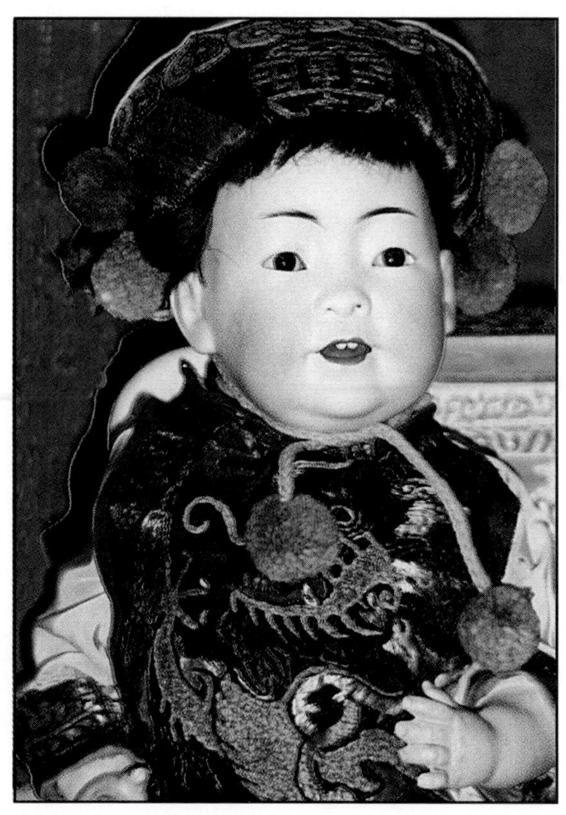

Kestner mold 243 Oriental. *Courtesy Sally Van Luven.*

13in (33cm) Kestner Oriental boy. *Courtesy Kate & Richard Smalley.*

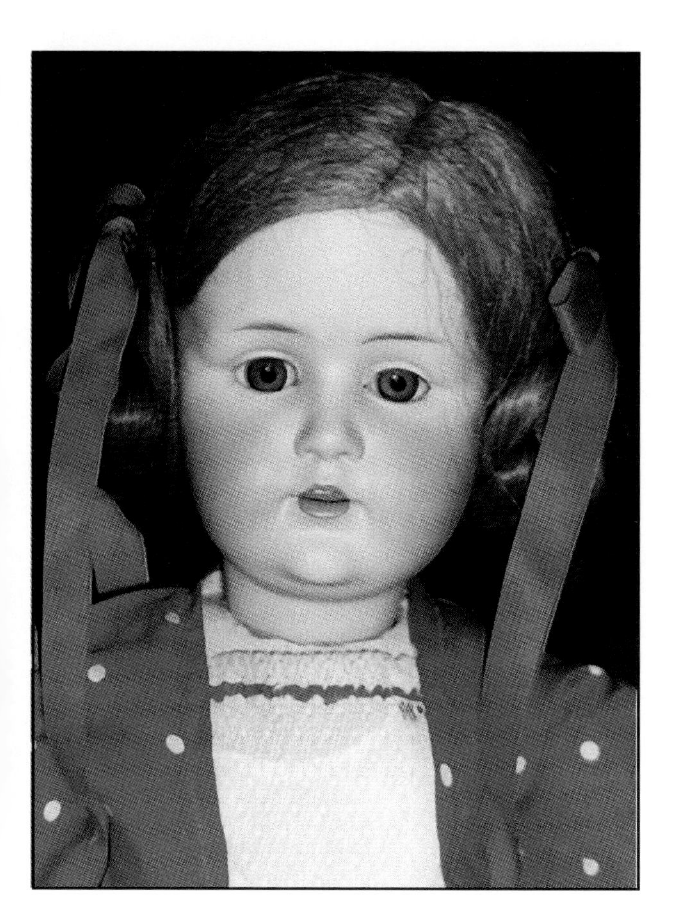

JDK 249. *Courtesy le Cheval de Bois.*

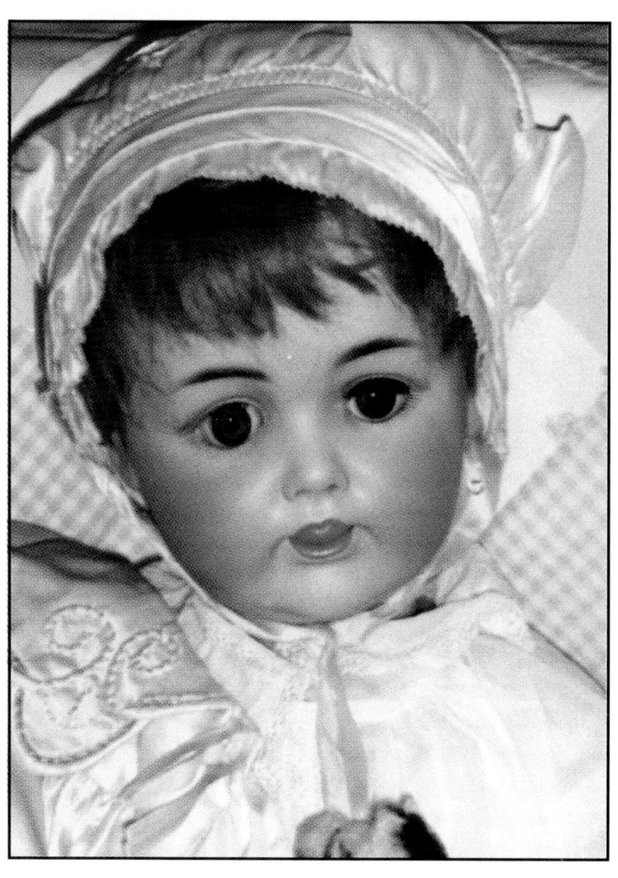

JDK 257. *Courtesy Ann Condron.*

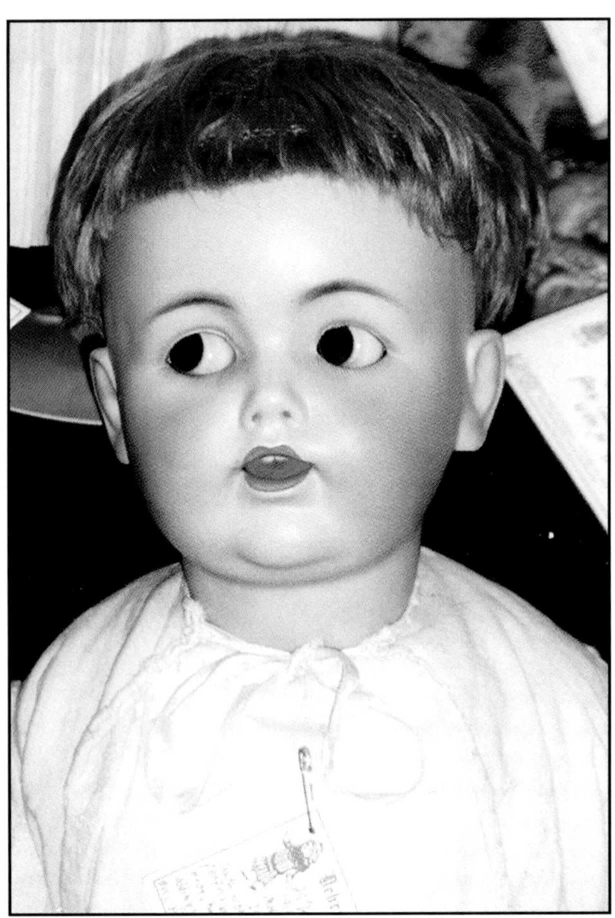

25in (64cm) Kestner mold 257. *Courtesy Debra's Dolls.*

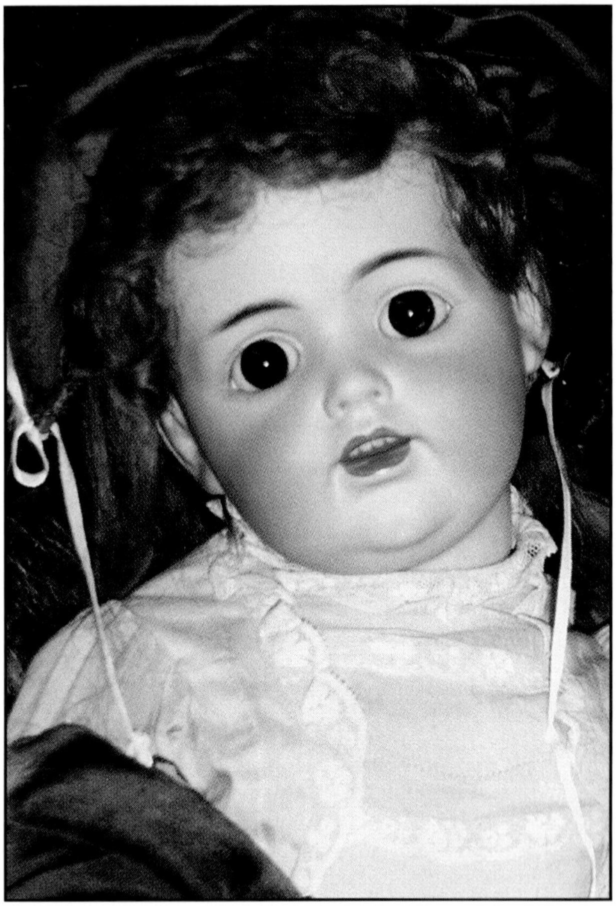

25in (64cm) JDK 260. *Courtesy Ann Condron.*

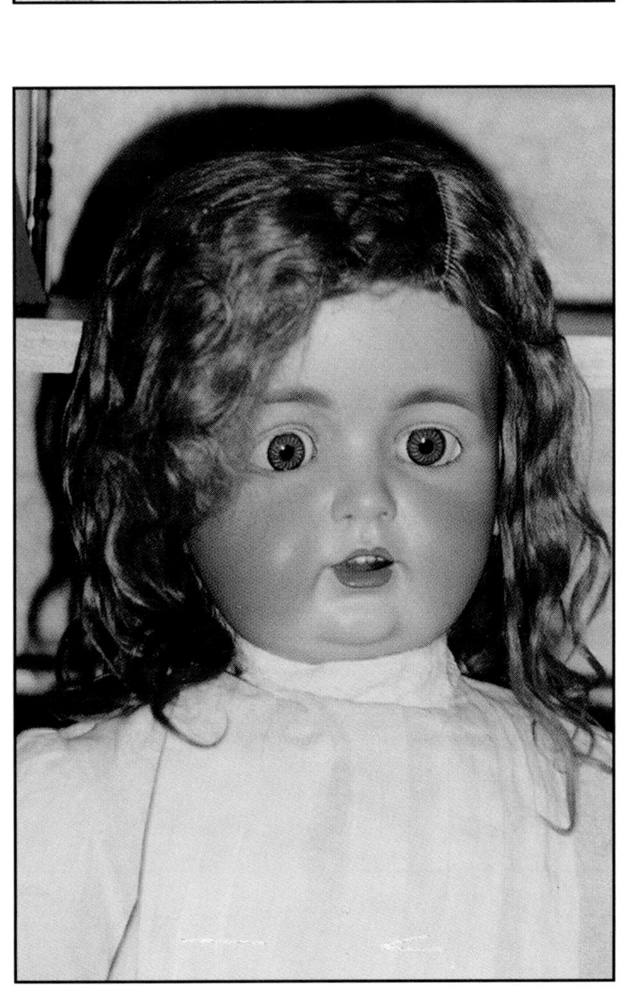

JDK 260. *Courtesy Esther Schwartz.*

Kestner mold 262. *Courtesy Unique Gallery*.

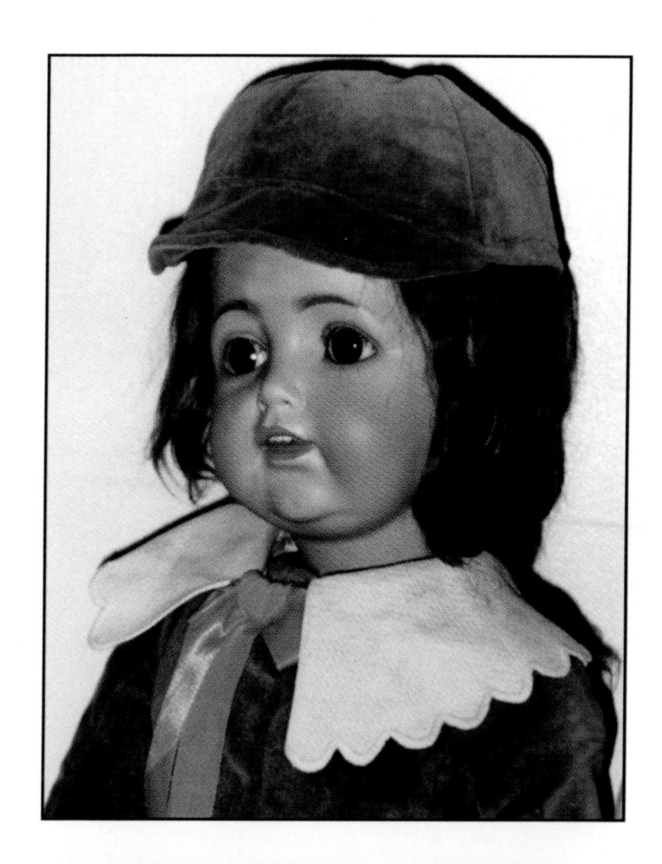

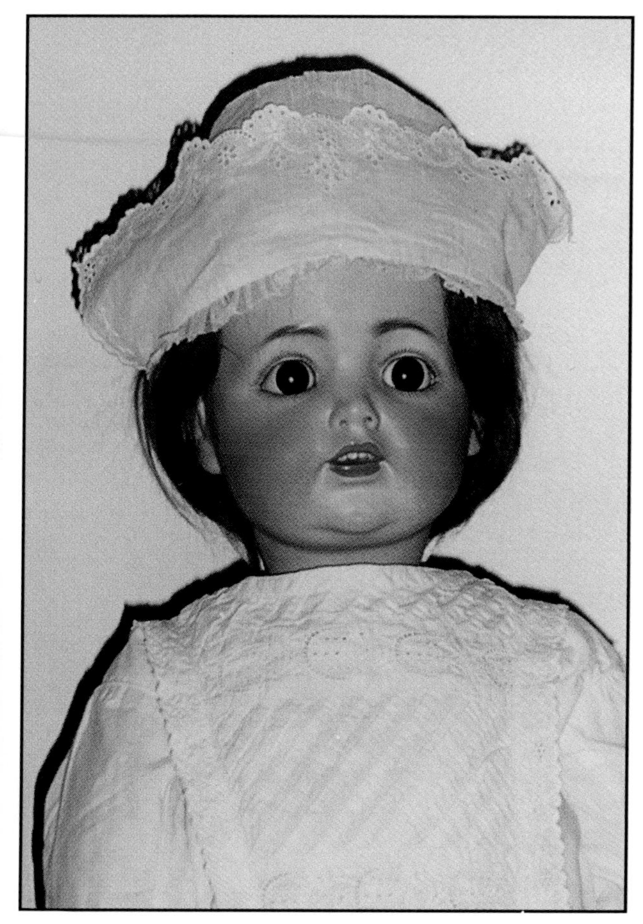

Kestner 262 dressed as a boy. *Courtesy Unique Gallery*.

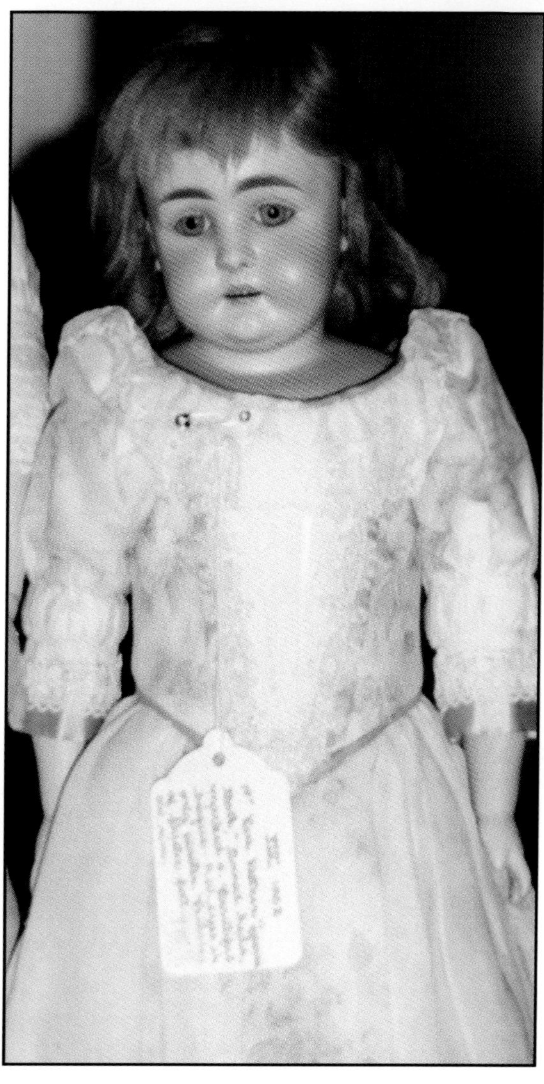

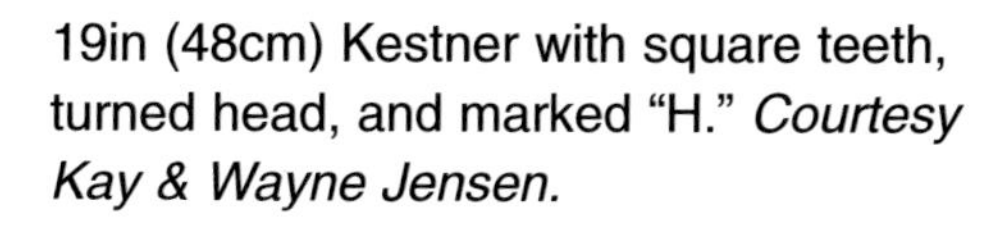

19in (48cm) Kestner with square teeth, turned head, and marked "H." *Courtesy Kay & Wayne Jensen.*

28½in (72cm) Kestner. *Courtesy Touch of Class.*

Kestner girl. *Courtesy Unique Gallery.*

Kestner all-bisque doll with molded boots. *Courtesy Ona's Toy Box.*

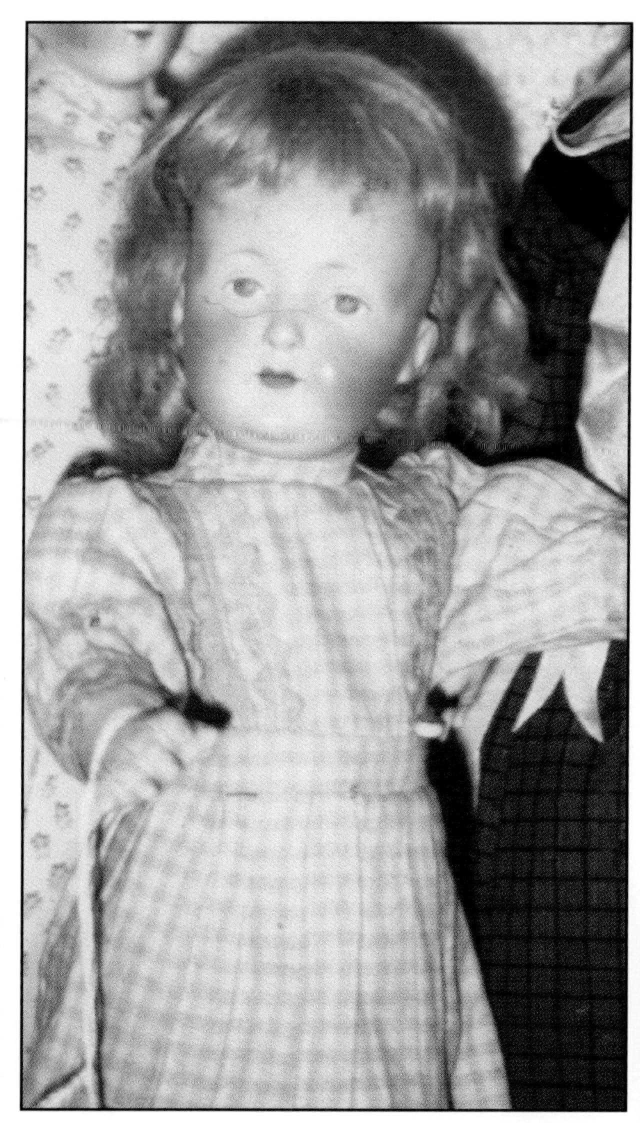

15in (38cm) Kestner mold 182 with painted eyes. *Courtesy Shelia Needle.*

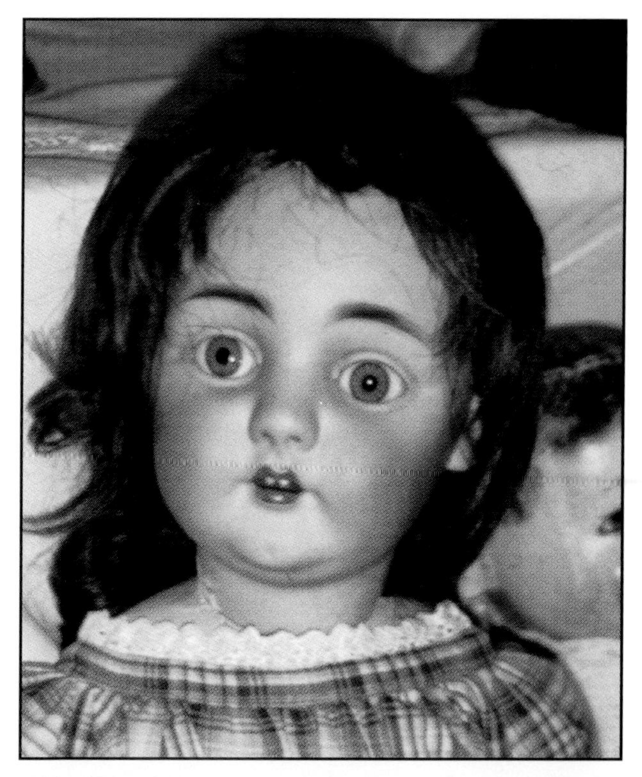

23in (58cm) Kestner 143 character. *Courtesy Geri Gentile.*

22-in (56cm) Kestner "Baby Jean." *Courtesy Dorothy Cohen.*

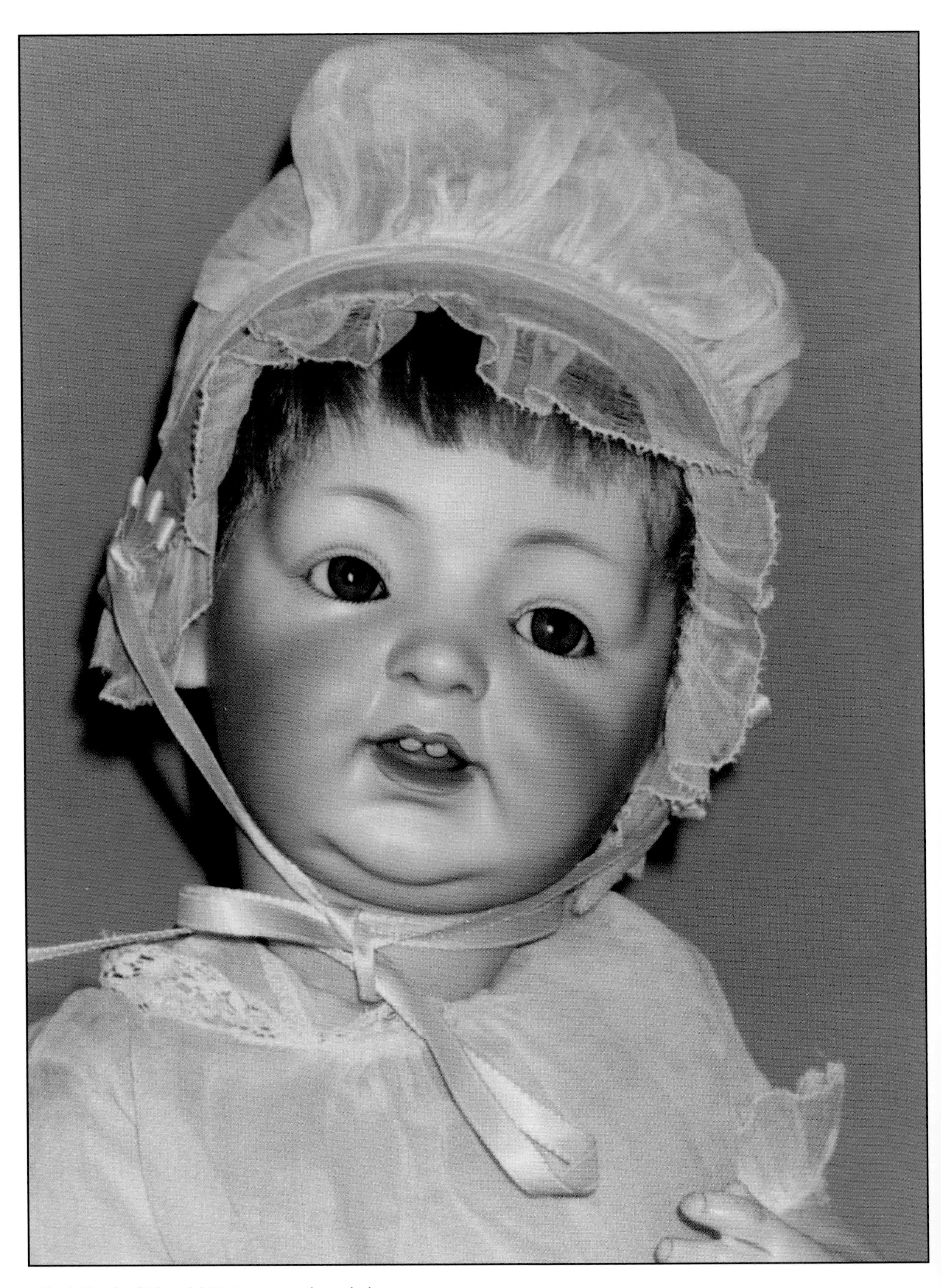

18in (46cm) JDK mold 226, two teeth and sleep eyes.

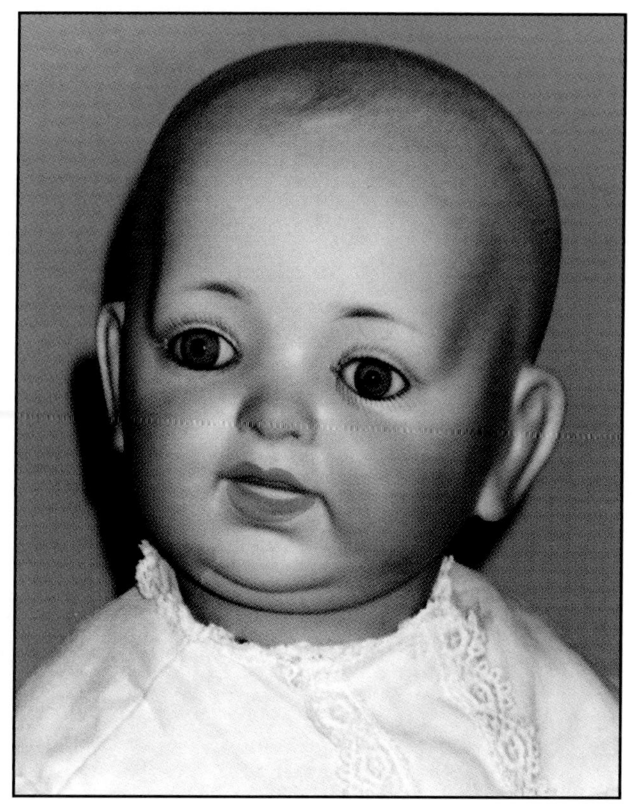

14in (36cm) JDK with no incised mold number, set eyes and open/closed mouth.

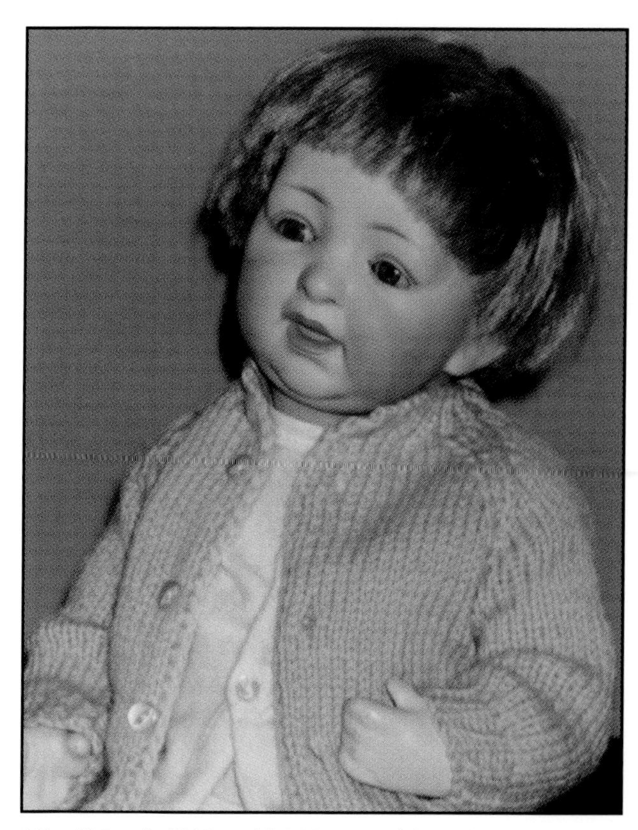

13in (33cm) JDK mold 211, open/closed mouth and sleep eyes.

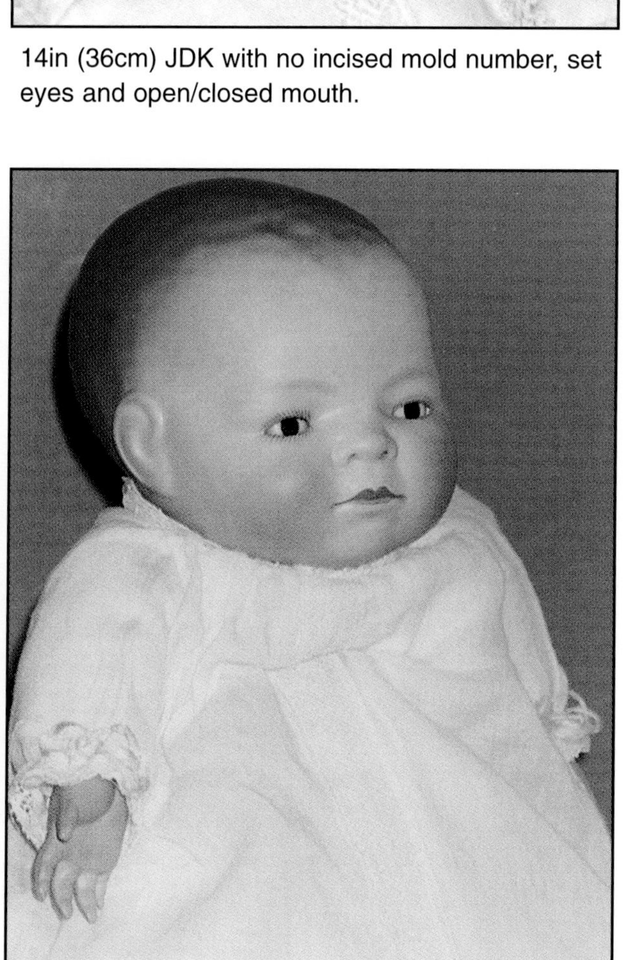

Many of the Bye-Lo babies were made by Kestner. This brown-eyes version is all-original.

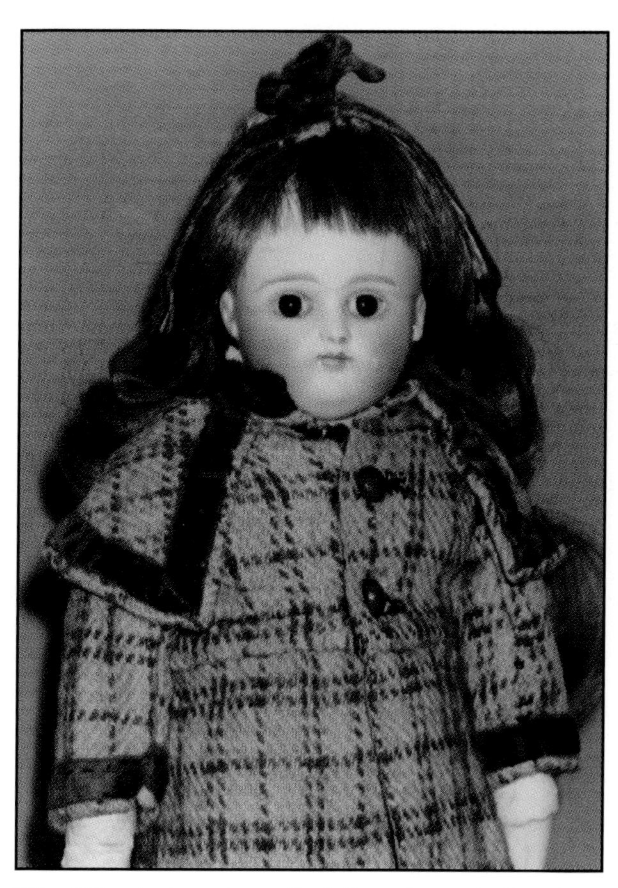

13in (33cm) closed mouth doll carries only the size number "2" but has Kestner characteristics.

14½in (37cm) size "D" or "8" is in her original chemise.

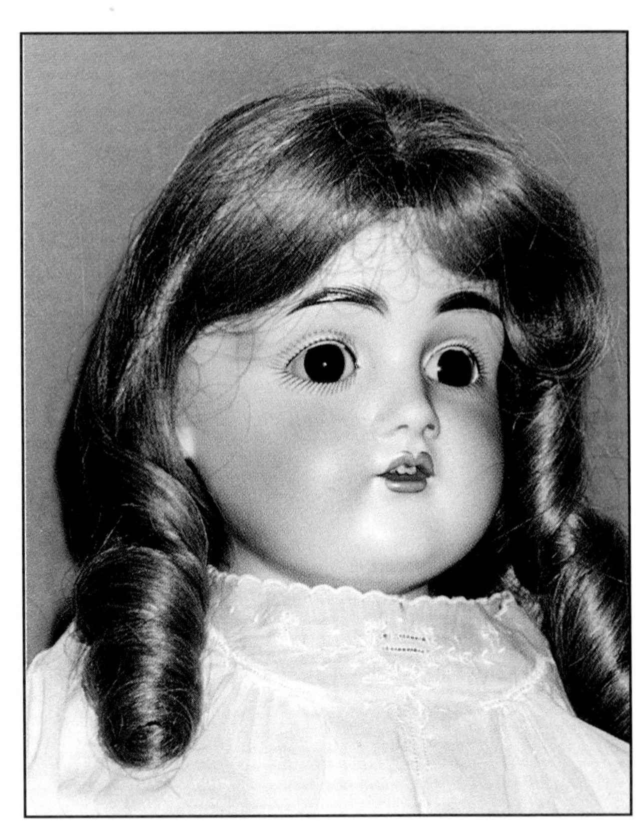

22in (56cm) Kestner 154, sleep eyes and open mouth.

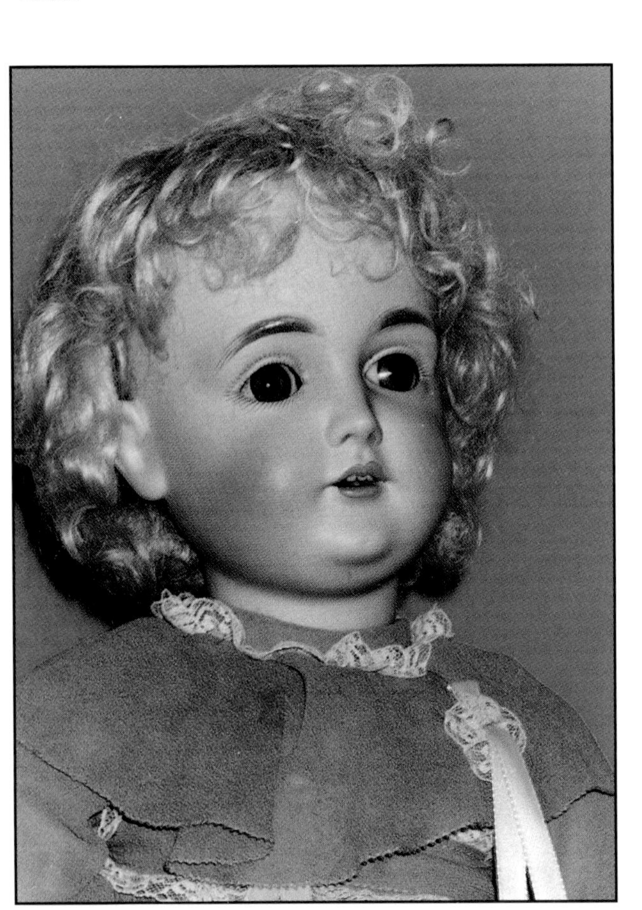

24in (61cm) mold 164 with molded brows.

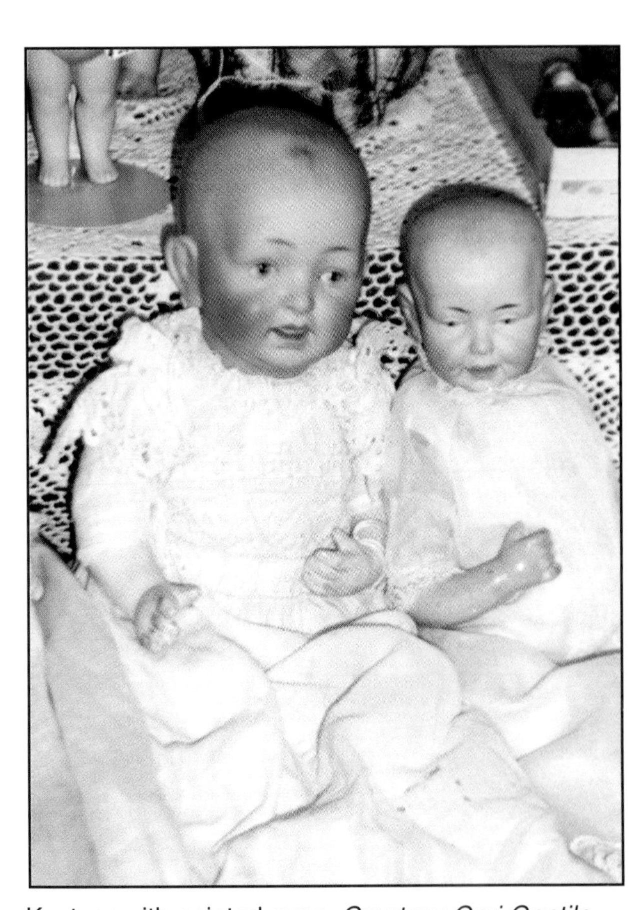

Kestner with painted eyes. *Courtesy Geri Gentile.*

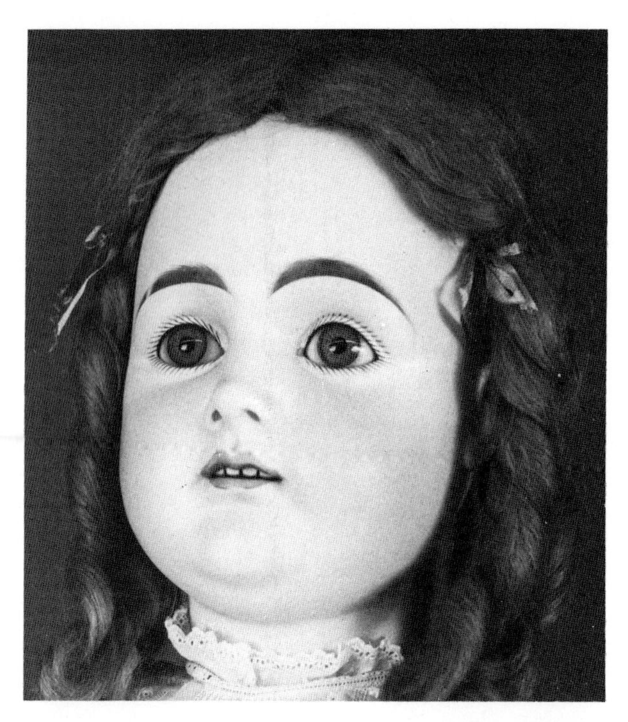

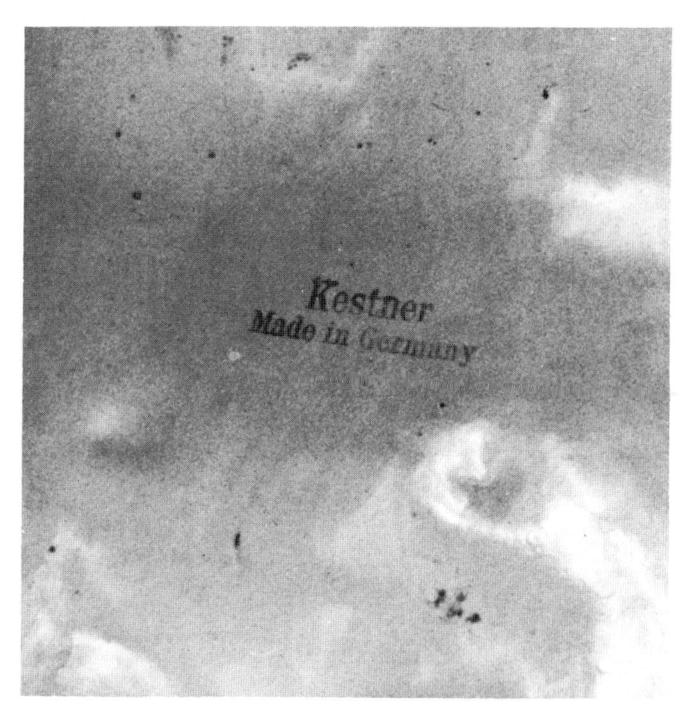

Illustrations 216 and 217. This gorgeous 38in (96.5cm) girl is incised on the back of her head with only a "17," which surely should be regarded as a size number. Her pale creamy bisque is of the finest smooth quality, with cheeks tinted a light pink. Her blonde eyebrows have quite a long arc to them, tapering to a tiny brush stroke. Her pale blue eyes look natural because of the threading blown into the glass. Her long mouth with nicely shaped lips barely open has four upper teeth set with a gap-toothed look. Her center-parted blonde human hair long curls are held back by satin ribbons on each side. Most interesting about her is the very heavy composition body with large separate ball joints at the thighs and a squared-off buttocks. Surprisingly, her body is stamped "Kestner//Made in Germany" in red; it is very unusual to find this mark. It is also very important to know that Kestner did indeed make this type of body, which very likely preceded the "Excelsior" body of 1892. Unfortunately, since this is such a large doll and we were not working in our own studio, we had no way to get a full view photograph of her body. *Emma Wedmore Collection.*

Illustration 217.

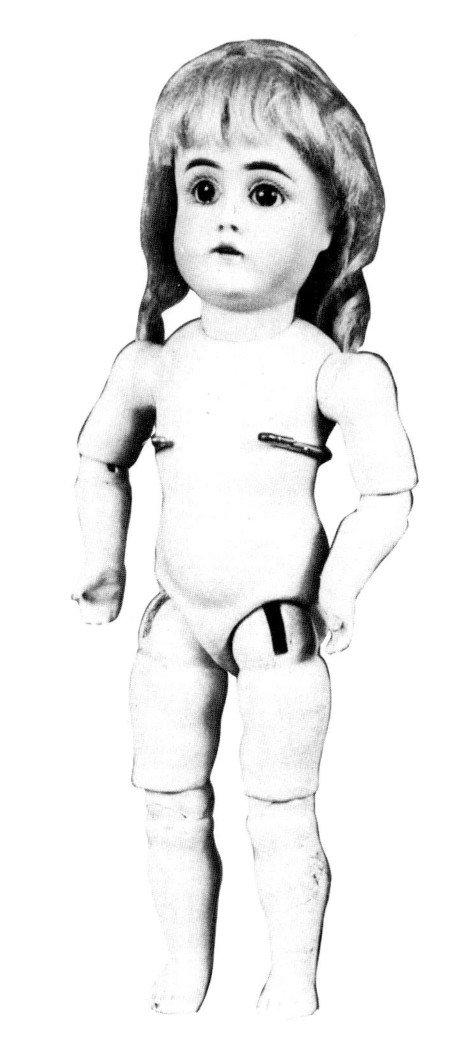

Illustration 218. This 12½in (31.8cm) girl certainly has a long face; the shape is quite different from any of those with mold numbers, yet she has the definite Kestner look. Her original blonde mohair wig covers a plaster dome. Her brown stroked eyebrows are long and tapered, matching her brown sleep eyes. Like the large doll in *Illustrations 216 and 217,* her painted lashes are short. Slightly parted lips show four upper teeth. Her tiny ears and fat neck are carry-overs from the closed-mouth dolls, as is her body with the unjointed wrists and cupped hands which appears to precede the "Excelsior" body of 1892. She is incised with the Kestner markings "6 made in B//Germany." The letter and number are in reverse of their normal positions. *Emma Wedmore Collection.*

Illustration 219. This is another of the long faced Kestners. Although her face has the same shape as the doll in *Illustration 218,* there are several characteristics which give her a slightly different look. Her eyebrows, although dark and stroked, are nearly flat on the underside; her brown sleep eyes have smaller cut sockets; and her mouth is smaller. The illustration does not show her four upper teeth. Her original blonde mohair wig and plaster dome are typically Kestner. She has exactly the same body as the doll in *Illustration 218.* Her Kestner style marking is "b made in 3//Germany." She is wearing a very nice old gold silk georgette dress with brown feather-stitching. *Elba Buehler.*

Illustration 220. Although marked only with a size number, this doll has been attributed to Kestner because of her characteristics. She has a blonde mohair wig over a plaster dome, and glossy blonde eyebrows in the Kestner style. Her fine quality bisque is pale and creamy; she has four molded-in square-cut upper teeth. About 16in (40.6cm) tall, she is on the jointed body with the straight wrists and cupped hands. *H & J Foulke.*

Illustration 221. This 18in (45.7cm) child is of the same mold as the doll in *Illustration 220.* Without the ties under her chin, one can note her very chubby cheeks. Again, she has quite a long face. Her glossy blonde eyebrows are typically Kestner, but have the flat underside like those on the doll in *Illustration 219.* Over her plaster dome is a replacement curly skin wig. She is on the early type body with straight wrists and is marked "made in Germany 9" with no letter size . *Maxine Salaman Collection.*

Illustration 222. Marked only "14," this 22in (55.9cm) child also has the oblong face. Although her cheeks are chubby, her long straight hair makes her face appear thinner. She has lovely brown blown glass eyes with amber flecks and glossy thick brown Kestner eyebrows. Her open mouth is actually a small slit with molded, almost buck, teeth and tiny thin lips. She is on a sturdy composition body. *Sue Bear.*

Illustration 223. This child is 25in (63.5cm) tall and marked only "15," her size number. Her face is shorter and cheeks are chubbier than the child shown in *Illustration 222.* She has an open mouth with full lips and two square-cut upper teeth. Her brown sleep eyes are accented by especially long painted lashes; her eyebrows are less curved and not so heavy. Her body of sturdy composition has straight wrists, but separated fingers. *Sue Bear.*

Illustration 225.

Illustration 224.

Illustrations 224 and 225. This small 10in (25.4cm) girl is incised only "4," her size number. Her original blonde mohair wig covers a plaster dome. Her blonde eyebrows are nicely tapered with flat undersides. Her distinctive feature, of course, is her mouth with her small lips and two large molded-in and square-cut upper teeth! Her jointed composition body is the early type with straight wrists and cupped hands. *Jane Alton Collection.*

Illustration 226. This 9½in (24.2cm) girl certainly seems to be related to the doll in *Illustrations 224 and 225.* She is incised only "3," her size number. Her plaster dome is covered by a blonde mohair wig. Her sleep eyes are a very deep gray color. She has the same puffy cheeks, fat neck, and tiny ears as her sister. However, her mouth is different since it has two upper teeth and one lower tooth! She is on a chunky jointed composition body like the one shown in *Illustration 196.* She is wearing what appears to be her original rose satin dress with tatted lace trim and matching bonnet. *Richard Wright Collection.*

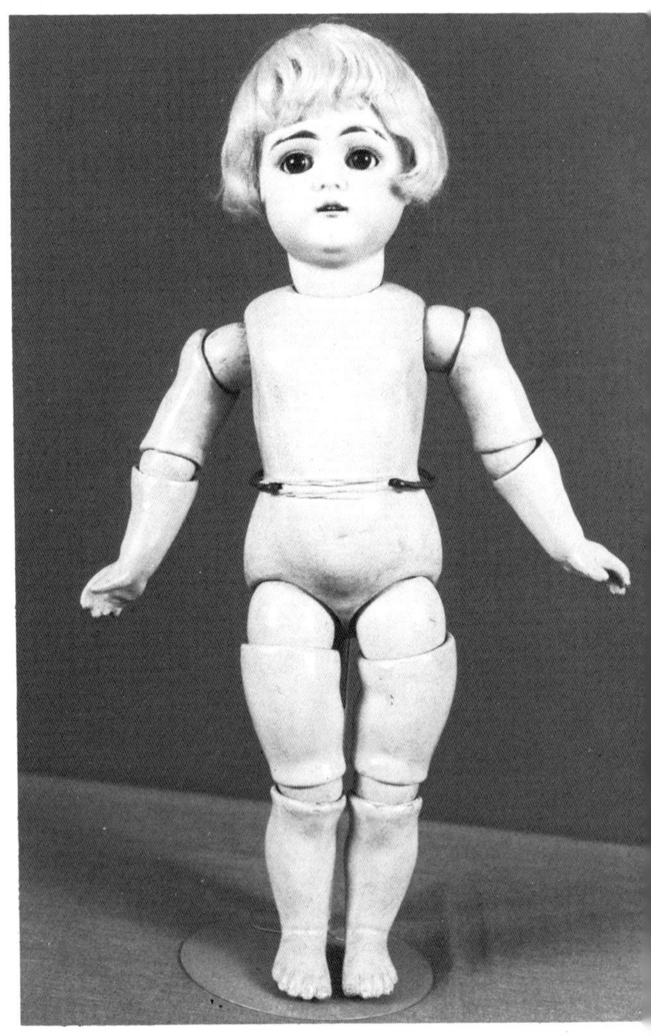

Illustration 227.

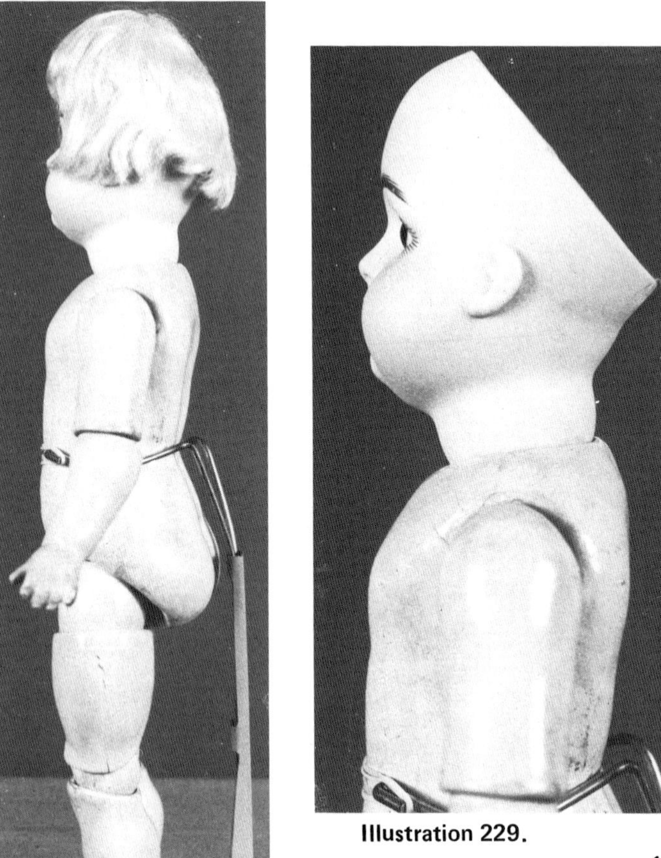

Illustration 229.

Illustration 228.

Illustrations 227, 228 and 229. This 16in (40.6cm) bo incised "Germany 10" way up at the crown openin appears to be a larger size of the two little girls in *Illustra tions 224 and 225* and *Illustration 226.* He has the sam oval face, tiny ears, chubby cheeks, fat neck, and doubl chin. His open mouth shows four upper molded and square cut teeth, although they are not as far to the front. His eye brows are in the Kestner style, being dark brown an glossy. His brown sleep eyes are outlined by closely painte lashes. He has the earlier type body seen on many of th closed-mouth dolls with the separate balls in the joints an especially the large ones at the thighs, like the doll with th marked body in *Illustration 217.* The side view shows h interesting profile, as well as the fat roll at the back of h head, his chubby neck, and his low cut crown, typical c this period doll. *Joanna Ott Collection.*

Turned Shoulder Heads

Kestner made a series of shoulder plate dolls with heads turned to the side using an alphabet sizing system. These are referred to in an 1888 advertisement from the Kestner factory which notes doll heads as being "sidewards directed" and being "quite impressive with their stationary or movable glass eyes, their inserted teeth and their magnificent mohair wigs." These apparently popular dolls were made over a period of perhaps 20 years. Companies tended to repeat an item that sold well. These heads definitely have the Kestner look. The eyebrows are arched and heavy; eyes are small with generous painted lashes; mouths are in the bowed style with upturned lips and four upper teeth; pates are of plaster. Some of them have molded eyebrows and real eyelashes. It seems that Kestner probably made these heads mainly to sell to other companies since they are found on a wide variety of kid and cloth bodies.

D

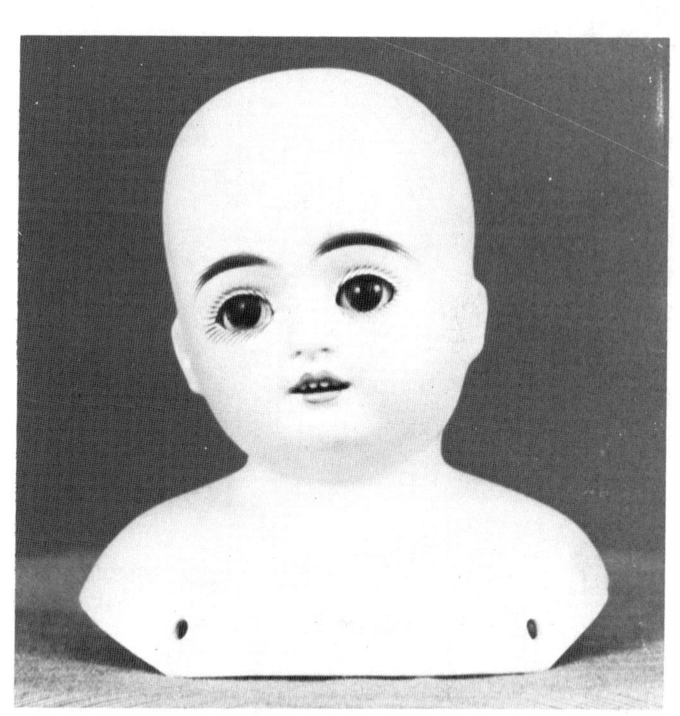

Illustration 230.

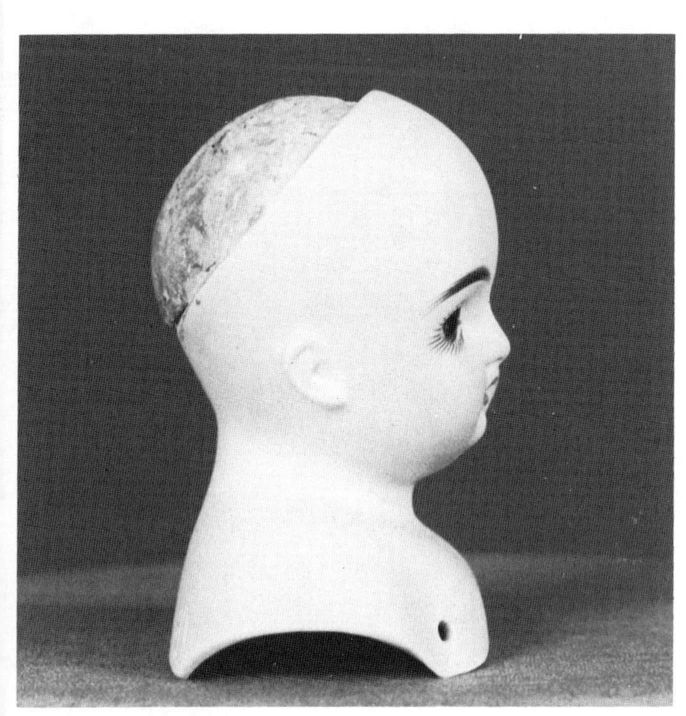

Illustration 231.

Illustrations 230, 231 and 232. This turned shoulder head is 4in (10.2cm) high and in an ascending scale is size D which is incised in a capital script letter at her crown opening and again at the bottom of her shoulder plate along with the words "made in Germany" which can be written in a variety of styles in different heads of this type. She has a lovely peaches and cream complexion with slightly tinted cheeks. Her eyebrows are typically Kestner, brown and glossy with many individual brush strokes. Detail of modeling is good for such a small face as her nose turns up at the end and she even has indentations between her nose and upper lip. The ears are very small, as are her brown sleep eyes. Her lips are a light orange, peaked and turned up at the corners. She still has her original plaster pate. *H & J Foulke.*

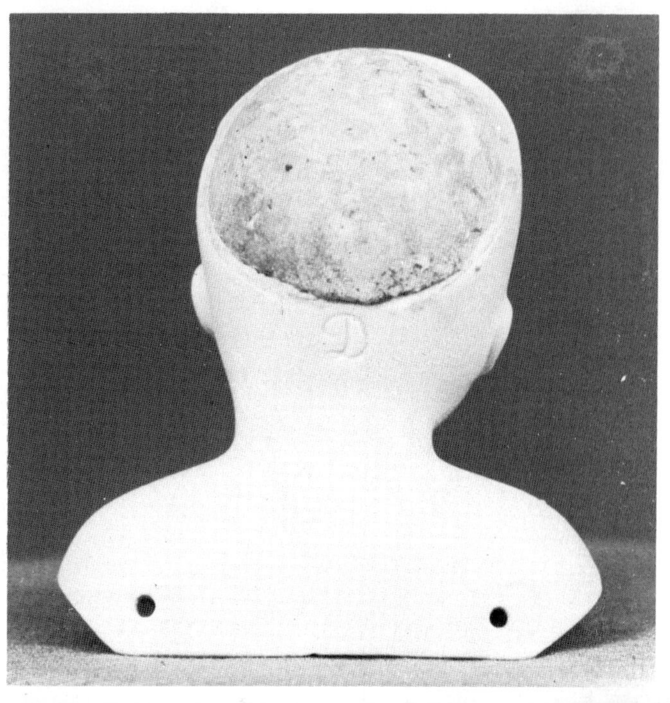

Illustration 232.

F

Illustration 233. This 16in (40.6cm) girl's head is turned quite a bit more than that of the one in *Illustration 230.* Otherwise, she has the same general look on her face, although her eyebrows are closer to her eyes and are less arched. She has the gray sleep eyes so often associated with Kestner. Her open mouth has four upper teeth. Her brown human hair wig matches her eyebrows which have many individual strokes; the inside edges of her eyelashes are quite heavy. She has a short stout neck and a chubby shoulder plate. Her gusseted kid body has bisque lower arms. She is incised with an "F." of the printed rather than script type. *Jane Alton Collection.*

H

Illustration 234. This 19in (48.3cm) doll is incised "H½" at her crown opening and "11½ made in Germany" across the bottom of her shoulder plate. She is on a gusseted kid body with bisque lower arms and hands. She has blue sleep eyes and her original gorgeous blonde curly mohair wig. Since this series is not usually found on the J.D.K. marked bodies, they were probably a less expensive line. In an 1899 catalog a good line of dolls similar to these and not having the Kestner crown label wholesaled for about 47¢ each (undressed but with shoes and stockings) which meant they would retail for less than $1.00. A marked Kestner doll of about the same size wholesaled for 75¢--50 percent higher. *J. C. Collection.*

Illustration 235. This girl is also incised "H," but she is 22in (55.9cm) tall. Part of this extra height is due to the fact that she has a different type of body, a cloth one with leather arms and sewn-on socks and boots. Also her head is turned much further to the side than that of the doll in *Illustration 234* and her plate is much deeper. Her human hair wig is a replacemant, but she has on a lovely old cotton dress. *Edna Black Collection.*

K

Illustrations 236 and 237. This girl, incised "K," is 25in (63.5cm) tall and again has a very turned head. She is shown twice: one, a head-on view for studying her face and second, a front-on view to show how much her head is turned. She is a lovely doll with brown sleep eyes, and open mouth with upturned lip, and her plaster dome under her original blonde mohair wig. She is on a jointed pink kid body with wooden upper arms and bisque lower arms. Her outfit is of pale green wool. *J. C. Collection.*

Illustration 237.

M

Illustration 238. This lovely 26in (66.0cm) turned shoulder head girl is incised "M." In her open mouth are four molded upper teeth; she has a stout neck and a dimple in her chin. Her features are beautifully painted. In addition to her dark and thick eyebrows, she has heavily painted eyelashes. She is dressed in appropriate old clothes and is on a kid body with hinged joints at the hips, knees, and elbows. *Joanna Ott Collection.*

O

Illustrations 239 and 240. This doll is 28in (71.1cm) tall and is incised "O made in Germany." She has the same face as the other dolls, but being so large she looks a little different and has more prominent ears. Her auburn mohair wig is original and styled like many of those shown in a 1904 wholesale catalog. Her white kid body has been preserved in fine condition. The upper arms are kid over wood with a rivet elbow joint and finely molded bisque lower arms. She has a *Ne Plus Ultra* riveted hip joint and a *Universal* knee joint. Her lower legs are also of kid with seamed and stitched feet. In 1904 the wholesale price of a doll like this without clothes but wearing shoes and socks was $5.00 which means her retail price would be $8.00 to $10.00. A handsome sum in those days indeed! *Edna Black Collection.*

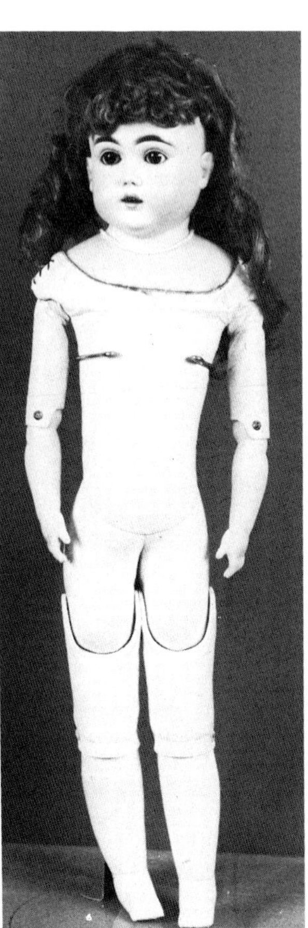

Illustration 239.

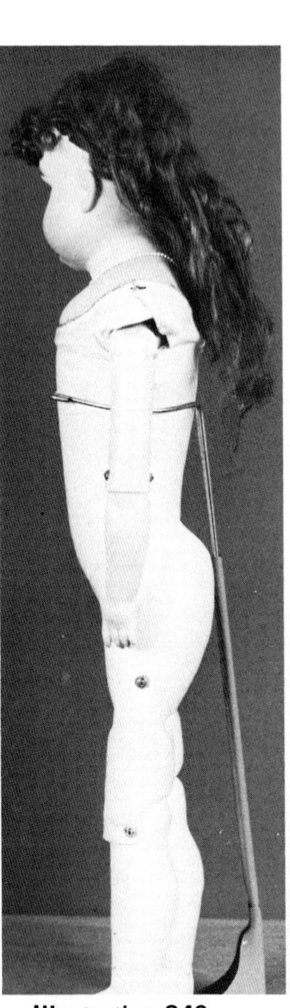

Illustration 240.

Illustration 241. Oddly enough this little pair of 12in (30.5cm) twins represents both the straight and turned shoulder head dolls. Neither has a mold number, but both appear to be Kestner dolls. They have single stroke brown eyebrows and brown glass eyes; mouths are open with molded upper teeth. Their blonde mohair wigs are in the Rembrandt style. They are wearing their original matching dresses, one pink and one blue. Kestner dolls about this size were advertised in 1904 to sell for 25¢ each, undressed but with imitation stockings and buckled shoes. The faces were "drawn from life." *Jane Alton Collection.*

Character Children

Introduction

The conception and development of the character doll was a German movement. At the turn of the century, sociologists and psychologists began paying more attention to the role and training of children. This, of course, had a great influence upon the development of toys, including dolls. In the first decade of the 1900s, there was much talk by both educators and artists against the French doll as unnatural and too elaborate in style, and, of course, this extended to the German dolly-faced doll since she was an outgrowth of the French one. The demand was for a more real-looking and childlike doll. This idea spread from Munich where it was fostered by designers, such as Marion Kaulitz, who became famous for her Munich Art dolls. This movement, which was prompted by an awakened interest in children as real people in their own right, had a profound effect upon the design of dolls. Käthe Kruse was one of those artists responsible for the decisive breakthrough from the dolly-face to the character doll when she showed her models which were miniature portraits of her own children.

In February 1908 Samstag & Hilder Brothers advertised character dolls in *Playthings.* These were dolls which had been "modeled from living subjects under the direction of the most famous artist of Munich...represent[ing] the very latest development of doll making." In January 1909 Strobel & Wilken Co. ran the first ad in *Playthings* which actually pictured bisque character dolls. In the same issue George Borgfeldt advertised "dolls with human expressions modeled by well-known artists in Munich from living subjects." And so the character doll had arrived in the United States, but the reviews were mixed. The character dolls were considered by many customers to be innovative and arty; they appealed more to liberal than conservative people who liked to stick with the familiar. The new dolls were not an immediate success, and the dolly-face doll continued to dominate the market in sales as well as production.

In January 1910 George Borgfeldt advertised in *Playthings* character dolls by Kestner, as well as Kämmer & Reinhardt and Heinrich Handwerck. This puts Kestner into the market within the first year; apparently the Kestner dolls were already in hand at the time as the ad stated that "immediate wants supplied from stock." The dolls were advertised as "true to life" in a "large variety of natural facial expressions."

This lovely series of character children produced by the Kestner firm is certainly a highlight of their character doll production. For some reason, however, the number made must have been limited as they are rarely found on today's doll market compared to the large numbers of Kämmer & Reinhardt or Gebrüder Heubach character children available. Perhaps this line of dolls just did not sell well and was discontinued in favor of the 200 series, which, judging from the numbers of dolls available, was an extremely popular line.

The marking on the heads of this series has been confusing to many collectors, since they have simply a number; the molds which have been found range from numbers 177 to 190 with the exception of 188 which has not been located to date. There are also several numbers in the 200s. On several models Kestner used his usual sizing system, but not his initials as part of the mark. In fact, some of the dolls do not even have the mold number and must be identified by comparing the face with a marked doll. The bisque on the dolls is of very fine quality, a grade which has begun to be taken for granted in a Kestner doll. The decoration is extremely well-done with soft looking many-stroked eyebrows and even tiny painted eyelashes and a red accent line above the irises of the painted eyes. The head openings are covered by plaster domes and the wigs, when known to be original, are blonde or brown mohair, many of the little girls having coiled braids in the popular German style. The heads are of the socket variety, and most of the bodies of these character children, when known to be original, have a pink finish more like the bodies of K*R dolls than the usual yellow finish ball-jointed Kestner ones with the "Excelsior" or "Germany" stamp.

LEFT: Illustration 242.

177

This wonderful boy incised only "177" appears to be the first of their character child series. What a great face he has! He is the only one of the series to turn up with molded hair. Further, he is the only known example of this mold number. His blonde hair is nicely molded around his face with a little curl on the forehead and combed back on the side; he has a molded crown with hair waved and comb marks down the back. His eyebrows are formed from four blonde brush strokes. The eye socket is molded, as are upper and lower eyelids. Eyes are painted blue with a black lash line, but do not have the tiny painted upper lashes most dolls of the series have. He has the same face as mold 178, but with molded hair instead of a wig. He is 11in (27.9cm) tall on a jointed composition body. *Courtesy of Jane Walker.*

178

Illustration 243.

Illustration 244

Illustrations 242, 243, and 244. Mold 178 has the same face as 177, but it is a wigged rather than molded hair model. The face certainly has a very appealing look with distinctiveness being expressed through the eyes and the mouth. The eye socket is molded as are the upper and lower lids. The boy has brown eyes with a white dot highlight; the girl's eyes are just a little smaller than those of the boy although they appear to be quite a bit smaller than his because they are light blue and do not show up as well in the illustration. They have enameled tiny upper lashes over the colored part of the eye and lovely blonde, lightly feathered eyebrows. Their noses are rather sharp; they have open/closed mouths with full protruding upper lips. Some dolls of this mold have a slight white space between their lips which gives a hint of teeth. Both of these dolls have blonde mohair wigs over their plaster domes. These dolls were purchased in London, England and are in completely original fancy costumes. The boy has a suit of gray satin with bead and lace trim; his coat and tri-cornered hat are of black velvet with satin lining. The girl's dress is pink satin with an ecru lace overlay which has pink ribbon insertion. The dolls are 12in (30.5cm) tall. This model has also been found with glass eyes. *Richard Wright Collection.*

179

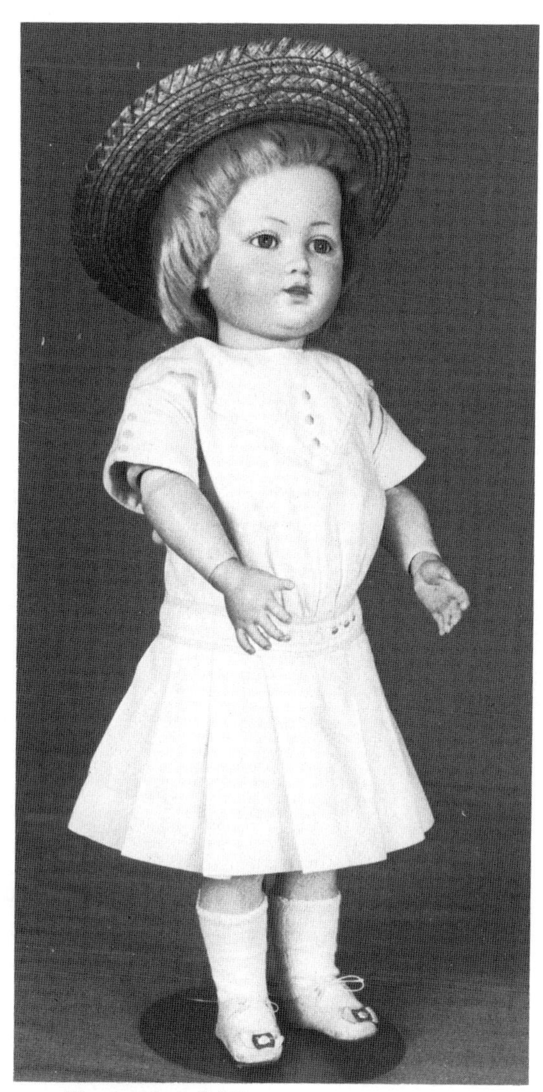

Illustration 246.

Illustrations 245 and 246. This doll, incised with mold number "179," is 16in (40.6cm) tall. Most of the dolls in this series have painted eyes, so a glass-eyed model is rare indeed. The face is quite a bit chubbier than that of mold 178, with nice puffy cheeks. There is very good molding detail around the eyes with long painted upper and lower lashes. The short, light, and delicate eyebrows are similar to those on the dolls in *Illustrations 242, 243, and 244.* The blue glass eyes have lovely threading. The open/closed mouth has a molded tongue and just a hint of upper teeth. The bobbed blonde mohair wig covers a plaster dome. The head is on a jointed composition body, unusual because of its celluloid hands. The doll is wearing a very nice old white cotton dress with lowered waistline and pleated skirt; the trim is light blue. *Jane Alton Collection.*

Illustration 247. Strange as it may seem, here is another 179 which is obviously a part of this series, but an entirely different face from the 179 with glass eyes. This doll has a longer, less chubby, but still full face. Her eyes are deeply molded with an incised red eye line; they are quite large and blue, but as noted with other dolls in the series, the blue eyes do not show up as large in the illustrations as they really are. Above the blue irises are the tiny painted lashes. Her mouth is the pouty type, small and definitely downturned. She is 15in (38.1cm) tall on a ball-jointed body with the red stamped and boxed "Germany" seen on so many of the Kestner dolls. *Jane Alton Collection.*

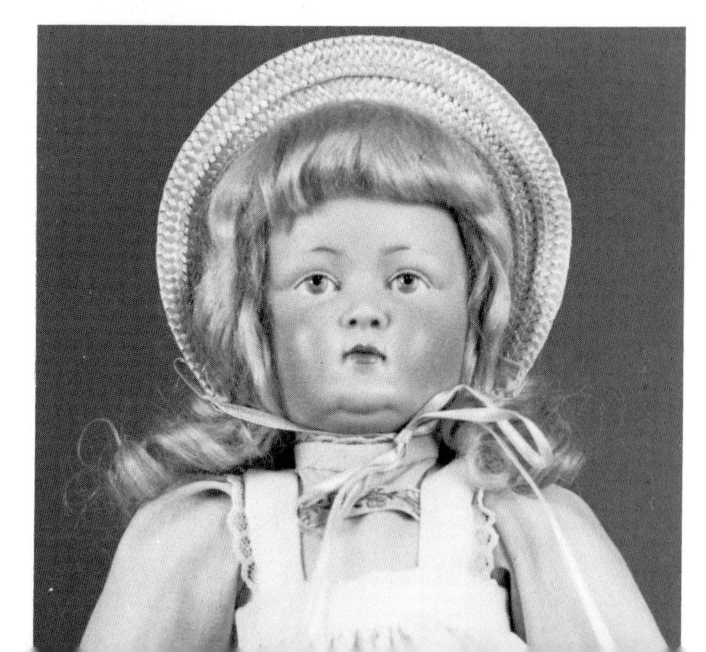

180

Illustration 248.

Illustration 248 and 249. This girl with mold number "180" is from the character set shown in *Illustration 278*. Her dark blonde mohair wig has coiled braids above the ears with her original turquoise ribbons. Her eyes are molded, with upper and lower molded lids also, but the eye socket is flat, not done in the intaglio style of the Gebrüder Heubach dolls. Her eyes are painted gray with a white dot highlight, a red line above the eye, and tiny short painted lashes above only the colored part of the eye. She has an open/closed mouth with four molded upper teeth and a tongue, which makes her look as though she is just ready to begin singing. She has the usual plaster dome. Her original underwear consists of a one-piece combination with turquoise ribbon trim and a full slip of fabric and trim which matches her original dress of white cotton with embroidered edges used for the bottom of the skirt and sleeves. Lace trims her collar and sleeves. *Richard Wright Collection.*

Illustration 249.

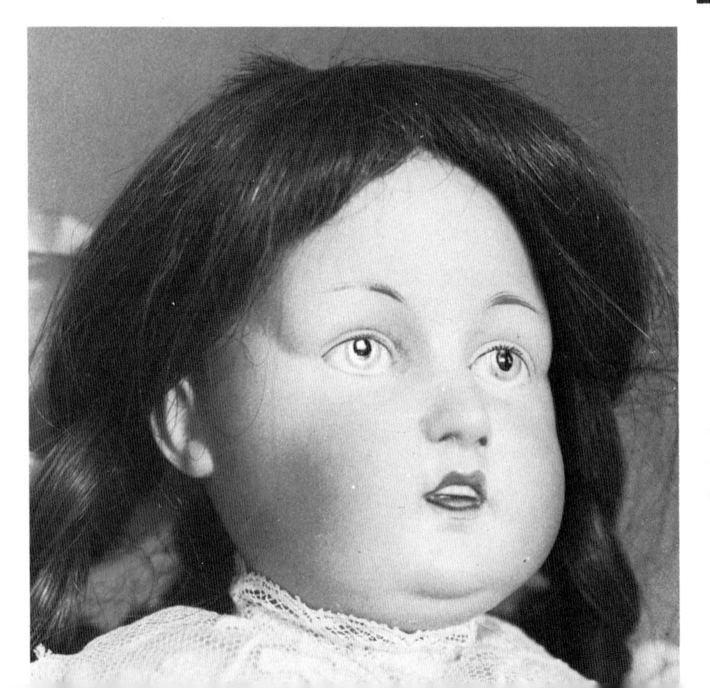

Illustration 250. This is another example of mold number 180, also a girl in the 18in (45.7cm) size. The side view shows the modeling of her ear as well as the molding of her side cheek line. Her eyes are a very vivid blue color. This model has also been found with glass eyes and a few even have real fur eyebrows. *Richard Wright Collection.*

181

Illustration 251. This delightful boy is one of the largest of the series, being 21in (53.3cm) tall. His face is full which gives him a healthy, sturdy look. His painted eyes are light blue with the same molded socket and lids as those of the other faces of the series. His black pupil is slightly indented, and he has the white highlight, red eye line, and tiny painted upper lashes. His open/closed mouth shows four molded upper teeth. His short blonde mohair wig covers a plaster dome. He has been appropriately redressed. *Maxine Salaman Collection.*

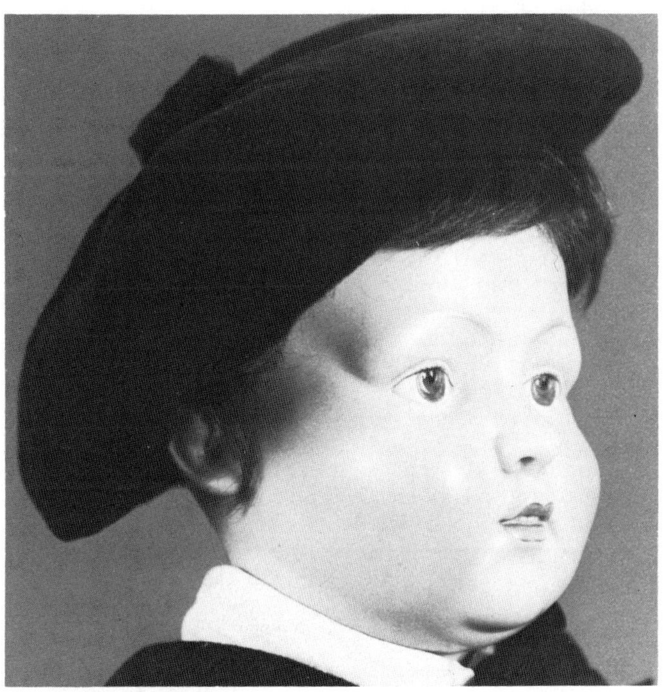

Illustration 252

Illustrations 252 and 253. This 21in (53.3cm) boy is unmarked, but he appears to be the same mold as the boy in *Illustration 251*, 181. His brown painted eyes are a little larger and also have the slightly indented black pupil. His blonde eyebrows are particularly natural looking as is his nose with rounded nostrils. The side view is particularly good to show his eye molding, and his interesting cheek line. His open-closed mouth has excellent decoration, the upper lip being shaped like a bow with upturned corners, lips slightly parted to show four molded-in upper teeth. His brown mohair wig complements his eyes. His jointed composition body has the red "Germany" body stamp as well as a label from the G. A. Schwarz toy store in Philadelphia, Pennsylvania. (This Gustave A. Schwarz was a brother of the better known F. A. O. Schwarz of New York, New York.) He is wearing an appropriate navy blue wool boy's suit. *Elba Buehler.*

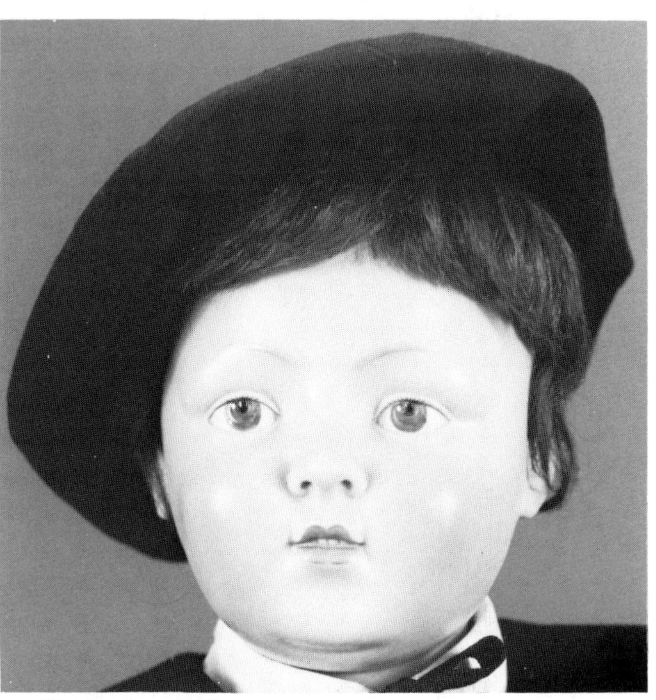

Illustration 253

182

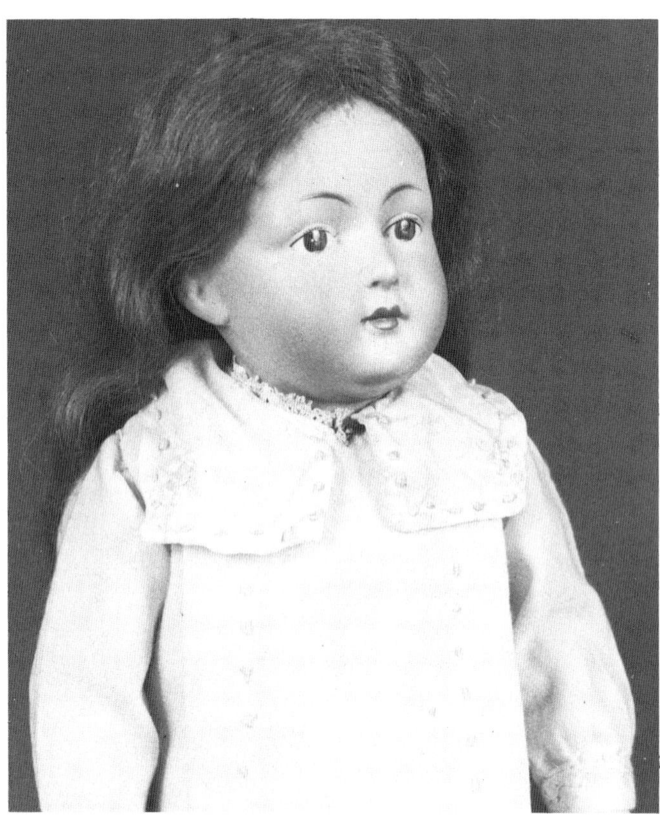

Illustration 254.

Illustrations 254 and 255. Mold number 182 presents a face with a very serious, yet serene expression. Her eyes are larger in proportion to the size of her face, and are different in that a black instead of red eye line was used. Her brown eyebrows give her a more dramatic look than many of the other characters which have pale eyebrows; also her long brown mohair wig emphasizes her dark eyes and eyebrows. Her mouth is given an entirely different treatment in that it is completely closed and rather small; the painting on the lower lip does not completely fill out the molded portion. Her handmade dress is of white silk with feather-stitching trim at the hem and on the cuffs. Her collar and front pleats are trimmed with French knots. On a good quality composition body, she is 15in (38.1cm) tall. *Richard Wright Collection.*

183

Illustration 256. This 15in (38.1cm) girl incised "183" appears to have the same upper face as mold 182. In fact, one wonders if the same child might have posed with varying facial expressions for several of these dolls. However, her face is much less serious, her mouth has lips slightly parted with molded teeth. Her large brown eyes have tiny painted upper lashes and her eyebrows are light. Her brown mohair wig with braids coiled above the ears appears original, but she has been redressed. This model has also been found with glass eyes. *Clendenien Collection.*

184

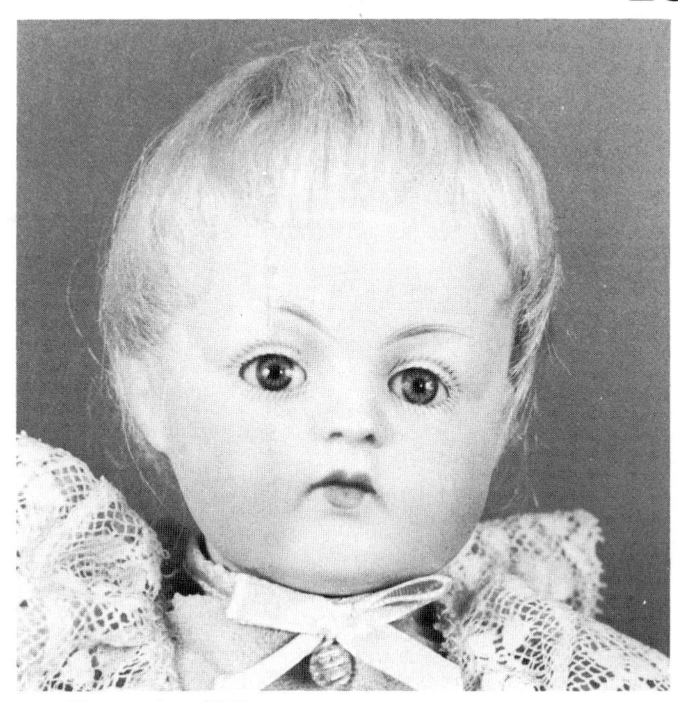

Illustration 257.

Illustrations 257 and 258. The face of this pert little character, mold number 184, is enlivened by his blue glass eyes. His face has particularly good modeling with cute plump cheeks and indentations at his temples which look very natural. His mouth is tiny and molded with a definite downturned upper lip. His lower lip is painted in an elongated fashion which does not even fill out all of the space allotted for it. He has long painted eyelashes and light fly-away eyebrows. His short mohair wig covers a plaster dome. His blue suit with wide lace trim makes him a fancy fellow indeed; he is 11in (27.9cm) tall on an excellent jointed composition body with a pink finish. *Rosemary Dent Collection.*

Illustration 258.

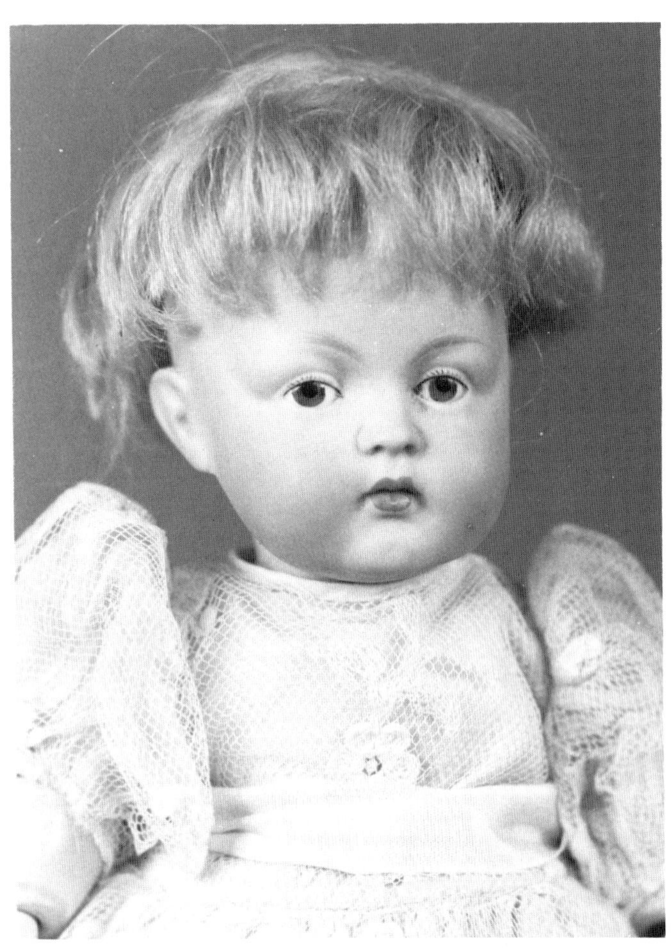

Illustration 259.

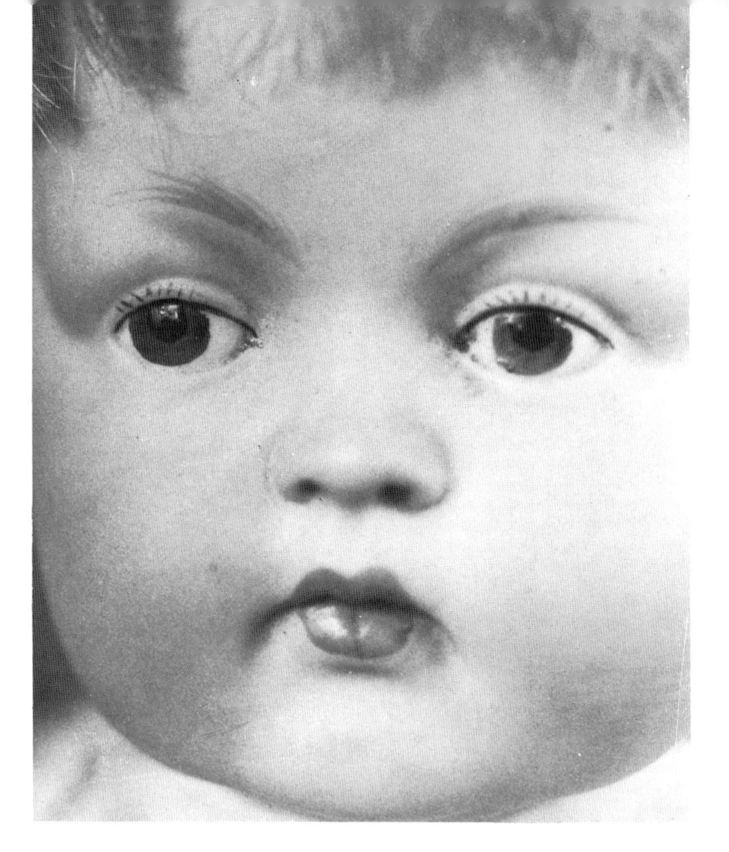

Illustrations 259 and 260. This 11in (27.9cm) example of mold 184 has brown painted eyes with very good molding and decorating detail, such as molded eye sockets, black eye line, red corner eye dots, and tiny painted lashes above the irises. Her eyebrows are very light and one is more upswept than the other. Her downturned mouth is small with a definite molding crease on the lower lip. Her chubby chin has a center molded dimple. She has her original blonde mohair wig over a plaster pate. Her pink composition body is of excellent quality and has the red "K & W" stamp of the König & Wernicke company of Waltershausen which apparently also used Kestner heads, a fact which is not surprising since both companies were in the same town. The close view gives a marvelous look at the eye and mouth detail. *H & J Foulke.*

185

Illustrations 261 and 262. This very appealing little girl is an example of mold number 185, showing marvelous detail in the facial modeling. The light has hit her right eye at exactly the proper angle to illustrate the molded eye socket and upper and lower lids which have been mentioned before. She has a molded indentation at her temples, hollows under her eyes, and tiny cheek dimples. Her open/closed mouth shows a wide smile and four molded teeth. Her large brown painted eyes are looking slightly to the left and are emphasized by her tiny painted upper lashes and red eye line. She is 18in (45.7cm) tall. *Sheila Needle Collection. Photograph by Morton Needle.*

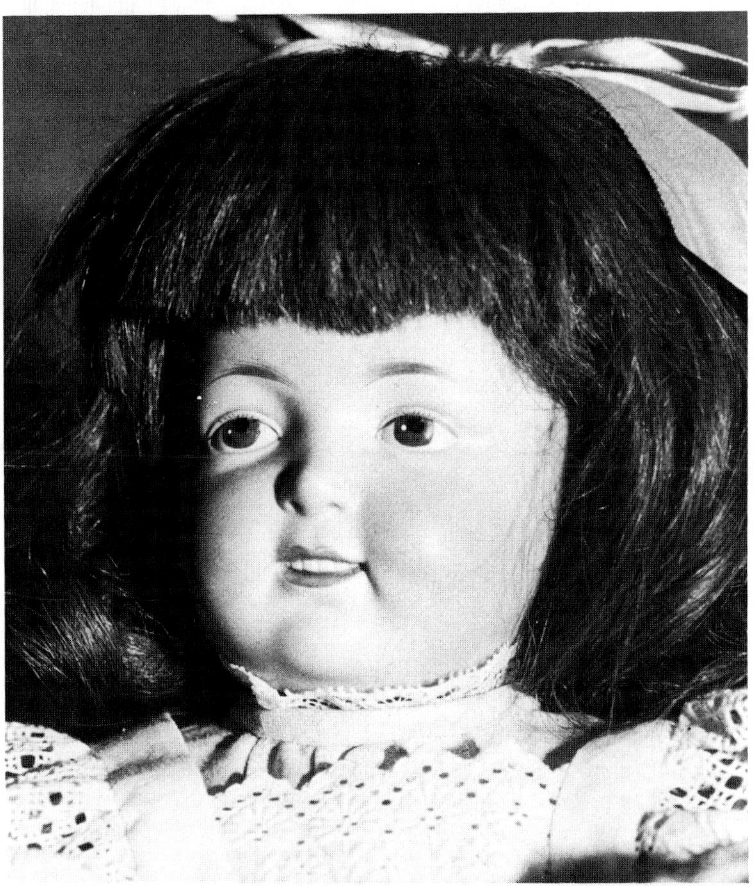

Illustration 262.

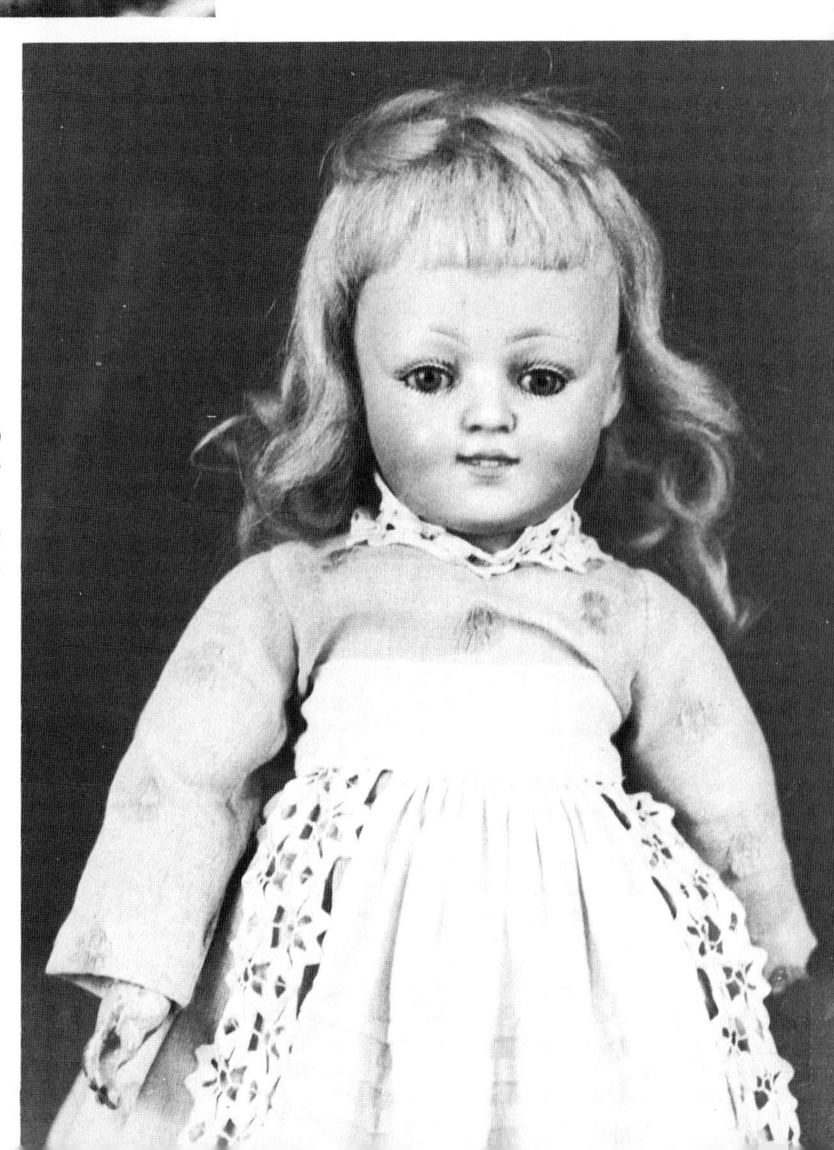

ustration 263. This darling little 11½in (29.2cm) rsion of the 185 mold has sparkling blue-gray glass es with delicately painted lashes which make her a ry pert little girl. She has tiny cheek dimples just like r larger sister. Her open/closed mouth has upturned rners which allow four molded teeth to show. Blonde oked eyebrows match her blonde mohair wig. *Edna ack Collection.*

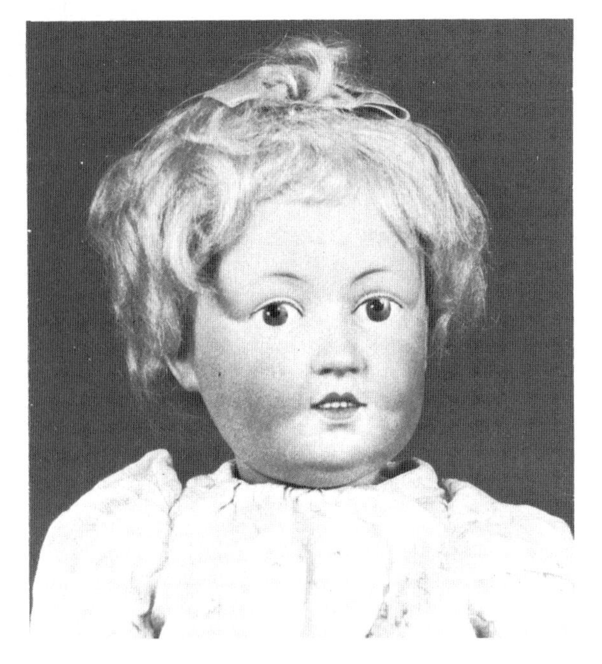

186

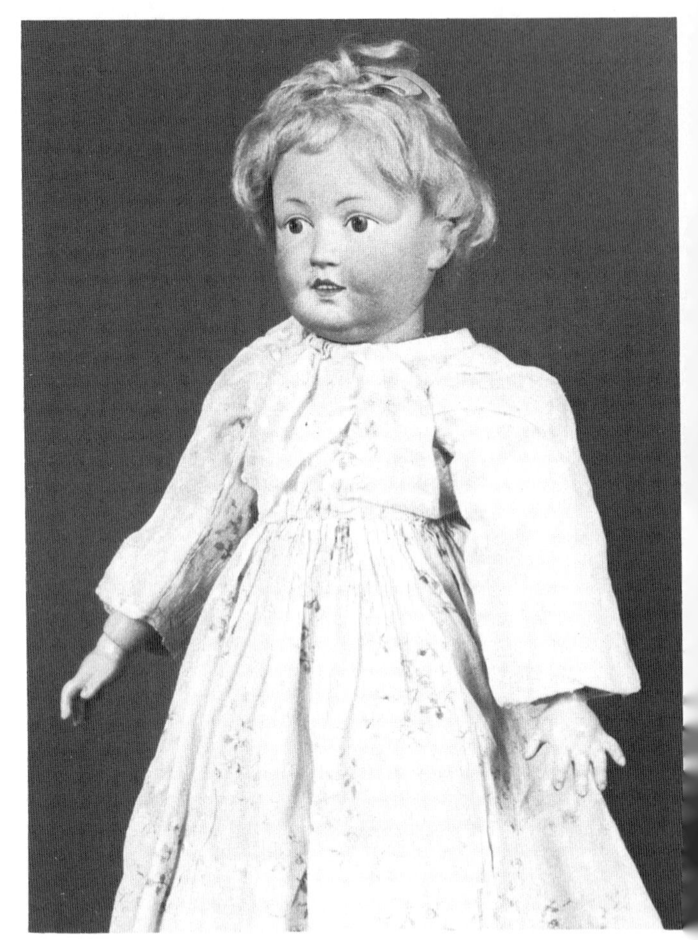

Illustrations 264 and 265. This is one of the heads from the character set shown in *Illustration 278* made into a complete doll, mold number "186" written in black on her front neck and the back of her head. Her short mohair wig is styled in soft curls all over her head and has a blue satin bow on the top. Her face has excellent eye and cheek molding, but her lower face is outstanding. Her open/closed mouth shows six molded upper teeth, and she is smiling, but not so broadly as number 185. She has a definite lower cheek line extending from each corner of her mouth and a prominent indentation under her lower lip. She is one of the most outstanding faces in this series. This model has also been found with glass eyes. *Richard Wright Collection.*

Illustration 265.

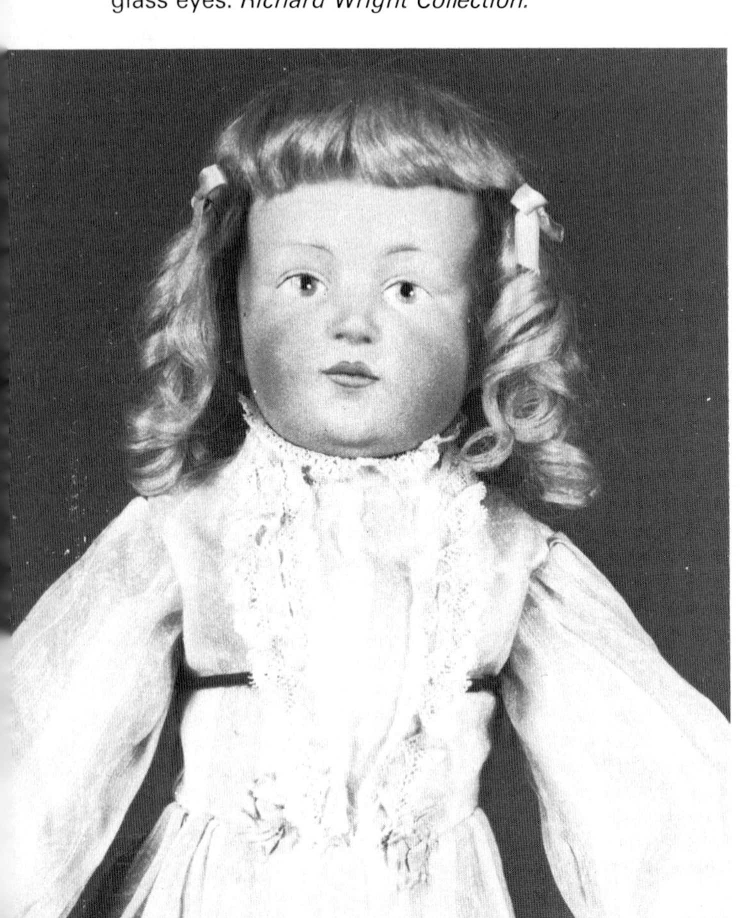

187

Illustration 266. The favorite of many who have seen these dolls is mold number 187, a child with a very solemn expression. Again, comparing the modeling through the top of the face of the last few dolls, one wonders whether the same child posed in varying moods for the artist who designed these dolls, as the eye molding remains virtually the same. This girl is 18in (45.7cm) tall with a beautiful shade of gray-blue painted eyes which look straight ahead. Again, her cheeks are very full and she has a plump double chin. Her closed lips are full with a darker red line dividing them. Her curly light blonde mohair wig is typical of the Kestner company. *Jane Alton Collection.*

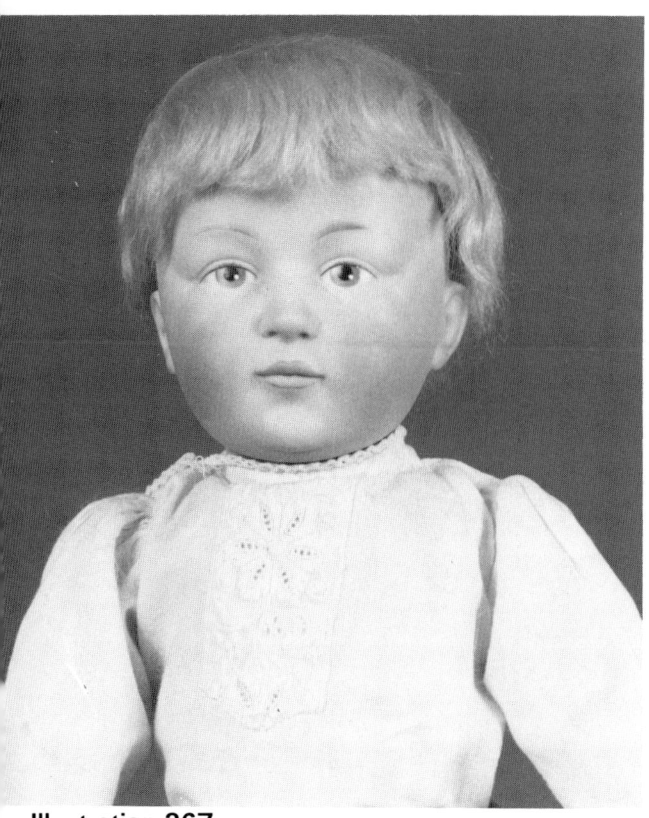

Illustration 267.

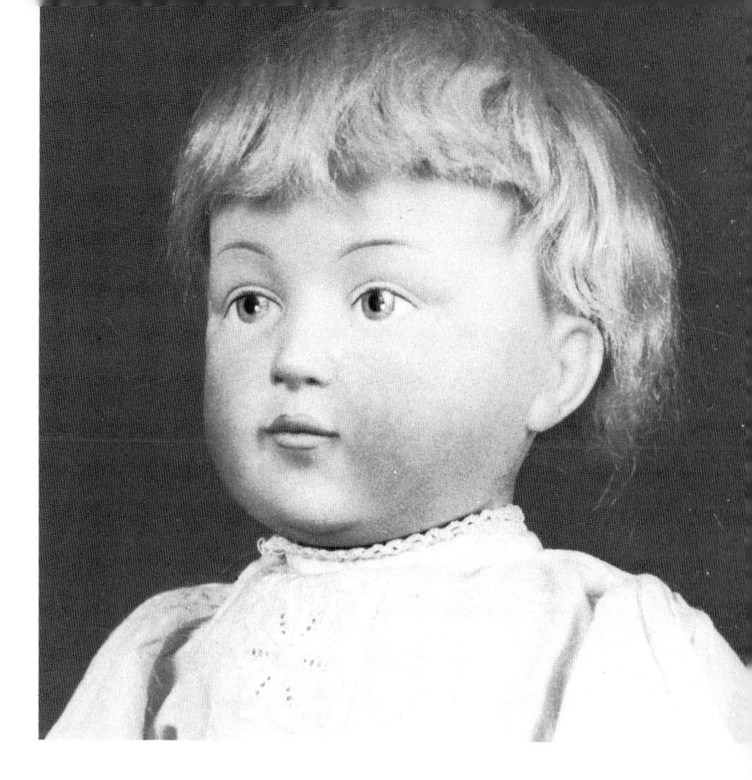

Illustrations 267 and 268. Here is the same mold number 187 in the same 18in (45.7cm) size given treatment as a boy character. His face appears to be fuller than that of the girl, but that is simply because of the shorter hair. The artist, however, has painted his mouth in a different fashion, as the lips are longer and less full. His short mohair wig is original, and he has been made up from the character doll set shown in *Illustration 278. Richard Wright Collection.*

---

188 This number has not been found to date.

189

Illustration 269. This character is a special doll to collectors who favor those with glass eyes. Hers are of the familiar Kestner gray and they do add a sparkle to the face and are appealing, but are actually much less artistic than the lovely Kestner painted eyes. In spite of her closed mouth, she is not really solemn-faced because she has two lovely dimples on either side of her mouth. She has stroked blonde eyebrows to match her original blonde mohair wig with coiled braids. Her excellent quality pink composition body has the red stamped "Germany" in a box with a size number as is often found on Kestner dolls. In addition she has the "Made for//G. A. Schwarz//Philadelphia" paper label found on other Kestner characters of this series. She is wearing a most luxurious white lawn dress with pink ribbon trim and is 16in (40.6cm) tall. *Richard Wright Collection.*

Illustration 270. This doll measures 15½in (39.4cm), but her face is exactly the same size as the doll in *Illustration 269* and she has the same mold number "189." These glass-eyed pouties are truly outstanding Kestner dolls. She has long thin lips with a contrasting red lip line separating them. Her eyebrows are blonde to match her blonde mohair wig. She is on the pink composition body and also has the "Made for//G. A. Schwarz//Philadelphia" sticker. *Jane Alton Collection.*

190

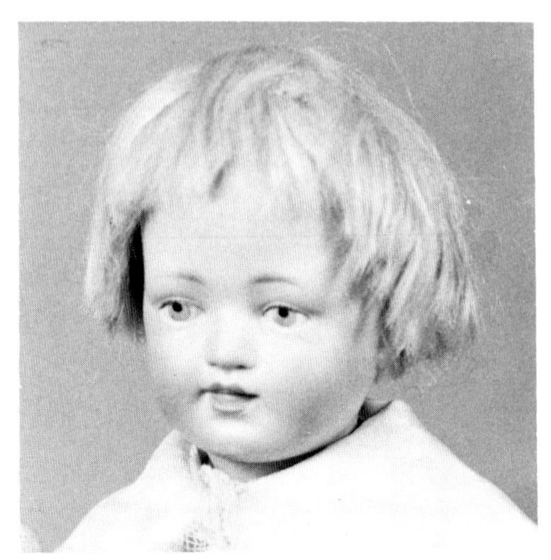

Illustration 272.

Illustrations 271 and 272. Just 12in (30.5cm) tall, is this sweet little boy incised "190." Different from the other characters shown, he has a complete Kestner-type mark "made in//b Germany 3//190." He has pale blue painted eyes without the usual white highlight, but with the tiny enameled upper lashes over the colored part of his eye. He is very similar to the 178 mold, but his mouth is slightly wider with more of a hint of a tongue. *Ruth Noden Collection.*

206

Illustration 273. Quite a few numbers are skipped to the next character child which is 206. She is also 12in (30.5cm) tall, and is wearing a very lovely original costume from Lapland. Her white apron and sleeve tops are covered with red embroidery. Her headdress is red and gold with braid and bead trim. Her outstanding facial feature is her tiny brown glass eyes with painted lashes and blonde eyebrows. Her tiny mouth has good molding detail and she has a dimple in her chin.

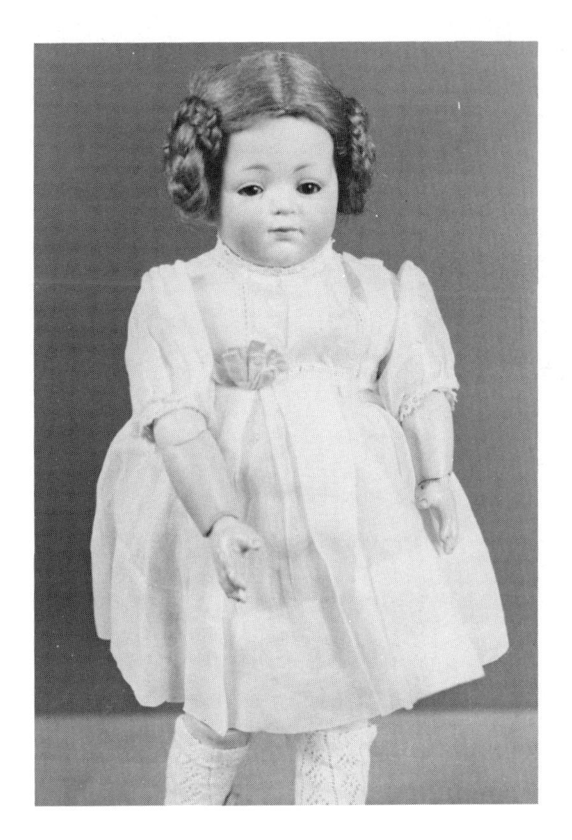

Illustration 274. This sweet example of mold 206 is 11½in (29.2cm) tall on a fully-jointed composition body. Her tiny glass sleeping eyes are enhanced by painted upper and lower eyelashes. Her closed mouth is pouty. Her mohair wig features braids coiled at the sides, a very traditional European style. This mold has also been found on a 19in (48.3cm) size. *Esther Schwartz Collection.*

208

Illustration 275. This little girl, also 12in (30.5cm) tall, is definitely a sister to the 206 doll but she in incised "208." She also has the tiny brown glass eyes and a very high cut crown in both front and back. The molding detail around her mouth has upper and lower lip as well as chin and lower cheek detail. She is on a pink composition body and is wearing a lovely, possibly original, white cotton dress with pink flower design and ribbon insertion at waist and sleeves. Her brown mohair wig with braids covers a plaster dome. Since she had to be repaired, her eyebrow painting should be disregarded, as the restorer did not know how to paint them properly. *Pearl Morley Collection.*

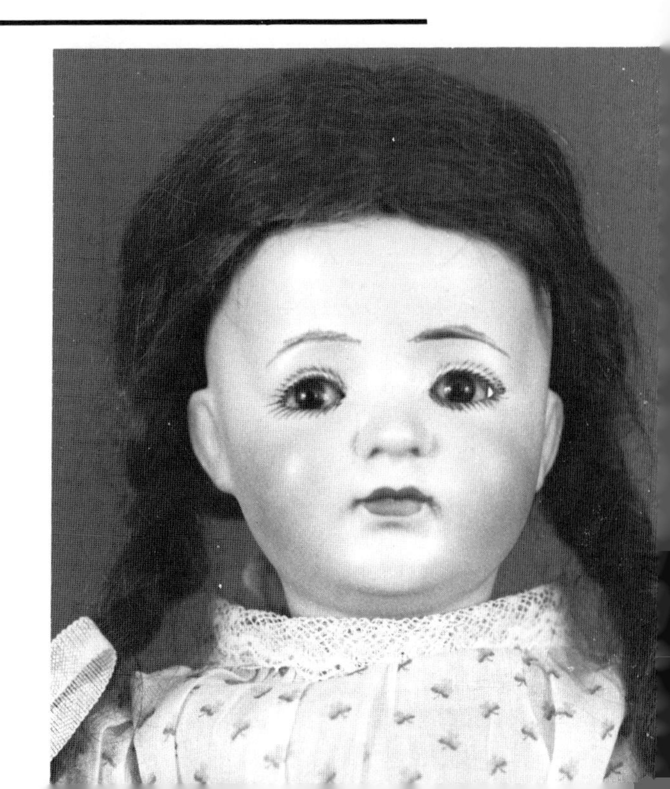

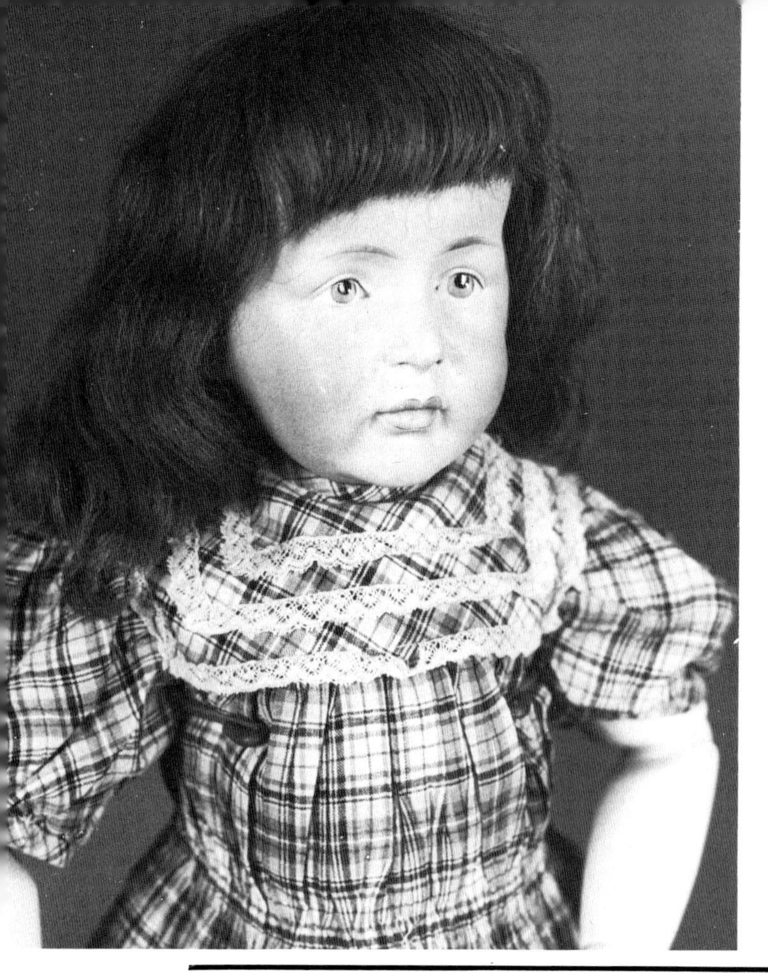

Illustration 276. This expressive girl at 23in (58.4cm) tall, is the largest of the Kestner character children and is an example of mold 208 with painted gray-blue eyes. In this lovely large size it is much easier to see the detail around the cheeks and mouth of her well-modeled face, her wide pug nose, her double chin, and her substantial pouty lips. Her glossy molded eyebrows are light brown. A brown human hair wig covers her plaster dome. Her eyes are outstanding for the painting detail characteristic of this series including dots in the eye corners, tiny glossy painted upper lashes, and painted red eye line. Also of note are the molded eye sockets and upper and lower lids, details more easily seen in this large size. Her complete mark is "K made in 14//Germany//208." *Richard Wright Collection.*

212

Illustration 277. This marvelous little 12in (30.5cm) character girl is so chubby-cheeked that she looks like she has mumps! She is incised "212," and fits in well with this character series. She has the tiny glass eyes typical of the Kestner characters. With her downturned upper lip, there is no doubt that she is pouting. She has lightly stroked eyebrows although hers are a shade darker than those of some of the other characters. Her blonde mohair wig covers a plaster dome. *Becky Roberts Lowe Collection.*

Wonder Doll

Both Kestner and Kämmer & Reinhardt made boxed character doll sets which contained one complete doll and three interchangeable heads, one of which was a conventional dolly face and three of which were character faces. The composition of the Kestner sets varied as to which character heads were included. There seem to be more of the Kestner sets known than the Kämmer & Reinhardt ones. Unfortunately, the number of boxed sets intact is dwindling; with the high prices the character faces are commanding, people are putting the extra heads on bodies and selling them as individual dolls.

Character doll sets were advertised in *Playthings* in January 1910. One set is pictured, but it is impossible to tell which company made it, as the ad mentions both Kestner and Kämmer & Reinhardt. The picture is really a line drawing and the artist made several of the faces look alike, although they all have different wigs. Anyway, the ad puts Kestner into the character doll market within a year of its introduction, perhaps indicating that the firm was better at imitation than innovation. In Germany Kestner advertised this set as "Wunderpuppe" (Wonder Doll).

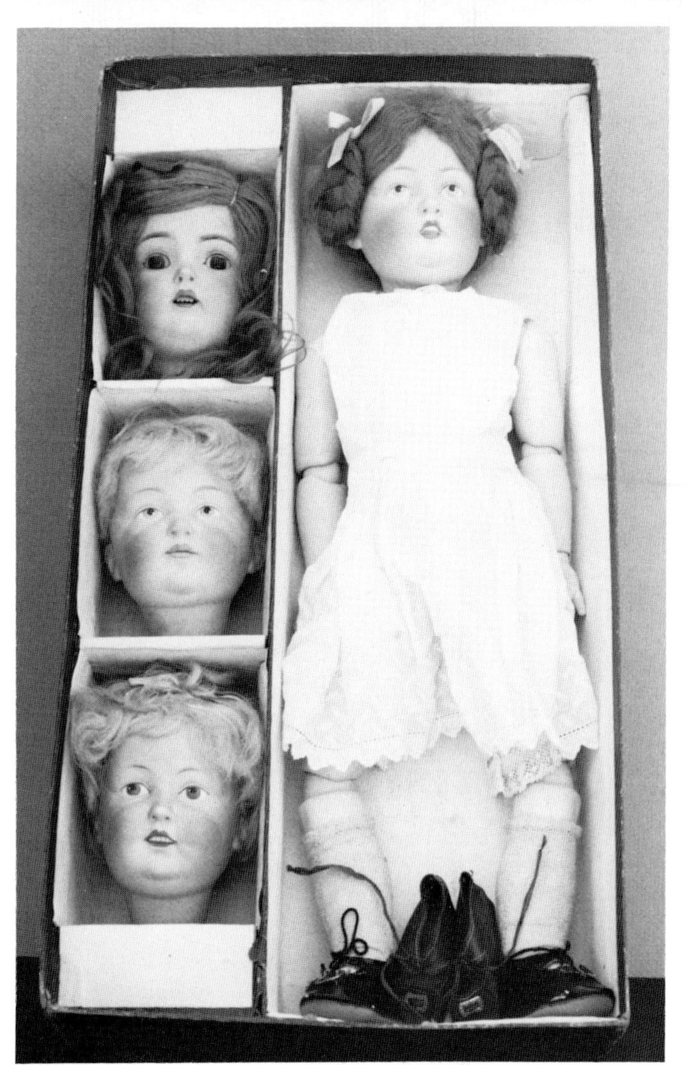

Illustration 278. The complete doll in this set is incised "180;" she is on an excellent quality pink-toned jointed composition body and is 18in (45.7cm) tall. She is wearing her original underwear consisting of a one-piece combination with turquoise ribbon trim and a full slip of fabric and trim which matches her original dress which is shown in *Illustrations 248* and *249*. Also original are her turquoise lace hose and her black leatherette shoes. In the box is an additional pair of brown leather shoes.

The dolly face in the set is the standard mold number 171, which was a very popular Kestner number. She has brown sleep eyes, a dark blonde mohair wig, side-parted with curls, and a red bow. Her eyebrows are of the molded and painted type; her open mouth has four upper teeth. The character head in the center section is mold number "187," so indicated on the front neck in black pencil and incised in the plaster dome. The short blonde mohair wig enables him to be made up as a boy; he is shown complete in *Illustrations 267* and *268*. The head at the bottom of the row is mold number "186," indicated in black pencil on both the front and back of the neck. The doll is shown complete in *Illustrations 264 and 265. Richard Wright Collection.*

Illustration 279. The end of the maroon box which contains the character doll set with the Kestner paper label showing not only the Kestner name, but the famous Kestner crown, although this one is slightly different from both the 1895 and 1915 trademarked ones. *Richard Wright Collection.*

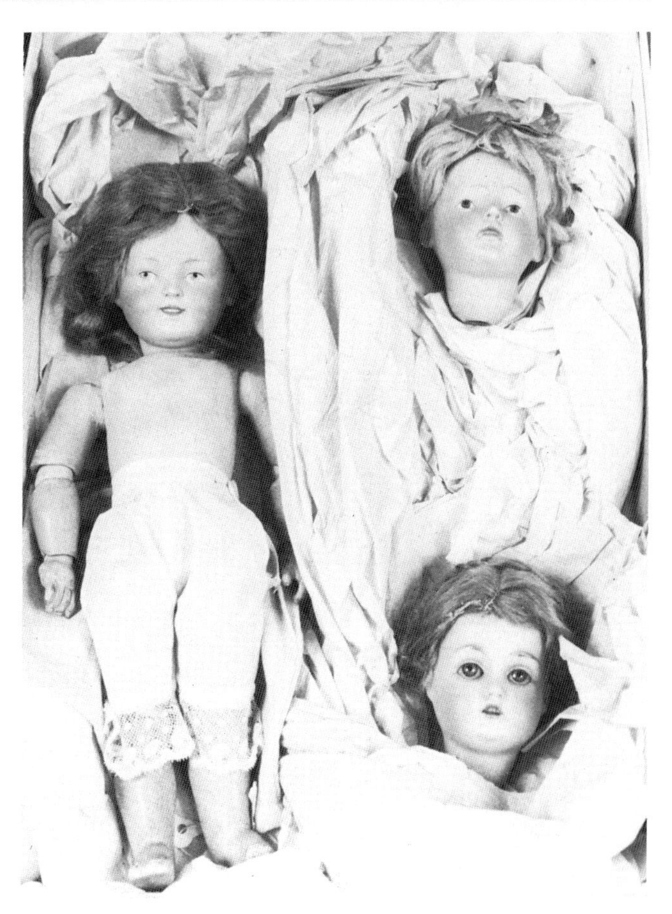

Illustration 280. These are three heads from an 11in (27.9cm) to 12in (30.5cm) character doll set. The head on the complete doll is mold number 185; the head at the top right is incised "178;" and the dolly face is mold number 174, a number which does not turn up very often and which is usually then only found in small sizes. All of the boxed character sets which I have seen have been either the 18in (45.7cm) or the 11in (27.9cm) to 12in (30.5cm) sizes. *Jane Alton Collection.*

241

It is hard to know exactly where to put this doll as she is really in a category of her own which is later character children. She is definitely not of the series beginning with 178 and finishing with 212. She has more of a relationship to *Hilda* and the 260 mold which is more often a small child rather than a baby. In fact there is a doll in the last Kestner catalog from about 1932, which is reprinted in *Die Deutsche Puppen-industrie 1815-1940*, which is very much like her. The 241 doll has a marvelously expressive face. Her eyes sparkle and she looks as though she is just ready to start speaking. She has a much more mature face than any of the other dolls in this series.

Illustration 281. This 241 mold is definitely a favorite with Kestner collectors, and a very difficult one to find. For such an appealing doll, it is hard to understand why so few were made. Apparently it just came on the market at the wrong time. Judging from its number it was made just before or about the same time as the *Hilda* doll. Perhaps the customers were clamoring for babies at that time, around 1914, and she just did not take hold. Her brown sleep eyes are small in proportion to the size of her face, yet not so small as the eyes of the earlier character dolls. She has lightly stroked eyebrows on a slightly molded ridge and lightly painted eyelashes. Her open mouth has a slightly drawn up upper lip, four upper inset teeth, and a separate tongue. She is 18in (45.7cm) tall on a very heavy jointed composition body. *Esther Schwartz Collection.*

Illustration 282. This is such a marvelous illustration of the 241 doll that little need be said about her. All of the reasons why this doll is so sought after are shown here. *Becky Roberts Lowe Collection.*

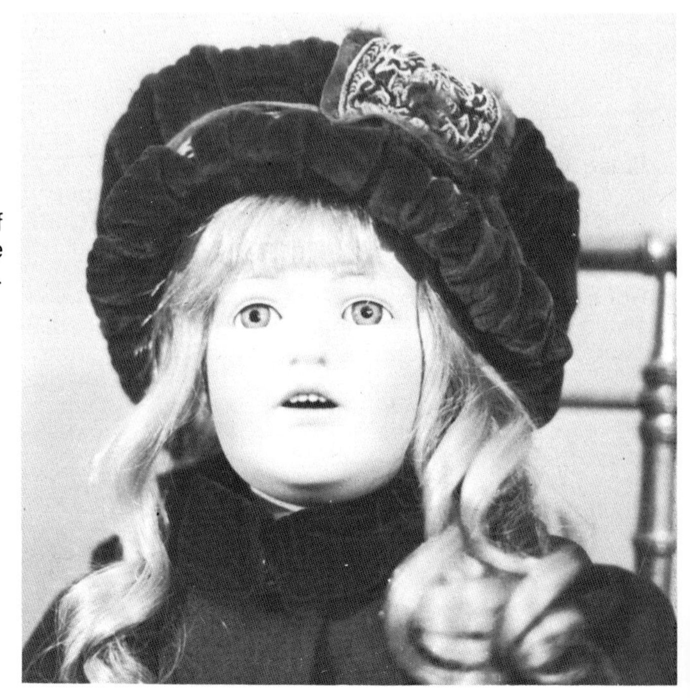

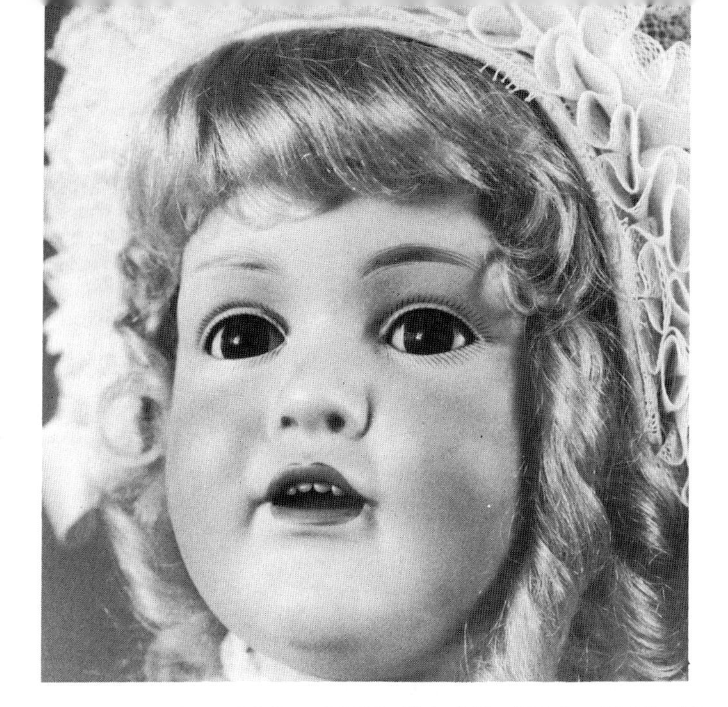

Illustrations 283 and 284. This 23in (58.4cm) 241 girl has a blonde mohair wig styled with bangs and finger curls which match her light blonde eyebrows. Her small brown eyes are set off by lightly painted eyelashes as well as real upper ones. Her mouth is very similar to that of the *Hilda* doll, but the upper lip is not so raised and the whole mouth is wider and includes four teeth, to contrast to *Hilda's* two, as well as a tongue. She is all dressed up in a white eyelet dress of old fabric and a white ruffled bonnet. *Mary Lou Rubright Collection.*

Illustration 284.

249

Illustration 285. This sweet character child, mold 249, is a very difficult number to find. She is definitely another version of the 241 child with a different mouth expression, which is much smaller in width and not so wide open. Her lips are tighter; she has four upper teeth. Her eyes are tiny, with both real upper eyelashes and painted upper and lower ones. Like the 241 she has a slightly molded eyebrow ridge, but not molded eyebrows. Her light brown eyebrows go well with her new brown human hair wig. Her coloring is natural with blushed cheeks. She is marked "a½ made in 4½//Germany.//249.//J.D.K." She is 13½in (34.3cm) tall on an excellent jointed composition body. *Virginia Yeatman Collection.*

Character Babies
200 Series Character Babies

Although offering an extensive line of Kestner dolly faces in their 1910 catalog, Butler Brothers showed only one character baby, which appears to be a K ★ R 100 Baby. It would seem that if Kestner had character babies at this time, at least one would have been offered. He was probably just behind Kämmer & Reinhardt in production of the babies, but not quite early enough to hit a 1910 catalog for which merchandise would have been chosen by at least the middle of 1909. The first ad for Kestner character babies appeared in *Playthings,* January 1910, where they were said to be "true to life" and presented in a "large variety of natural facial expressions." K ★ R character dolls were also advertised by Borgfeldt in the same ad; the dolls pictured appear to be K ★ R molds 100 and 101 and not Kestner dolls which perhaps were not quite ready when the ad was prepared. Of course, it does take some time for dolls to filter down from the manufacturer to the store catalogs.

If Kestner made his molds in order, then the first character baby widely produced would have been number 211. A doll which appears to be of this mold was advertised in 1912 as "Natural Baby." A George Borgfeldt ad in *Playthings* of January 1913 called Kestner character babies "The Great Hit of the Past Holidays. They are so cunning and life-like that no little girl is satisfied until she possesses one." The dolls were noted as having real skin wigs. Borgfeldt further stated that sales for 1912 of Kestner character babies went so far beyond their expectations that they were unable to satisfy the demand. The two dolls shown are photographs and appear to be mold 211 with skin wigs. One baby has the cardboard crown tag as shown in *Illustration 12.*

The majority of the Kestner baby heads are on excellent quality composition bodies with curved arms and legs. See *Illustrations 351, 352 and 353.* The finish is shiny and has a rosy tone which gives the body a natural look. The torsos show anatomical detail through the distended bellies and shoulder blades. The upper arms have chubby fat rolls; the hands are posed with outlined fingernails. It is in the modeling of the legs that Kestner is superb — they have rolls, creases and crevices. The toes are intricately done with joints defined and nails outlined. Although the bodies are usually not marked, they can be distinguished by their modeling detail and hand position — the left hand curves in, the right one out. A Kestner baby nearly always hits himself in the face when he raises his left hand! A few Kestner babies are found on kid baby bodies with rivet joints.

209 and 210

Molds 209 and 210 are shoulder head character babies. Both have lightly molded and painted hair, very large ears and open/closed mouth with no teeth. They appear to be another version of the standard Kestner bald-headed baby shown in *Illustration 351.* Heads are mounted on kid baby bodies with the J.D.K. crown label and bisque lower arms. For a photograph of the 210 mold, see Ciesliks' *German Doll Encyclopedia 1800-1939,* page 150. The 209 mold seems to have a longer and more narrow mouth than the 210. Neither doll carries the J.D.K. initials incised in the bisque but both use the Kestner letter/number size system.

Illustrations 286 and 287. This 20in (50.8cm) long baby is without a doubt one of the finest examples of mold number 211. He has the creamy bisque for which the Kestner firm is noted, beautifully decorated by their talented artists. He has his original blonde mohair wig over his plaster dome. His bent-limb baby body is of the finest composition with a glossy natural color finish. The hands are in the typical Kestner style, the right curving out, the left curving in. This particluar doll is an example of the 211 mold with the open/closed mouth design; probably fewer have this type of mouth than the open one with teeth. However, the most outstanding feature of this doll is his completely original outfit. He was purchased just this way from Hamley's Toy Store in London, England. He is wearing voluminous undergarments topped by the elaborate christening gown with front inset of two types of lace. His ivory silk coat and hat complete the costume. *Richard Wright Collection.*

211

Mold 211 is the first Kestner head to carry the "J.D.K." initials (Johannes Daniel Kestner, although Adolph Kestner, his grandson, was head of the firm at this time.) A sample of this new mark is shown in *Illustration 292.*

There is no doubt that in creating this mold, Kestner did an outstanding job. Collectors today seem to agree as this particular baby is an especially popular one. It is easy to see that the face was molded from that of a real child. There is excellent modeling through the temples with indentations beside the outer corner of each eye. There are also small hollows under the eyes and chubby cheeks with extra ridges around the mouth including creases at the sides and under the lower lip in addition to a well-defined philtrum. Decoration of these heads is lovely with softly tinted cheeks, attractive curved eyebrows showing brush marks, and finely painted eyelashes. For some reason I find myself referring to dolls of this mold as boys, but I can find no proof that they were intended to be such.

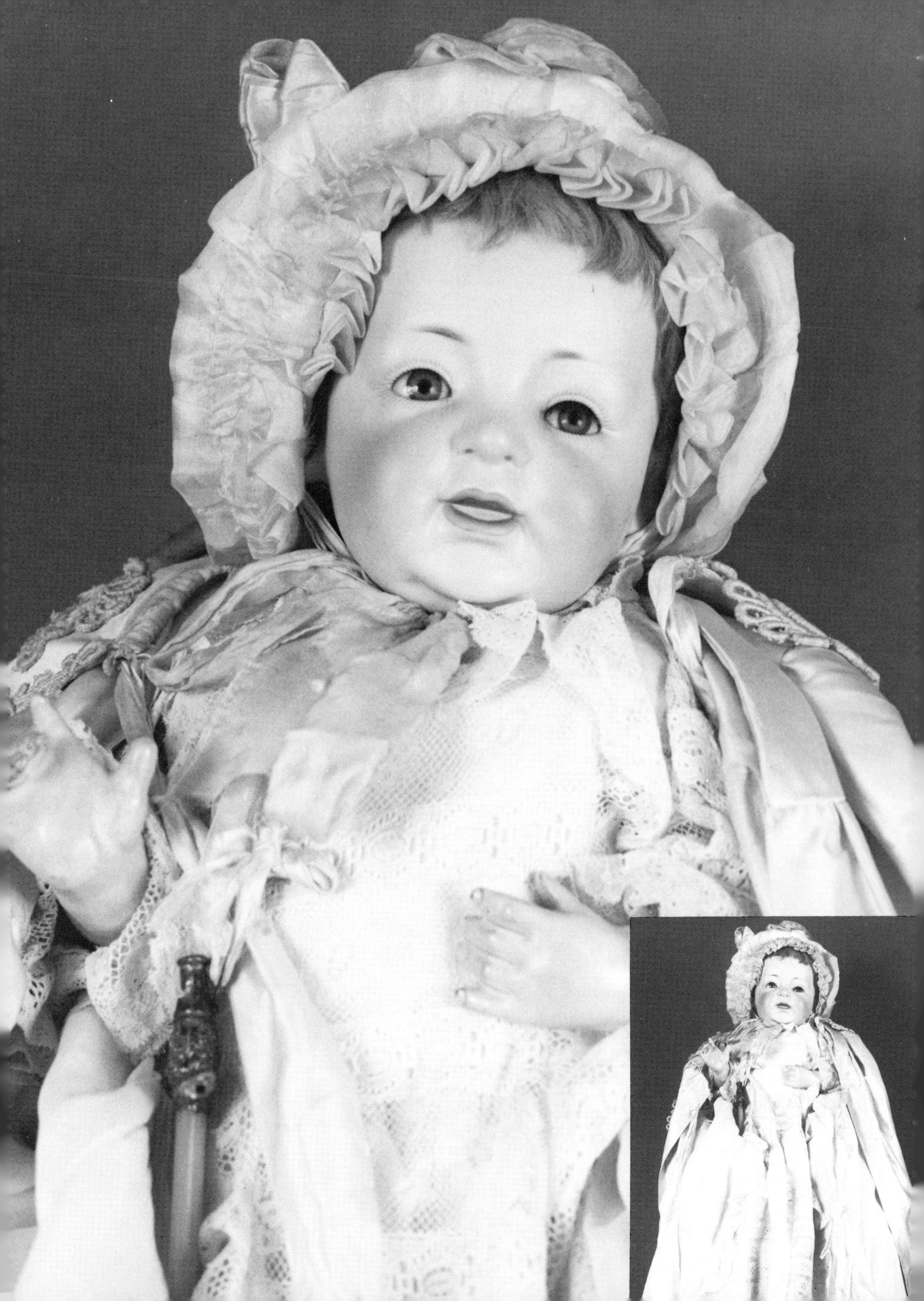

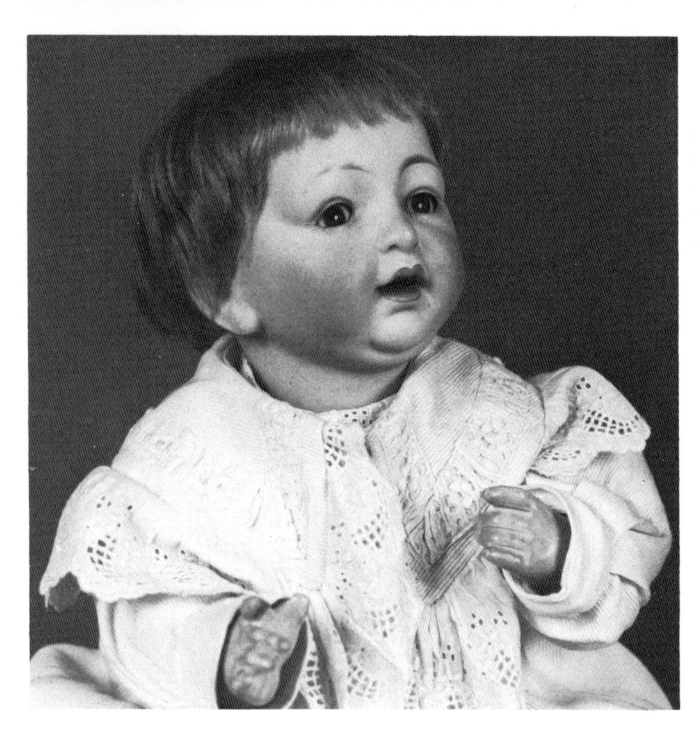

Illustration 288. This 15in (38.1cm) number 211 has the usual open mouth with molded gums and two small lower teeth set quite far back so that they really do not show in the illustration. He has sleep eyes and a light brown mohair wig. Also, the outline of his ear shows it to be fairly large especially when compared to those on dolly-faced dolls. *Richard Wright Collection.*

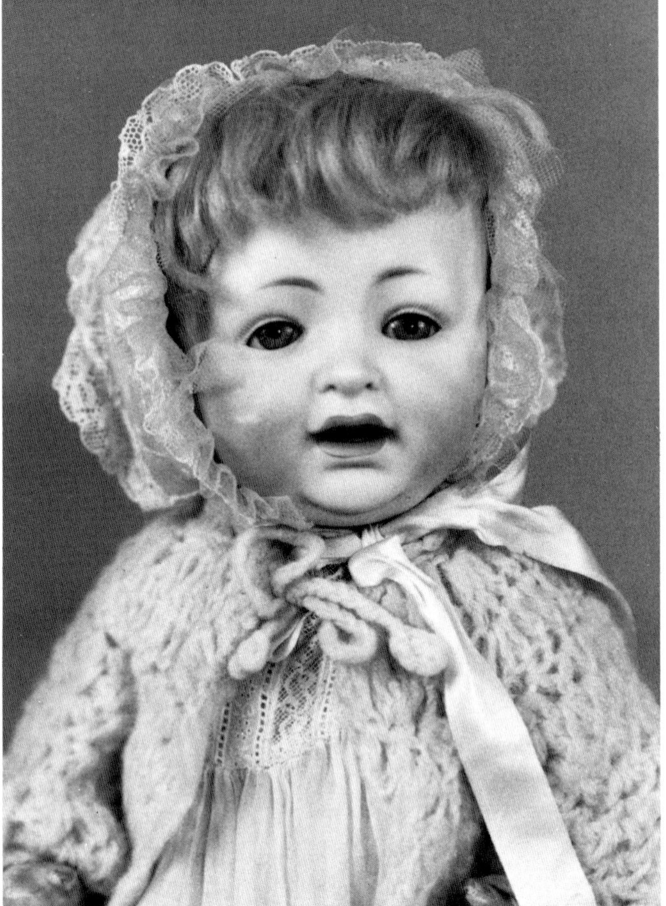

Illustration 289. Here is a 14in (35.6cm) example of mold number 211. The front-on view shows up his fat double chin. He is wearing nice old baby clothes and is holding his glass nursing bottle in a crocheted case tied to his hand with a ribbon drawstring. Most of these bottles are found capless as the nipples have nearly all disintegrated. *H & J Foulke.*

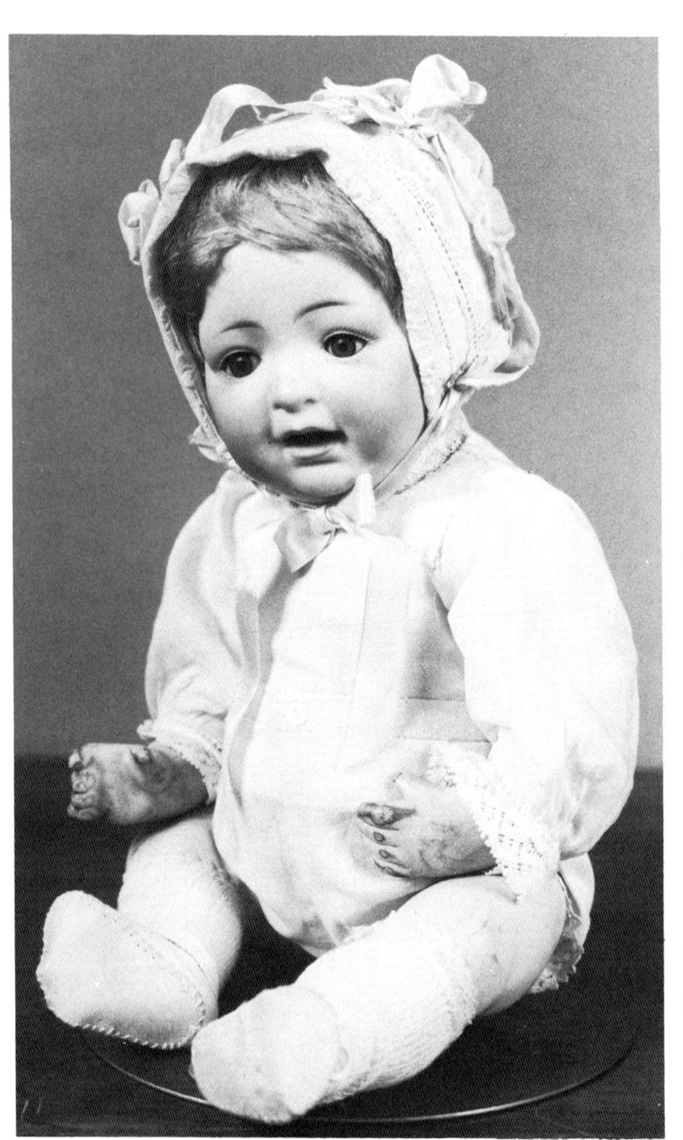

Illustration 290. This small 12in (30.5cm) size of the 211 mold does not have quite as much modeling detail around the mouth and eyes as the larger examples, but it is, nonetheless, a wonderfully cute baby. Even the small Kestner bodies have the right arm going out and the left arm turning in. His one-piece romper suit costume is an interesting departure from the usual long baby dresses put on these types of dolls. *J. C. Collection.*

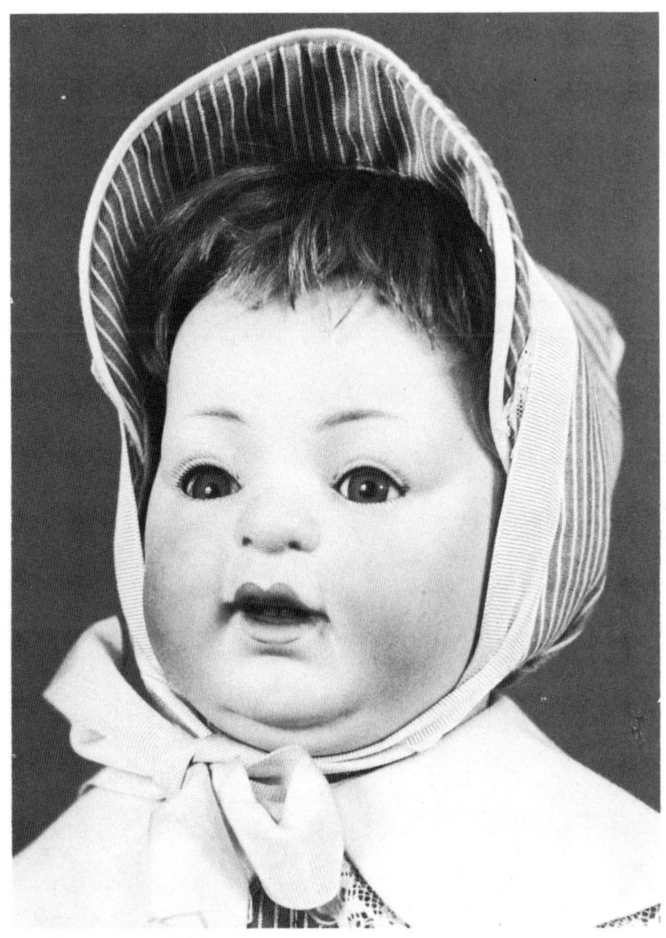

Illustration 291.

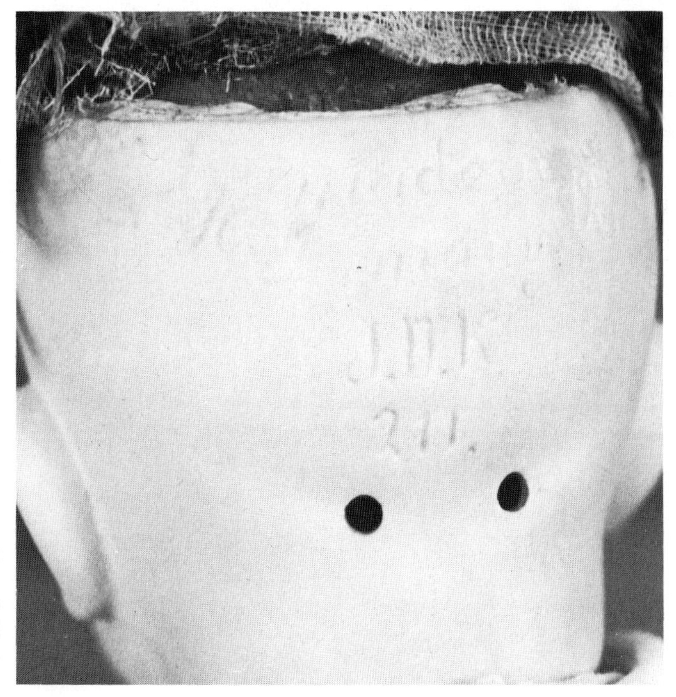

Illustrations 291 and 292. This 17in (43.2cm) 211 is another example of dressing the doll in a romper suit to make him look more like a little boy baby. This outfit is a particularly nice old one of blue cotton with white raised stripes and a matching cap which makes a very appealing doll. He has gray sleep eyes, a very nice original blonde mohair wig, and an open mouth with the two lower teeth. *H & J Foulke.*

Illustration 293. Although most of the Kestner character babies have the mohair wigs since they are fine and soft and more nearly resemble baby's hair than the coarser human hair ones, this 15in (38.1cm) mold 211 has his original animal skin wig, a type sometimes found on Kestner dolls. It is blonde in color and very silky to the touch, but many of these lovely wigs did not survive because the leather backing became brittle and crumbled. This is another example of the open/closed mouth with no teeth. He has blue sleep eyes, a shiny oily finish, and rosy cheeks. His clothes are original, and the owners feel that he was originally purchased in 1916. *Jimmy and Faye Rodolfos Collection.*

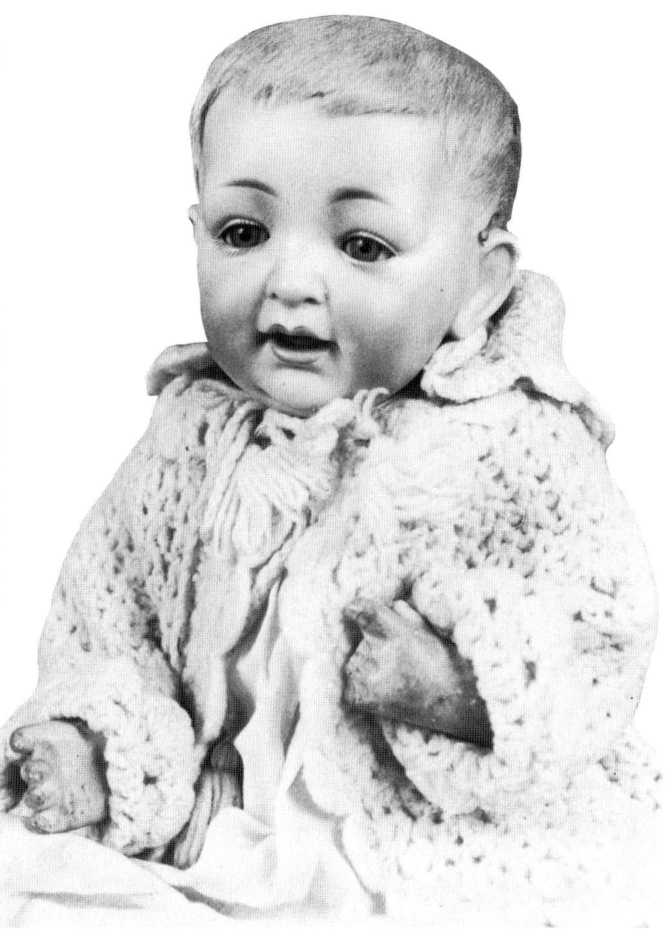

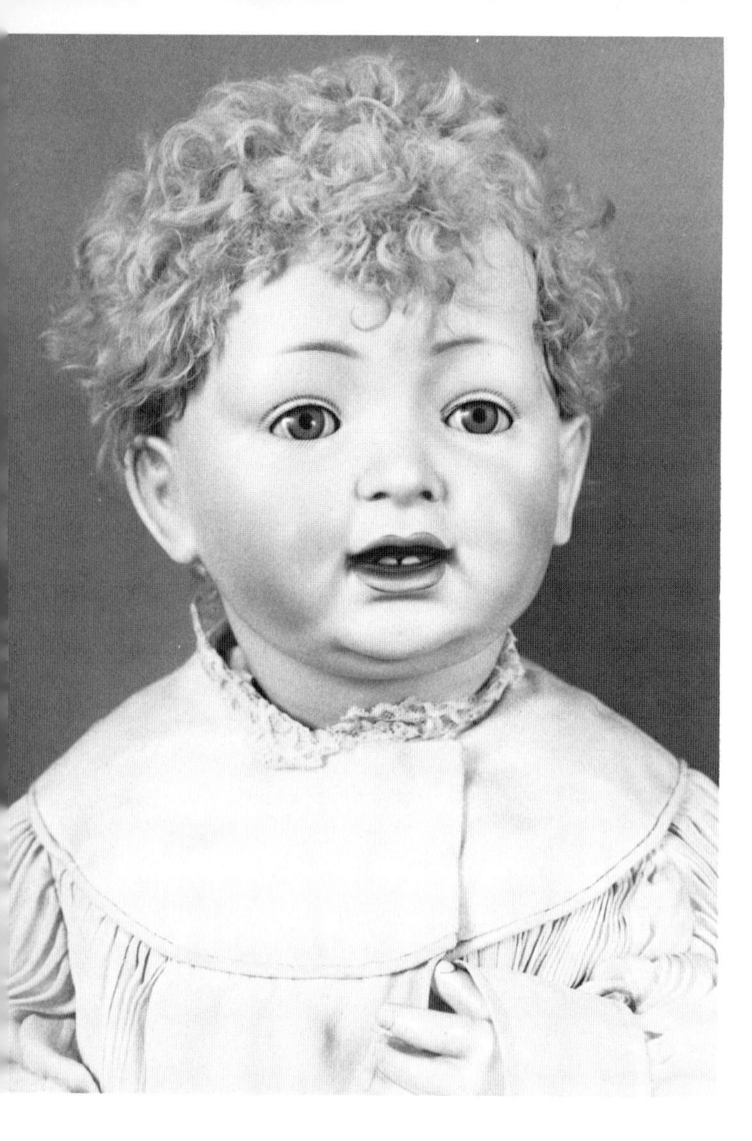

Illustration 294. As far as I can determine size Q 20 is the largest head that Kestner made, and there are not too many of them around. This large J.D.K. 211 baby measures 25in (63.5cm) and makes a larger than life model. Remember when computing size that these babies have a lot more girth than a regular child doll the same height, so there is a lot of doll here for your money. In fact, this doll is so large and heavy that it would have been rather difficult for a child to have played with it. His fuzzy blonde skin wig has been well preserved. Kestner babies with skin wigs were shown in a *Playthings* ad of January 1913. The modeling on his face is outstanding when one examines up close all of the ridges and contours which make him such a lifelike baby. His sleep eyes are blue and he has an open/closed mouth with two lower inset teeth and a molded tongue. *Esther Schwartz Collection.*

Illustration 295. This is another example of the Q 20 size, but this doll is 27in (68.6cm) tall on a fully-jointed chubby toddler body like that on the J.D.K. 220 doll shown in *Illustrations 299 and 300.* She has brown sleep eyes, blonde eyebrows, and very closely painted eyelashes. The illustration clearly shows the molding ridges around her mouth as well as the darker red shading lines along the top edge of her upper lip and the bottom edge of her lower lip. She has an open mouth with a tongue, molded gum ridges and two lower inset teeth. Her blonde human hair wig with long curls is a replacement. *Ruth Noden Collection.*

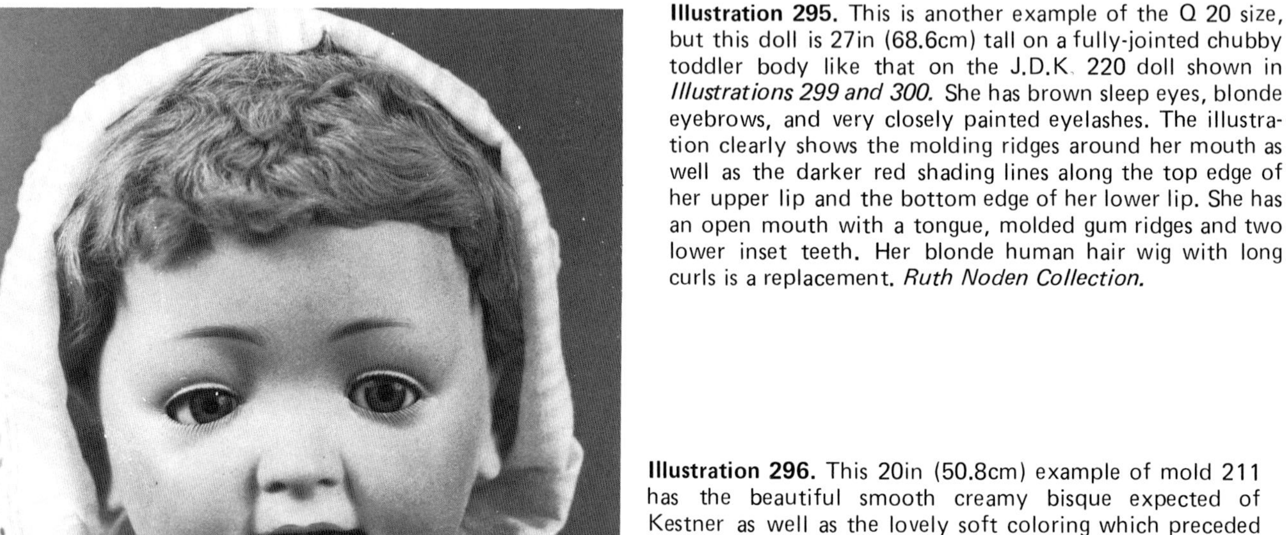

Illustration 296. This 20in (50.8cm) example of mold 211 has the beautiful smooth creamy bisque expected of Kestner as well as the lovely soft coloring which preceded the harsher rosy look of the later character dolls. His original blonde skin wig covers his plaster dome. He has a large number of painted lashes and a molded upper eyelid. *Ruth Noden Collection.*

220

Illustrations 297, 298, 299, 300 and 301. It only takes a quick glance to know that this fellow, mold number 220, is a relative of number 211, but he is much much harder to find, causing speculation as to why so few of this mold were made. The bisque on this doll is very smooth and flawless, his cheeks are rosy to the right degree, not heavily roughed like some of the later dolls. His eyebrows are light auburn, gently stroked on a molded ridge above his eye. His matching auburn wig is of human hair which is appropriate as he is a little child, not a baby. His ears are very large and done with much realistic detail. His face is in a broad smile with wide dimples on either side of his mouth. His lips are very full, and his mouth is open with a separate tongue and two inset upper teeth.

As is common with most of these very large dolls, his head is definitely out of proportion to his body. From the tip of his toe to the top of his head, he is 27in (68.6cm) tall. His ball-jointed body is an excellent example of the typical Kestner toddler body, which was actually a combination of the ball-jointed bodies of the girl dolls and the bent-limb bodies of the baby dolls. This body is quite chunky and chubby when compared to the girl body. Especially notable are the upper legs with the slant hip joint and the fat rolls. The lower legs show great detail in the modeling of the knees, calves, ankles, even the toes with nails and dimples. Character dolls on these types of toddler bodies are more desirable to most collectors than those on baby bodies.

"Herbie" is one of those dolls who just talks to people. Some are like that, just so real looking that they are absolutely irresistible. That is the way "Herbie" was to me. I saw him in a doll shop in 1974, but I did not take him home with me because he was so expensive; $550 for a Kestner character, other than *Hilda*, was certainly a top price in those days. But I found over the next few days that I could not forget that little boy doll sitting in that antique shop. A phone call assured me that he was still there, so off I went the following week to get him and he has been sitting in my living room ever since. People coming to my house for the first time do a double take when they see him because he looks so much like a real little boy sitting there in his chair. I have never regretted the day "Herbie" came to stay with us. My daughter named him after a character in a book we were reading -- a chubby little boy named "Herbie" and the name just seemed right.

All of the mold 220 dolls I have seen have been on toddler bodies. A doll which appears to be a 220 is shown in a Kestner advertisement of 1914. *Jan Foulke Collection.*

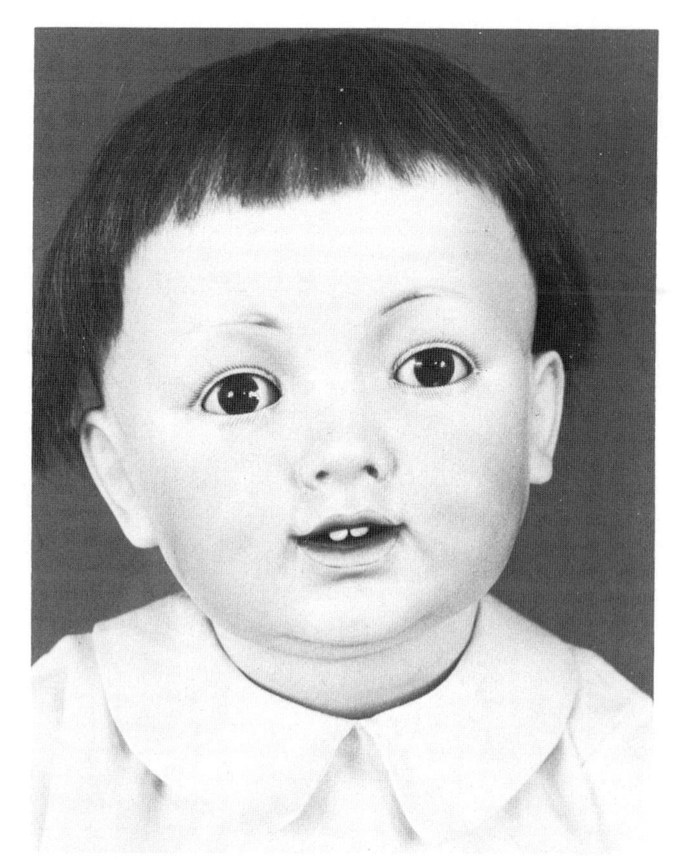

Illustration 297.

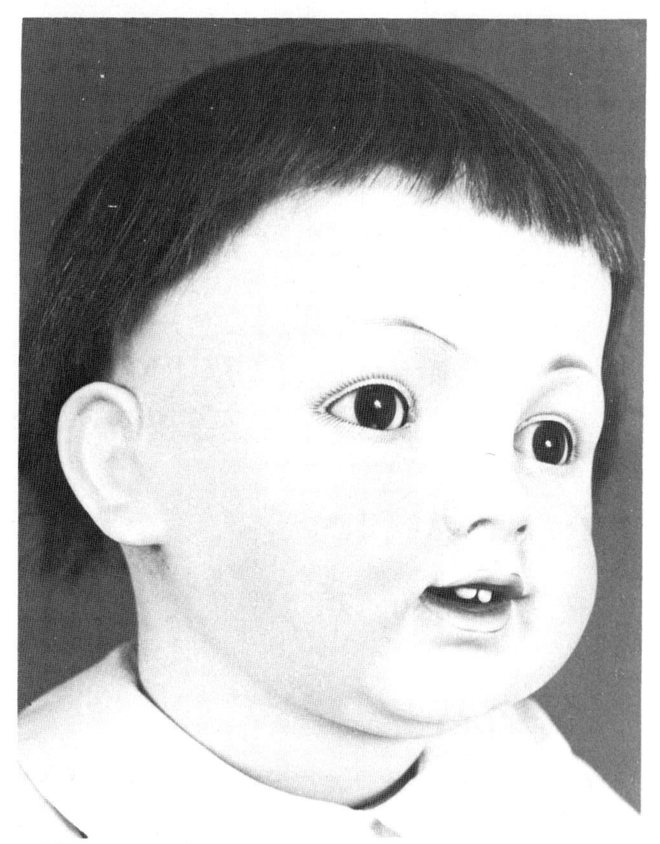

Illustration 298.

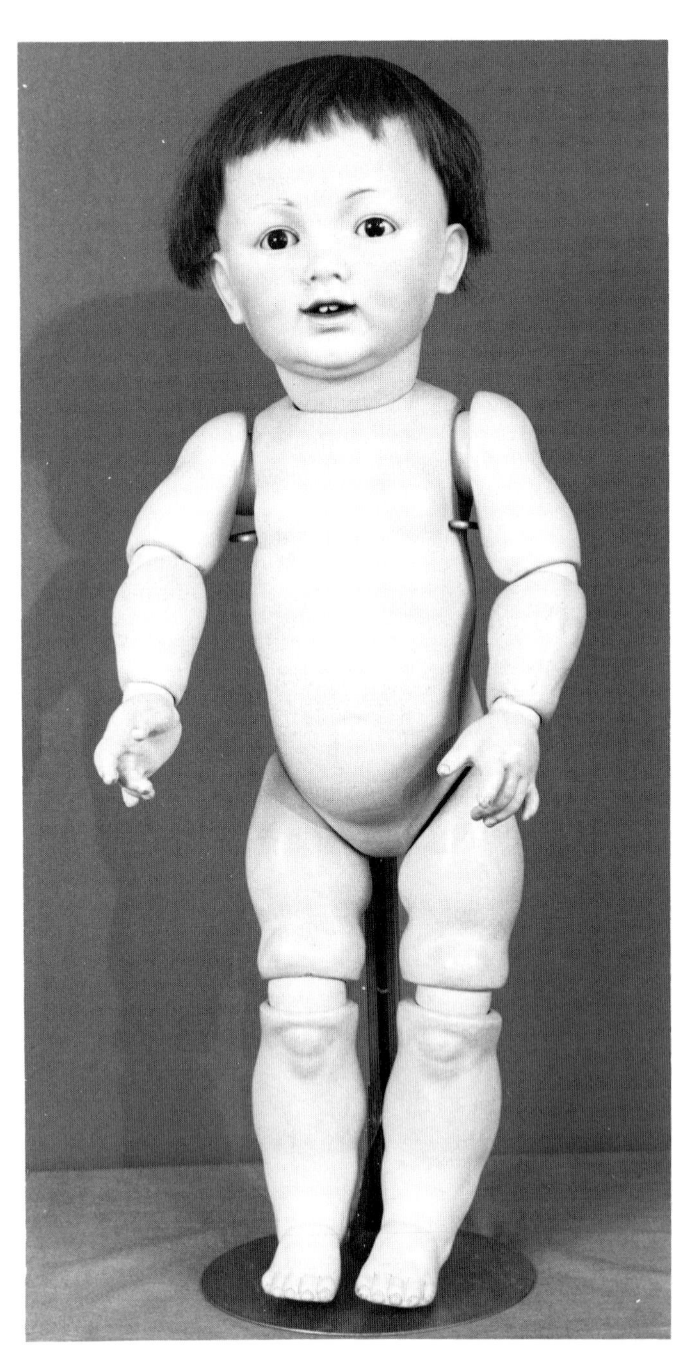

Illustration 299.

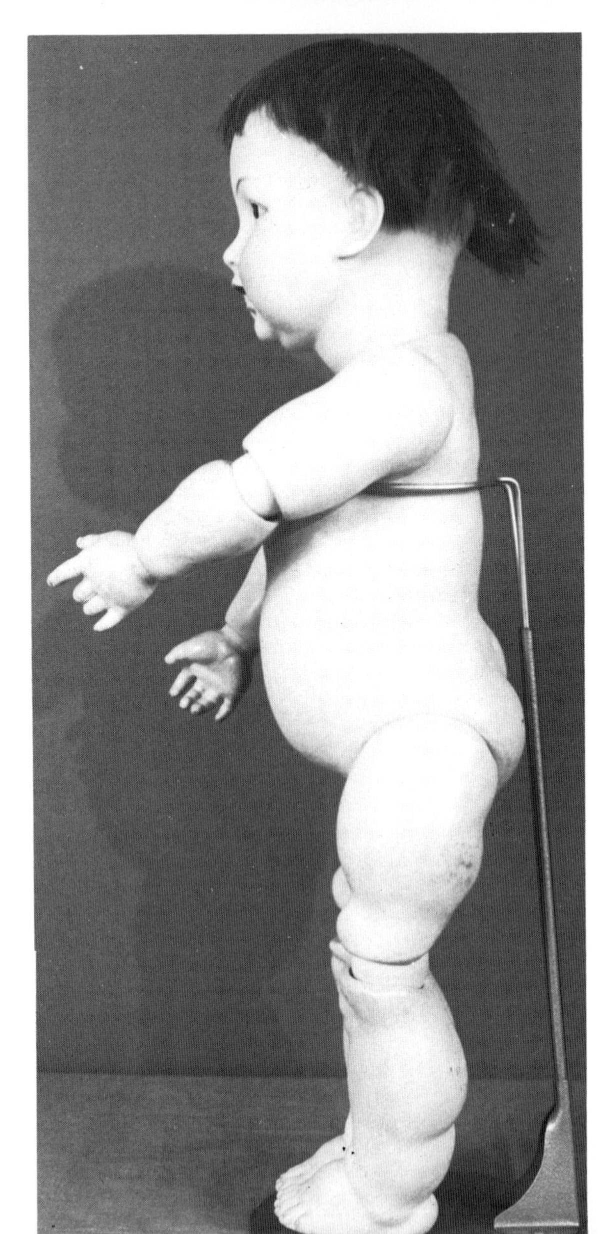

Illustration 300.

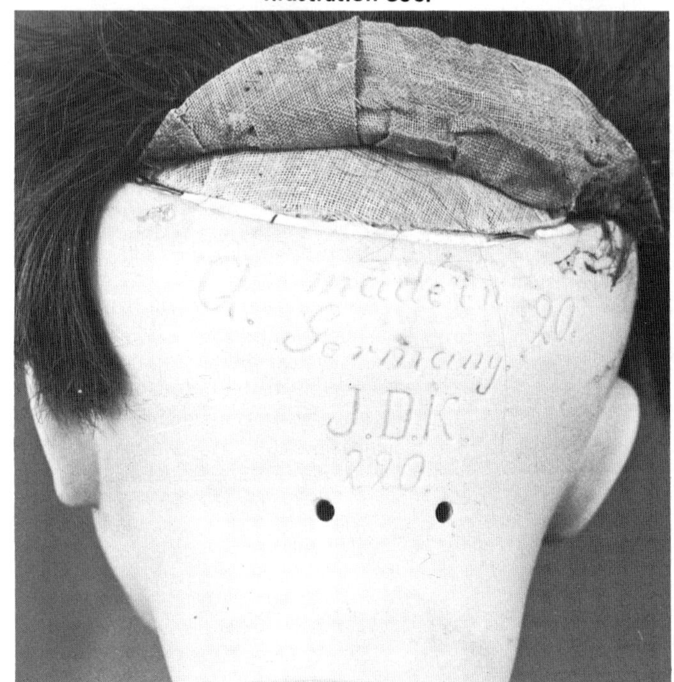

Illustration 301.

226

Illustration 302. This 18in (45.7cm) doll of mold number 226 is a good head-on view for studying his face. Since it is very similar to number 211, I often have difficulty distinguishing between the two unless I have them side by side. The eyes on the 226 are usually larger and the mouth is wider open with the two inset teeth at the top instead of the bottom, although I have owned several 226 dolls which had no teeth at all. Either all of them had lost their teeth or they really did come without any! This doll has the same lifelike molding around the eyes and mouth that the 211 has as well as the large ears of the 220. This sweet smiling face with such deep dimples is nearly irresistible. Over his plaster dome is a blonde mohair wig. Dolls from mold 226 are not as easy to find as those from 211, but they are more plentiful than the 220. *H & J Foulke.*

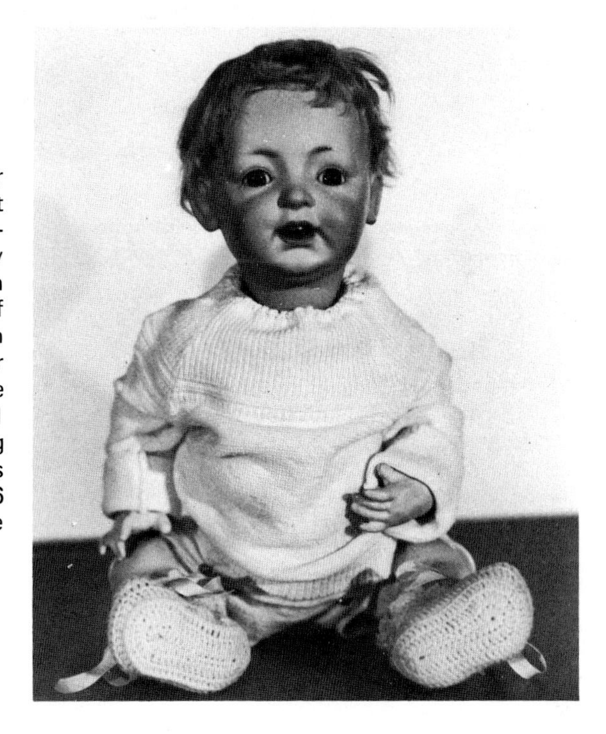

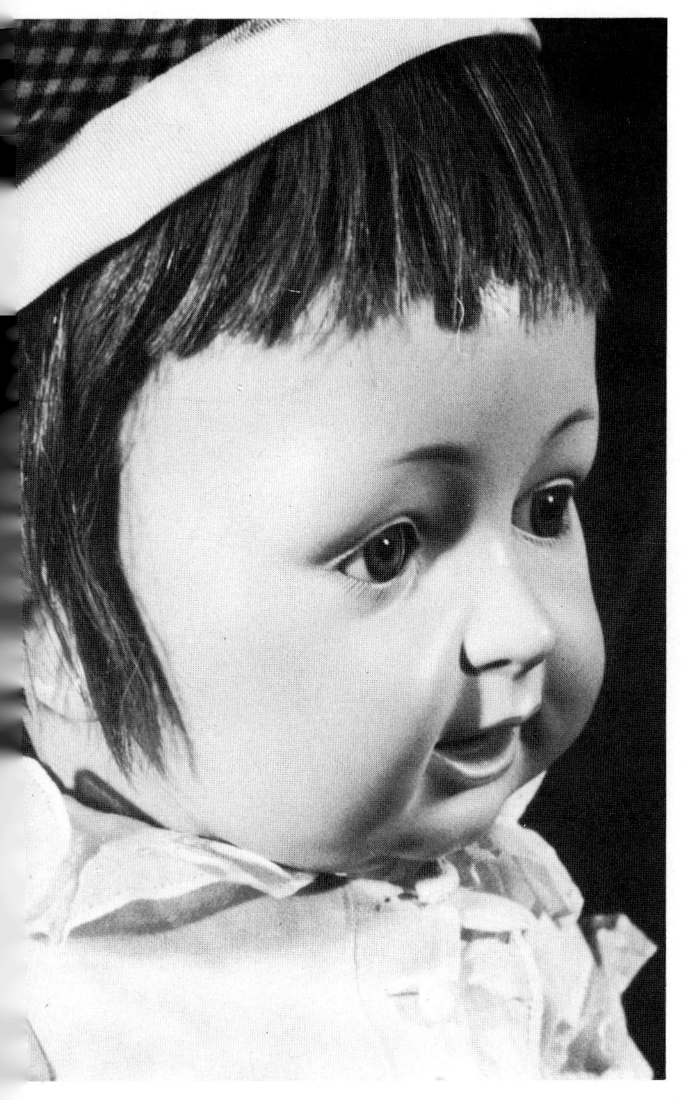

Illustration 303. This large 24in (61.0cm) 226 came from his original owner, but unfortunately all of his clothes had been lost. He was redressed in a romper suit and matching cap of old fabrics, a white linen blouse and blue and white checked short pants. Luckily he still has his original light brown human hair wig. The side view emphasizes his nose and cheek line. *H & J Foulke.*

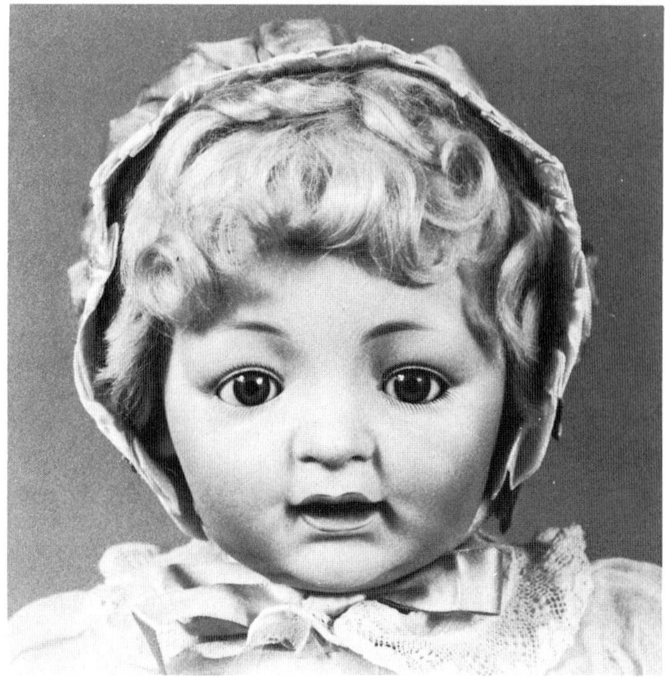

Illustrations 304 and 305. In this 18in (45.7cm) doll, the eyes are smaller but seem to be set farther apart than those of the doll in *Illustration 302.* Also her mouth is not as wide open, but still she has that unmistakable Kestner look. Her mohair wig is curled around her face, and her silk bonnet has ribbon trim. She has blue sleep eyes and the wispy multi-stroked eyebrows. Her inturned left hand indicates that she is on the proper body. *J. C. Collection.*

Illustration 304.

Illustration 306. This 17in (43.2cm) 226 is appealing not only because of his wonderful face, but also because of his cute old outfit made from blue and white striped cotton with matching cap. Again the romper suit effect makes the doll seem more like a little boy. His brown wig is a replacement, but is quite attractive and gives good contrast to the color of his face. He is incised "made in//H. Germany 12.// 226.//Z." *H & J Foulke.*

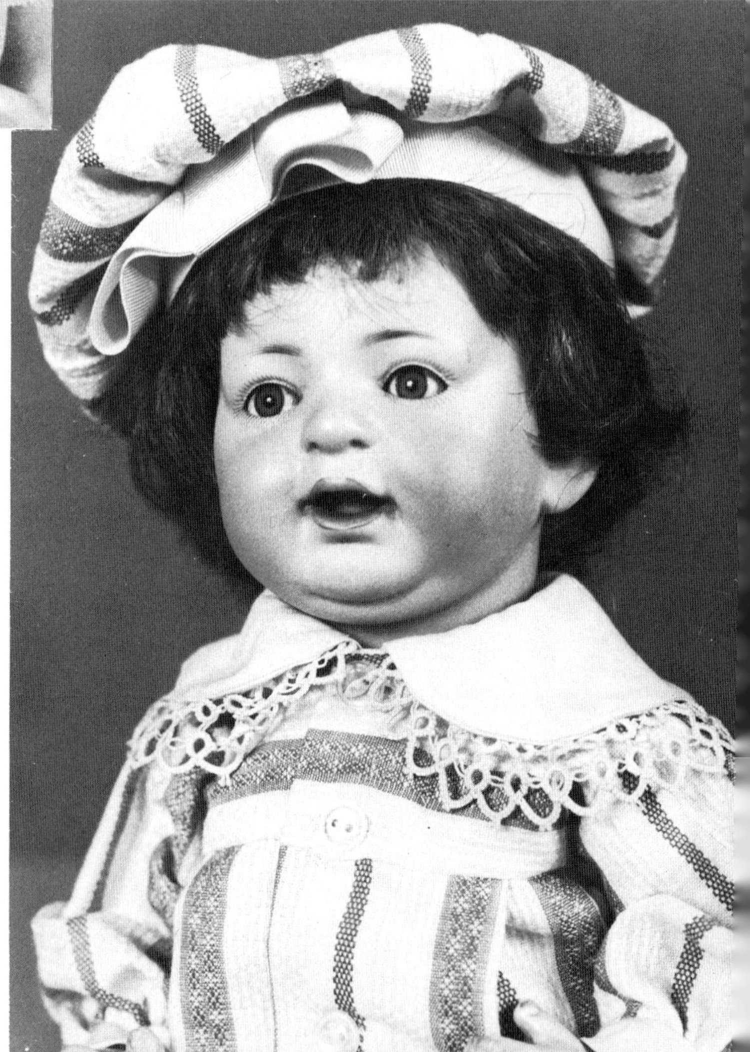

234

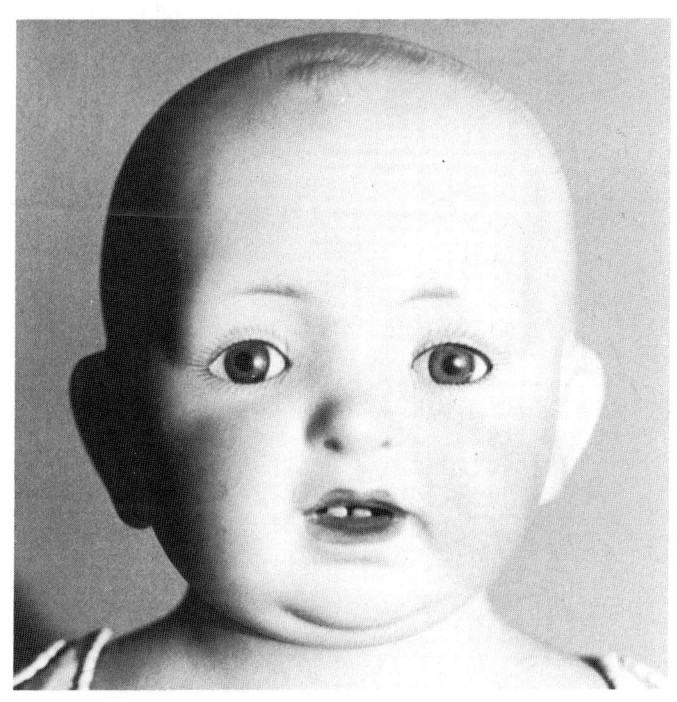

Illustration 308.

Illustrations 307, 308, 309 and 310. This is a very rare character baby which we have never before seen. Incised "J.D.K. 234," he is unusual because of the shoulder head and kid body construction and rare because of his solid dome head. The hair on his bald head is painted on in light strokes with no molding except for one curl in the center front. His ears are well molded and interesting since one is long and narrow and the other is shorter and sticks out quite a bit. He has chubby cheeks, a pronounced double chin and a cute round pug nose. His blue sleep eyes are set in small sockets with a lot of lightly painted lashes and glossy blonde eyebrows. He is pleasant faced, but not broadly smiling; his open mouth has an upper molded gum ridge and two lower teeth; his full lips are shaded in a darker red tone. The molded hollow in the center of his lower neck gives a realistic touch. His white kid body is in never-played-with condition and is rivet-jointed with lower arms and legs of composition. His left arm is bent in towards his body in the Kestner manner; the finger and toe molding is quite detailed with even the nails and joints outlined. He still retains his original J.D.K. crown label. *Gail Hiatt Collection.*

Illustration 310.

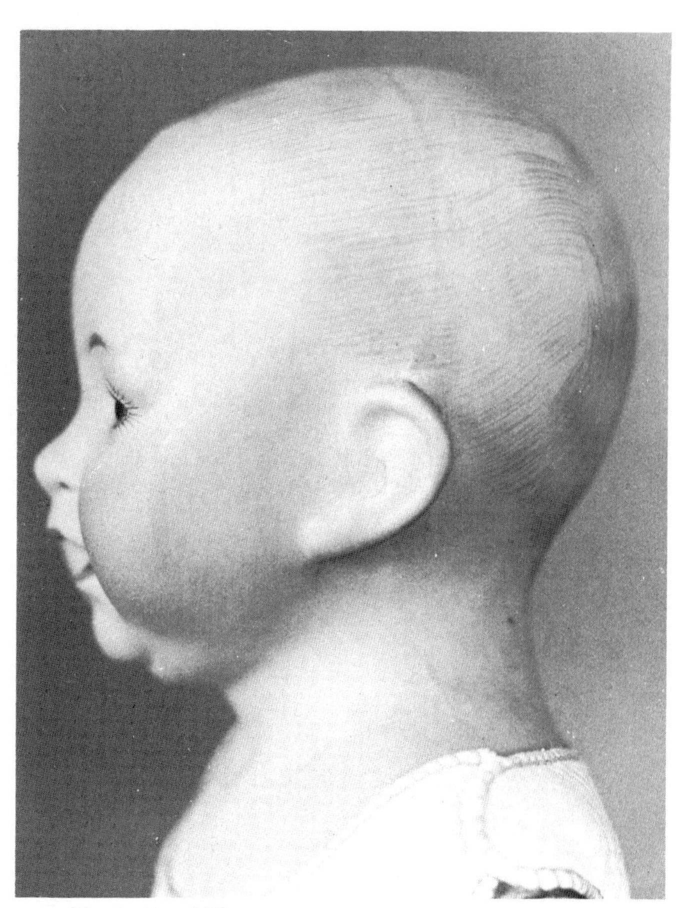

Illustration 309.

235

This mold is a shoulder head character with a faintly smiling face, sleep eyes, and an open mouth with two lower teeth; he is actually a wigged version of mold 234. It was not unusual for Kestner to make both wigged and solid dome variation of his faces.

236

This mold is a socket head character baby used on a composition baby body. He has glass sleeping eyes, mohair wig and open mouth with tongue. He is shown in Patricia Smith's *Antique Collector's Dolls*, Second Series, page 144, but I have never personally seen an example of this mold.

239

This is an absolutely wonderful character baby face, somewhat related to *Hilda*, but much more rare. Her mouth is fairly wide open with four upper teeth, so apparently she is meant to represent a slightly older child. She has glass sleeping eyes, a broad nose and feathered eyebrows. She has been found on either a toddler or baby body. *Old Curiosity Shop.*

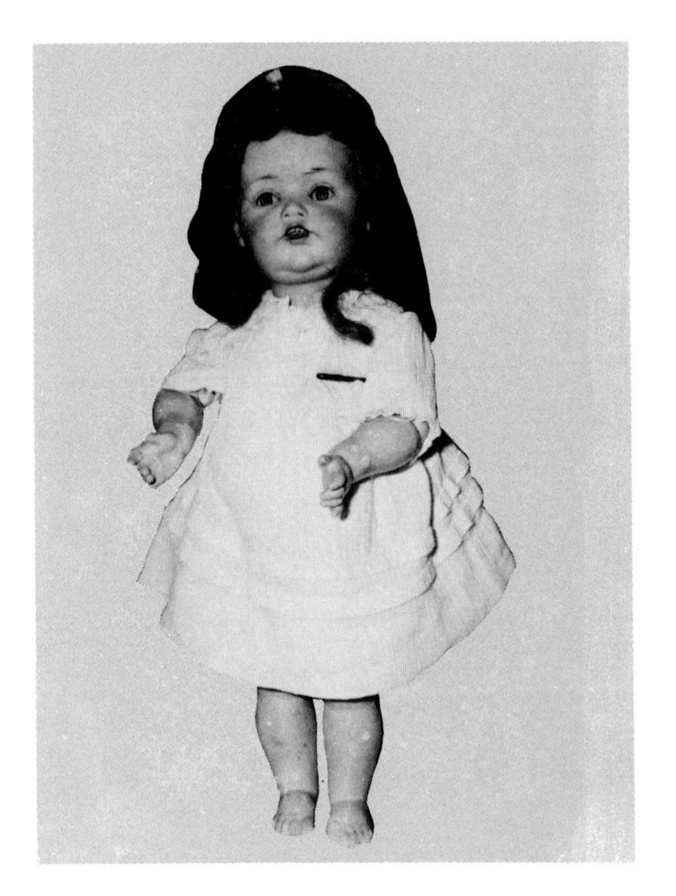

243

Without a doubt the J.D. K. Oriental baby, mold number 243, is quite a choice doll, and one which is avidly sought after by collectors. Not too many of these dolls come onto the sales market. Actually more German companies produced Oriental girls than babies. Armand Marseille produced an Oriental baby and I have seen several others by unidentified companies, but all of these babies lack the tremendous personality, charm, and appeal of the Kestner doll.

Illustration 311. This 16in (40.6cm) doll from mold 243 is made of the finest, smooth olive-toned bisque. His glass sleep eyes are very dark, as are his stroked eyebrows and black human hair wig. His eyes are shaped like half moons and have painted upper lashes only. His lips are slightly parted to form an open mouth which has two upper inset teeth and a separate tongue. His outfit is all original and has remained in excellent condition. It has colors of purple, melon, gold, blue, and turquoise in both silk and embroidered fabrics. *Richard Wright Collection.*

Illustration 312. This 14 in (35.6cm) 243 baby has replaced outer garments, but we have unbuttoned his coat to show his original brightly printed flowered chemise and his Kestner crown label. His eye sockets, cut nearly flat on the bottom, are rather wide spaced and small in comparison to the size of this face which is quite wide with plump lower cheeks. His slightly parted lips show two upper teeth and a separate tongue. His very dark sleep eyes, stroked eyebrows, and human hair wig are in contrast to his smooth pale ocher skin. *Mary Lou Rubright Collection.*

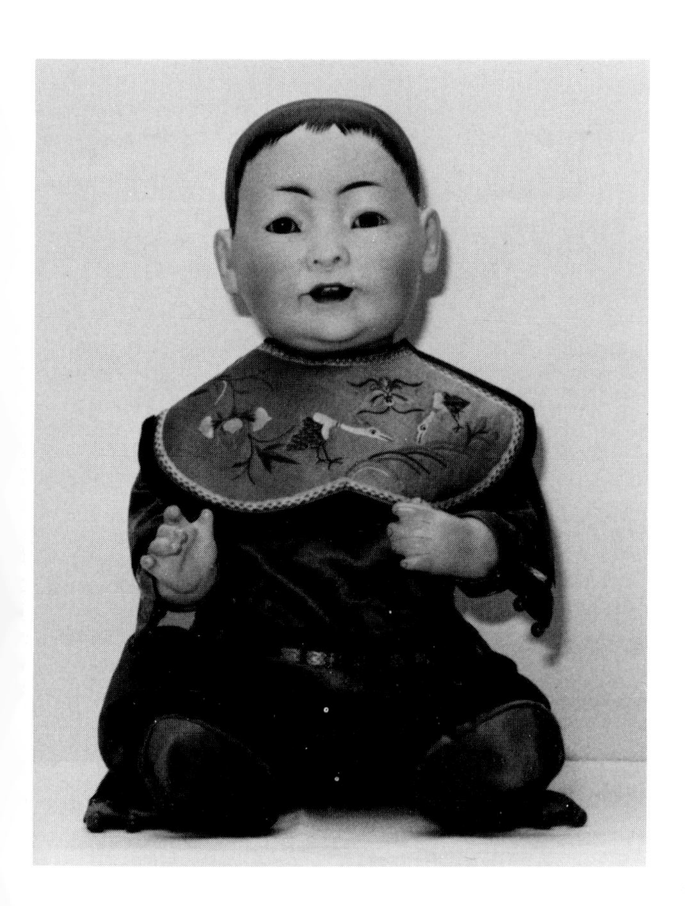

Illustration 313. This 15¾in (40.1cm) Oriental baby is a rarity because of his solid dome head with painted black hair. Brush strokes around the face add a realistic look as evidenced by comparing him to the wigged model in *Illustration 312.* I have never seen this model with a mold number but it is obviously a Kestner. Even the arms have the Kestner position. This doll is marked K. *Private Collection.*

Illustration 314.

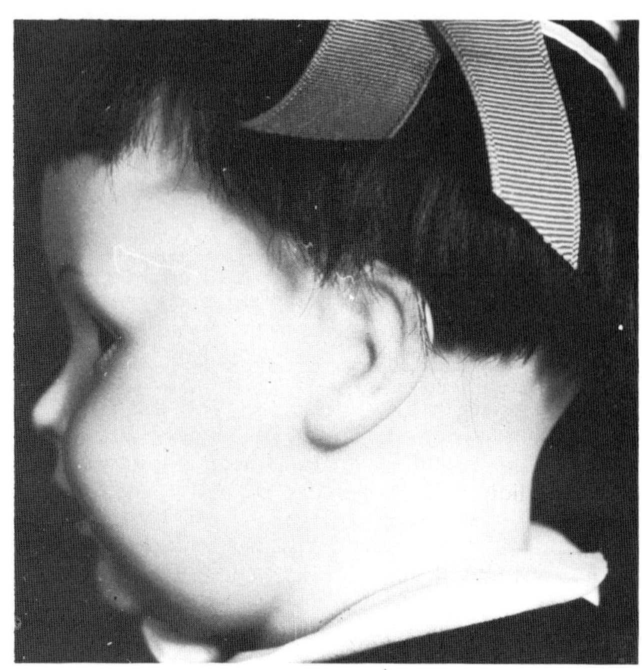

Illustrations 314 and 315. This baby, incised "J.D.K.-247," is one of my favorite of the Kestner characters. However, he is one that does not show up too often. Having been made right after the 245 *Hilda,* perhaps he was dropped in favor of her, for certainly there are a lot more *Hildas* around that 247s. His eyes are a little larger than on some of the characters and his cheeks are nice and plump. I also like his double chin. His mouth is especially interesting, though, in its elliptical shape. It does not create a too broadly smiling baby, just a very pleasant one. The profile shot shows his contours, even his jutting chin and well molded ear which shows much more detail than those on the bald head characters. He is 14in (35.6cm) tall on a Kestner bent-limb baby body. *H & J Foulke.*

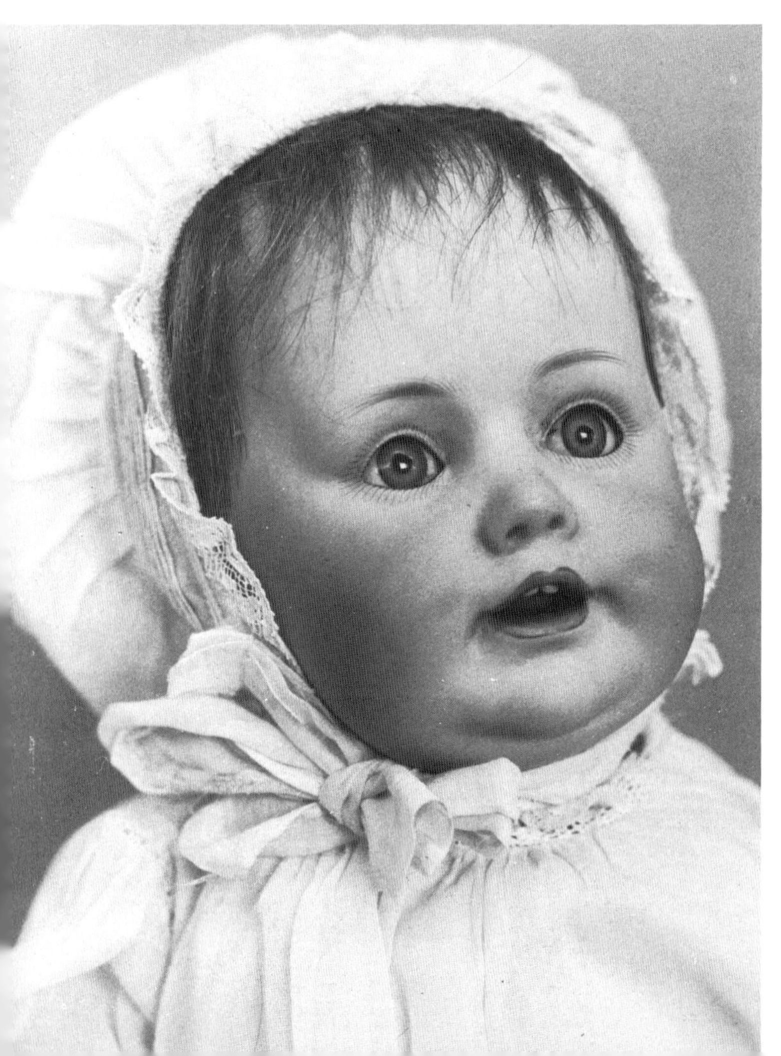

Illustration 316. This 16in (40.6cm) 247 baby remains in the family of the original owner. He has blue sleep eyes, long painted lashes, and light brown eyebrows with individual strokes. His mouth is open just a trifle wider than the other 247 shown, and his upper lip has a higher peak. He has two upper teeth and a separate tongue. This illustration shows to advantage his high dimples, one on either side of his upper lip, the fullness through his lower cheeks, and his chubby double chin. His dress and cap are of white lawn with lace trim and are original to him.

Hilda

Hilda—Certainly the magic name in German character babies is *Hilda*, a sweet little girl who is a favorite among doll collectors as well as noncollectors, many of whom choose her as their most preferred doll. *Hilda's* soft and subtle look which makes her such an appealing doll comes from a combination of factors which together contribute to her realistic look. The subtle coloring of her cheeks is peachy instead of harshly rosy. Her eyebrows are delicately feathered, and her soft, light colored mohair wig looks like real baby hair. Her sparkling eyes are accented by real hair eyelashes (if they have not disintegrated) and long but delicately painted upper and lower lashes. The modeling of her face is quite different from most character babies. Not nearly so wide-eyed as many, her eye sockets are longer and narrower, almost half moons. Her pug nose is rather broad. Her mouth molding is destinguished by her protruding upper lip which rises quite high into a peak. Inside are two tiny upper teeth and a separately molded tongue, some of which are on a spring. Surprisingly enough, this mold looses nothing in translation to smaller sizes, and the darling 10in (25.4cm) baby faces are just as lively as the larger ones. *Hilda* is shown in a January 1916 ad by George Borgfeldt in *Playthings* where she is dressed in a knitted romper and cap, and has a pacifier. This wigged version was shown with her cardboard crown label. At this time only a few dolls were coming in from Germany because of the war.

The mark on the back of her head does not have to say "Hilda" for her to be genuine, but the wigged version must be mold nubmer 245 or 237. The bald head version will sometimes have the number 1070; the wigged version sometimes also has the number 1070 which is the German registration number. Two examples of *Hilda's* marks are as follows:

Bald version:

J.D.K.
ges. gesch. 12 K 1070
Made in Germany

Wigged version:

M Made in 16
Germany
245.
J.D.K. jr.
1914
©
Hilda.

There are other variations as well.

It would certainly be convenient for doll collectors of today if all companies had put the date on all of their dolls as Kestner was obliging enough to do with the *Hilda*. Although the doll was doubtless made for many years, at least we know when it was first put into production. While she commands quite a high price, *Hilda* is not a rare doll, just a scarce one which is highly desirable to collectors. She also was made in a now very rare black version.

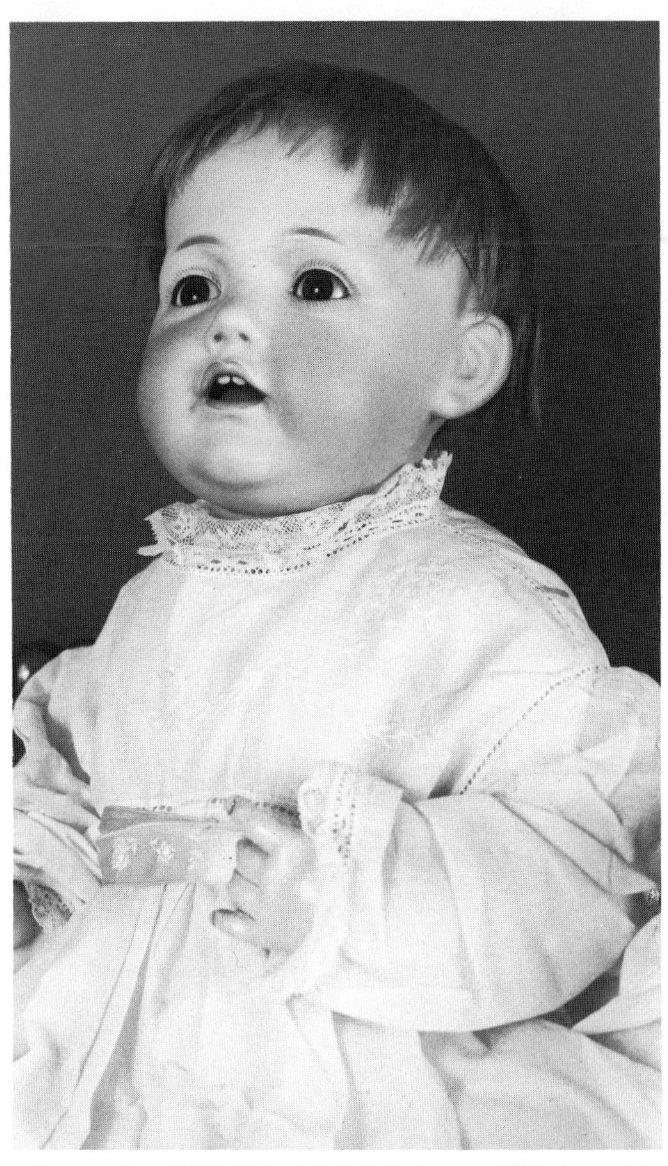

Illustration 317. This delightfully happy *Hilda* is mold number 245. She has her original light brown mohair wig with matching painted eyebrows and dark brown luminous eyes. This view is particularly appealing because it shows her chubby cheek line and the dimple in her chin. It also illustrates her distinctive mouth. A nice large doll, she is 20in (50.8cm) tall. *Richard Wright Collection*.

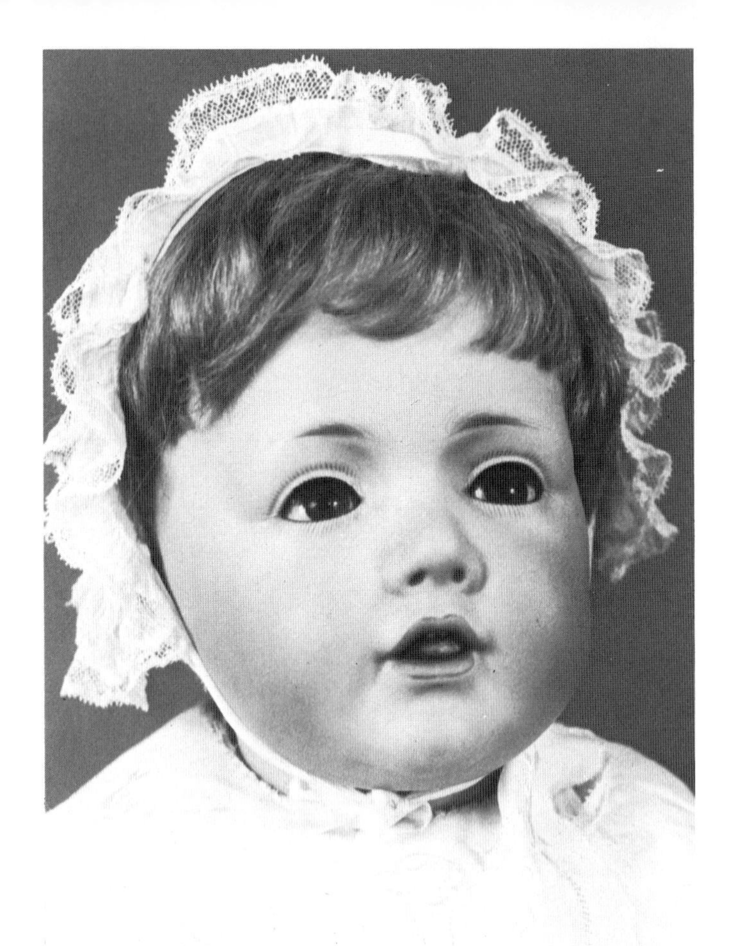

Illustration 318. This sweet 20in (50.8cm) *Hilda*, number 245, sitting in my living room in her rocking chair is a delight to all of our visitors. She is just a little different from her sister in that her tongue, which is on a tiny spring, sticks out quite a bit more and she has her original real eyelashes. *Jan Foulke Collection*.

Illustrations 319 and 320. Although the greatest majority of *Hildas* are on bent-limb baby bodies of the typical Kestner variety with the left hand curving in and the right hand curving out, occasionally one is found on a jointed toddler body, especially in the smaller sizes. These toddler *Hildas* are just simply irresistible! This 19in (48.3cm) example of mold number 245 has a blonde human hair wig, light eyebrows and contrasting dark brown eyes. Her lacy cap with satin ribbon trim and her fine white cotton dress with lace insertion, ruffles, and many tucks are appropriately made of old fabrics and contribute to her most appealing look. *Maxine Salaman Collection*.

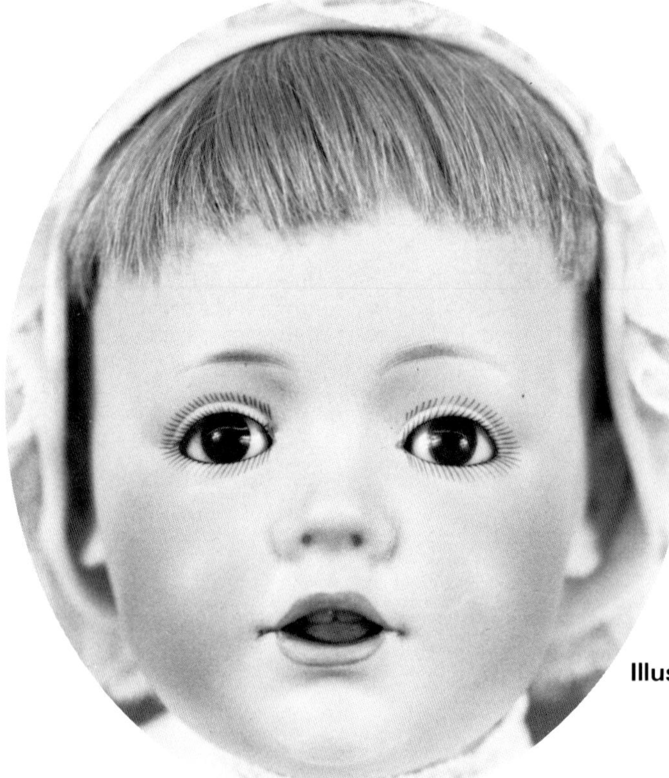

Illustration 319.

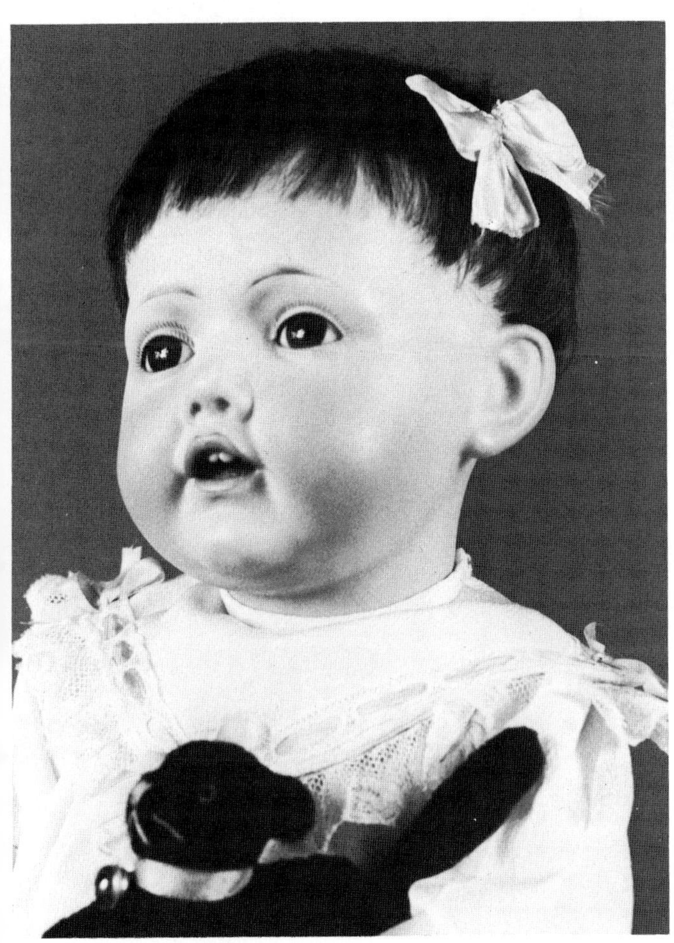

Illustration 321.

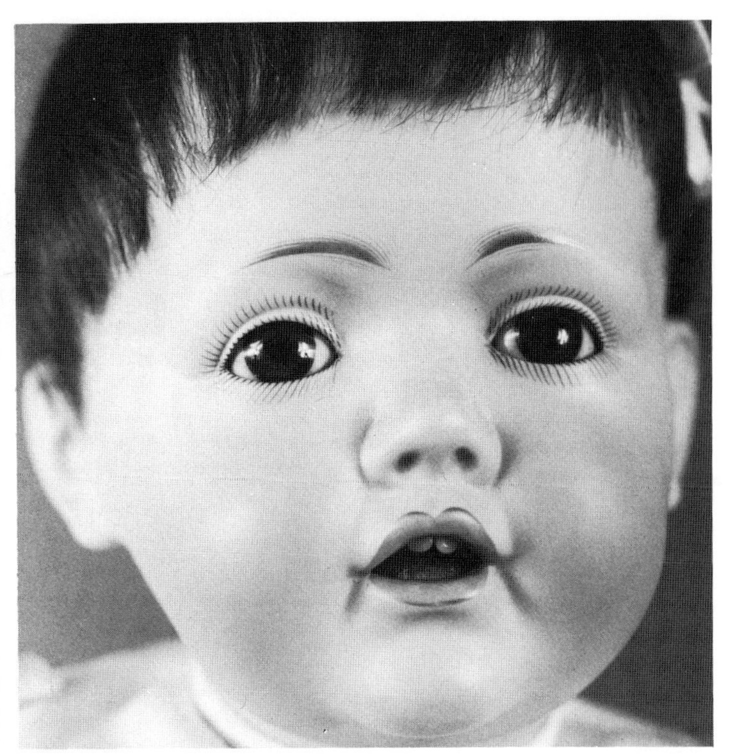

Illustrations 321 and 322. Here is another *Hilda* 245 toddler. She is 21in (53.3cm) tall. With her very dark brown mohair wig and sleep eyes she has an unforgettable face. The close-up view provides an excellent chance to study the molding around her mouth. Her full lips are painted with care including a darker red shading along the top of her upper lip and the bottom of her lower lip. The three-quarters view shows the molding detail in her rather large and realistic ear. *Mary Lou Rubright Collection.*

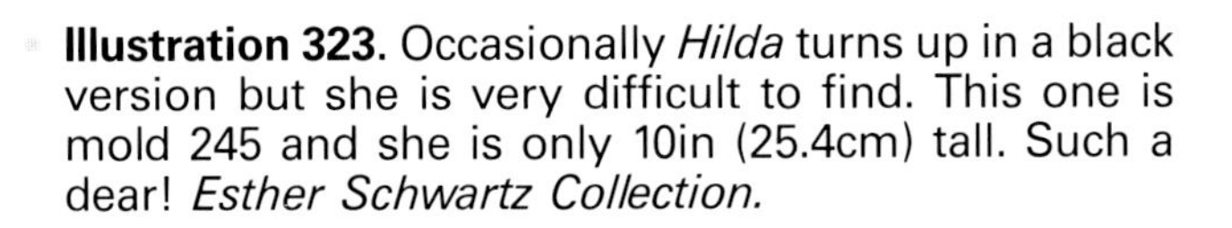

Illustration 323. Occasionally *Hilda* turns up in a black version but she is very difficult to find. This one is mold 245 and she is only 10in (25.4cm) tall. Such a dear! *Esther Schwartz Collection.*

Illustration 324. One wonders why Kestner made another mold with a different number for *Hilda* when it is obviously the same face. This one is number 237. Her complete mark is "M made in 16 // Germany // J.D.K. //237.//ges. gesch. N 1070". On her bent-limb Kestner baby body she measures between 21in (53.3cm) and 22in (55.9cm) tall. There is always some variance in measuring these baby bodies, as the *Hilda* in *Illustration 318* also has the size M 16 head, yet only measures 20in (50.8cm). This number 237 *Hilda* is unusual in that she has a cry box in her torso. She has a brown human hair wig and brown eyes. *Rosemary Hanline Collection. Photograph courtesy of the owner.*

Illustration 325. This 10½in (26.7cm) 237 *Hilda* just barely fills up your hand, but what a darling handful she is! Her light mohair wig makes her brown sleep eyes seem much more vivid. The mold loses nothing in the translation to the small size, and the decoration is done in proportion with feathered eyebrows and delicately painted eyelashes. Her left arm is bent in typical Kestner fashion. She is incised with the mold number "237," but not with the word *"Hilda." Jimmy and Faye Rodolfos Collection.*

Illustration 326. Another tiny little *Hilda*, this one is marked with both her name and "237." She is just 11in (27.9cm) tall. Her tiny sleep eyes are blue and her hair is brown with matching eyebrows. Even in this small size, her upper bowed lip is distinctive. Another marvelous handful of doll! *Maxine Look Collection.*

Illustration 327. This 22in (55.9cm) 237 *Hilda* is dressed in light blue velvet as a darling little boy. He has a blonde skin wig over his plaster dome, lightly stroked blonde eyebrows, and many lightly painted eyelashes surrounding his blue eyes. He has the two upper inset teeth, but does not have a prominent tongue like some of the other *Hildas* illustrated. *Mary Lou Rubright Collection.*

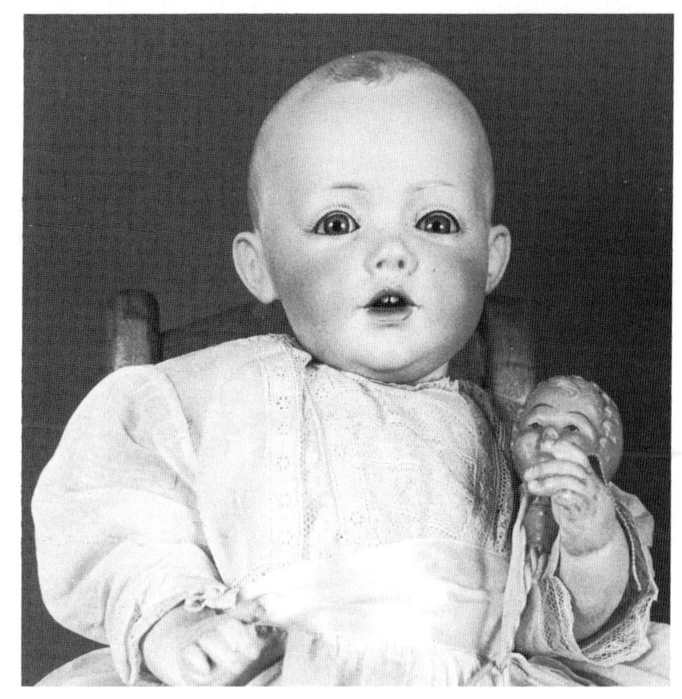

Illustration 328. Even *Hilda* lovers have favorites; many think the wigged ones are the sweetest, but just as many are partial to the bald version like this 18in (45.7cm) charmer. *Bald* is a good descriptive word for her since the sculptor has given her just a few molded locks on the top front of her head, and barely a hint of molded hair above her ears and at the nape of her neck. The bisque and decoration are particularly lovely on this doll; she has blue sleep eyes, dark blonde eyebrows, and painted hair composed of many wide and rather far apart brush strokes. This view is good for showing her large and intricately molded ear with outer fold and inner ridges in fine realistic detail. Also note some variation in the Kestner hand. Nearly all of the *Hildas* have a molded indentation on the left wrist whereas that of the 211 doll is smooth. *Richard Wright Collection.*

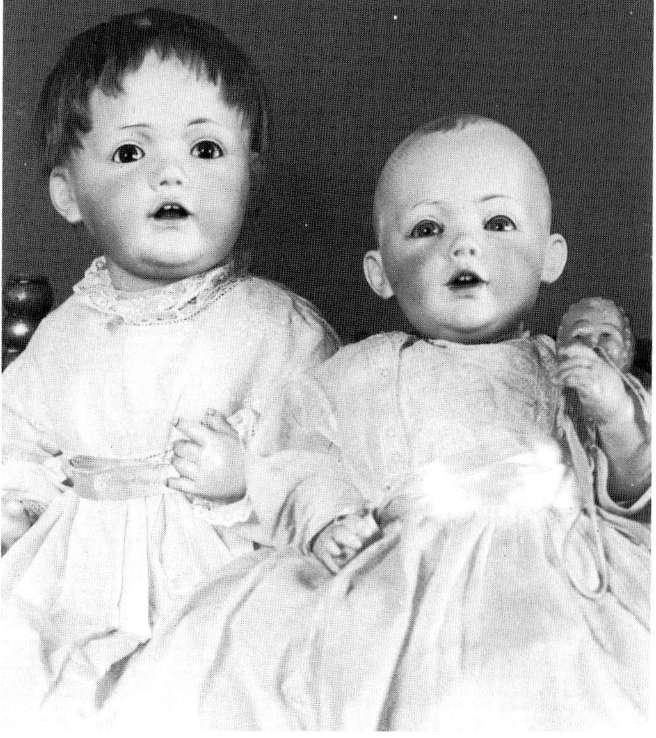

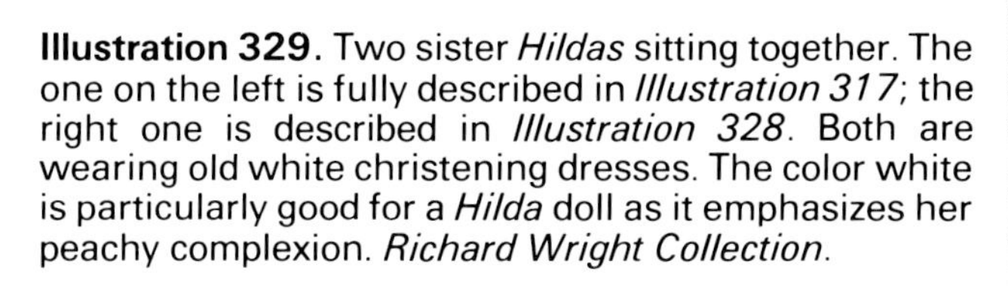

Illustration 329. Two sister *Hildas* sitting together. The one on the left is fully described in *Illustration 317*; the right one is described in *Illustration 328*. Both are wearing old white christening dresses. The color white is particularly good for a *Hilda* doll as it emphasizes her peachy complexion. *Richard Wright Collection.*

255

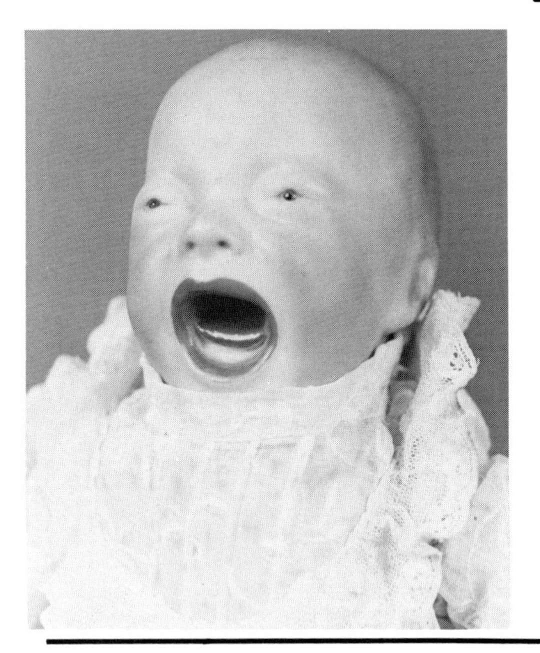

Illustration 330. The Ciesliks have attributed this baby to Kestner & Co. in their *German Doll Encyclopedia, 1800-1939*. Since it has a strange O.I.C. mark, it was apparently made as a special order Kestner poured for an, as yet, unidentified doll producer. The baby certainly is unusual and stirs up a lot of interest when he is displayed as he is obviously unhappy and somewhat uncomfortable — colic perhaps? wet diaper, or just hungry? He has a wide open molded mouth with tongue and tiny glass eyes. The flange neck is sewn onto a cloth body. The complexion is very ruddy as befits his fretful disposition. *Dr. Carole Jean Stoessel Zvonar Collection.*

257

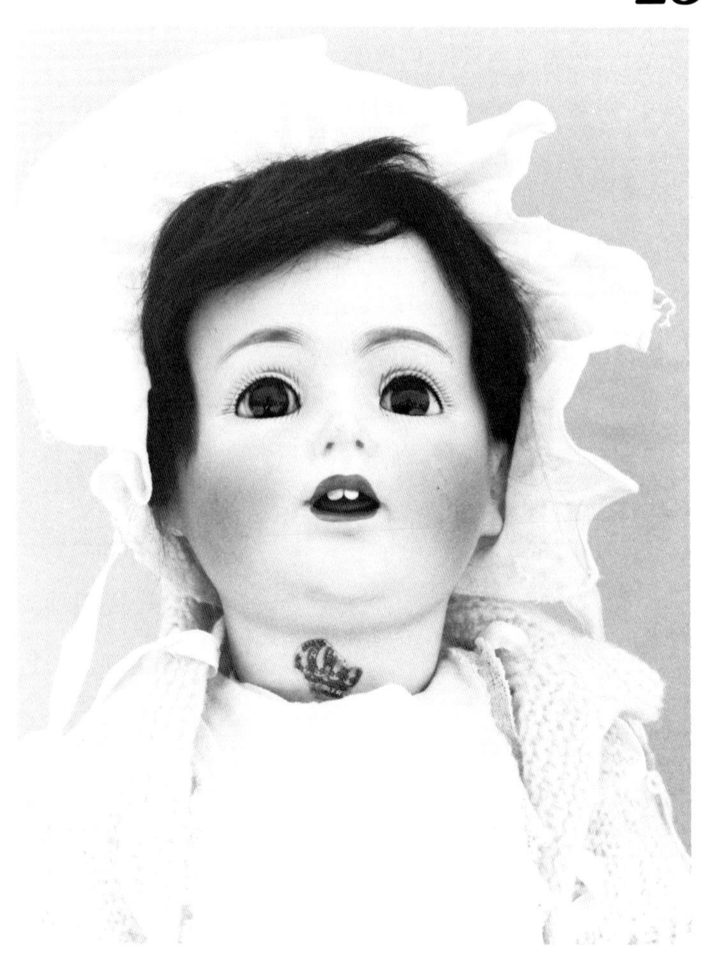

Character dolls which appear to be from mold number 257 are shown in the Kestner catalog reprinted in *Die Deutsche Puppenindustrie 1815-1940* by Georgine Anka and Ursula Gauder. Since the authors indicate that this was the last catalog of the Kestner firm, it appears that these were among the last dolls made by the Kestner factory. Although we are not sure of the date this mold was made, it was probably shortly after 1914, the date of the *Hilda* doll, so apparently this mold was in production for quite a few years. This also seems possible as this face is a fairly easy one to find. This 257 head is usually found on a bent-limb baby body or a five-piece toddler body, sometimes on other types as well.

Illustration 331. This marvelous clear view of the mold number 257 face shows a very pleasant baby but one which lacks the charm of the *Hilda* doll. This cute 18in (45.7cm) baby has very deep brown eyes with remains of real eyelashes and a dark brown mohair wig in bobbed style like those shown in the last Kestner catalog. The coloring on this doll is fairly high with quite rosy cheeks; this characteristic seems to have been the fashion in the later bisque dolls. Her fairly wide open mouth has two inset upper teeth and a separate tongue. This baby has the remains of a Kestner sticker on her throat. She is on a Kestner bent-limb baby body. Her head is incised *"Made in//Germany//J.D.K.//257.// Made in//Germany//41."* Most of these 257 and 260 dolls do not have the regular Kestner sizing on the heads. The number at the bottom of the head "41" on this doll and "46" on the one in *Illustration 334* is probably a centimeter size, although few heights actually work out precisely as there is much variation in measuring baby bodies.

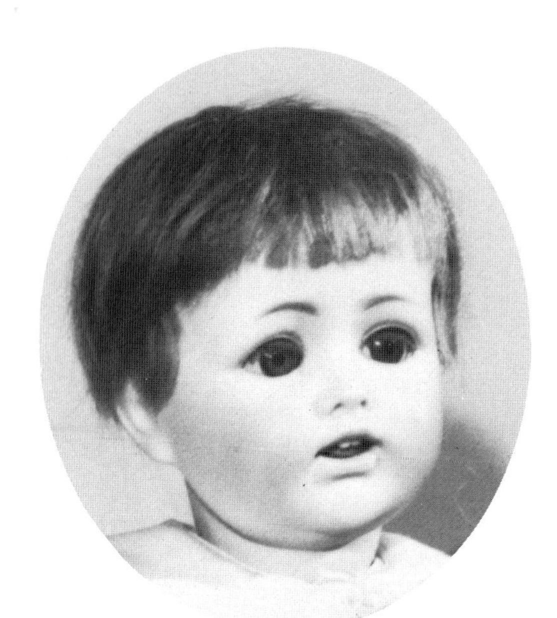

Illustration 332. This darling little boy is 27in (68.6cm) tall on a ball-jointed composition body. He is a delightful child with lovely bisque and coloring. His dark brown eyes have long lashes; his eyebrows are brown and so is his mohair wig.

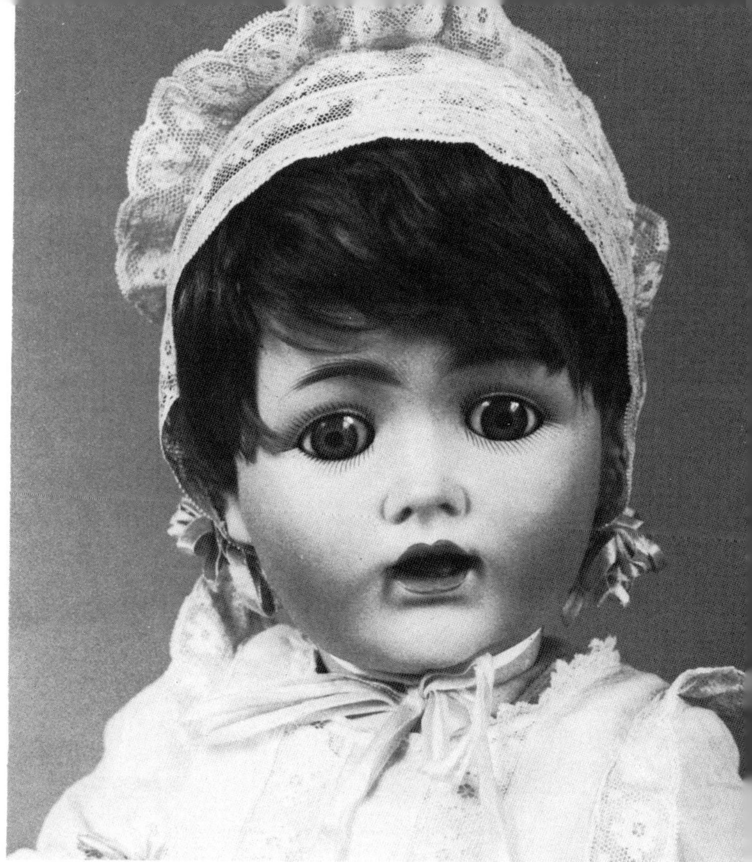

Illustration 333. This large 22in (55.9cm) 257 baby is typical of several shown in the last Kestner catalog which are dressed in long or short baby dresses with matching bonnets. Several of the babies have great huge bows in their hair. This particular baby has blue eyes. She also has two upper teeth which are difficult to see at this angle. *J. C. Collection*.

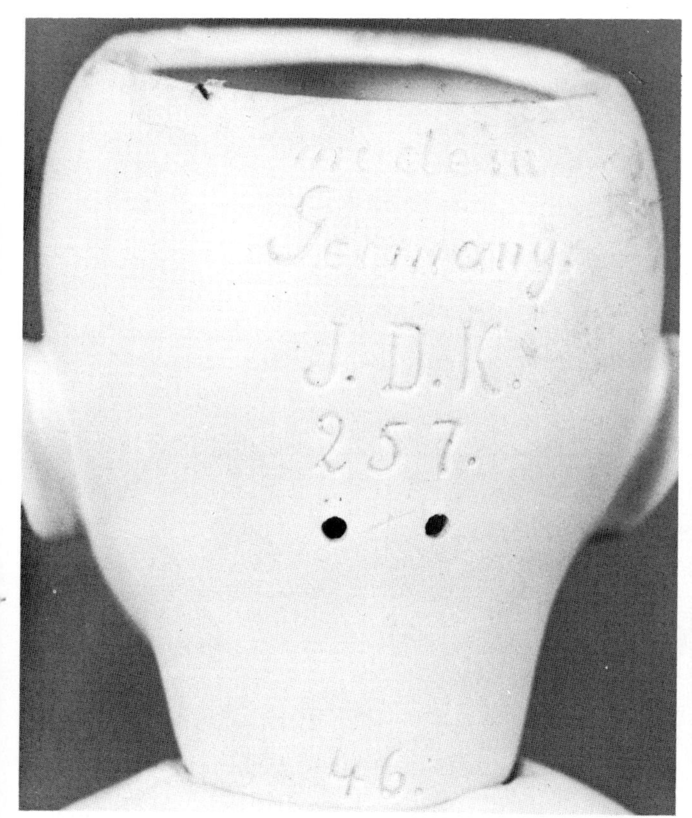

Illustrations 334 and 335. This 20in (50.8cm) example of mold 257 has lovely oily bisque with a beautiful peachy decoration, and is one of the nicest of this mold I have ever seen. He has large brown sleep eyes with very long lashes. At first glance you are sure that someone has added them, but a closer look shows that certainly they are original. As on most dolls of this mold the eyebrows are nicely stroked, but are quite a bit heavier than those on the 211 models. His original light brown mohair wig is wispy and very natural looking. *H & J Foulke*.

Illustration 335.

Illustration 336.

Illustration 337.

Illustration 336 and 337. This tiny 8½in (21.6cm) baby of mold 257 is a little gem. The face is darling, and even though small, retains the features of the larger dolls. The tiny brown sleep eyes have real lashes which contrast with his original blonde mohair wig. The small open mouth shows two upper teeth and a tongue. He is wearing his original gown with the cardboard Kestner crown label attached. *Jimmy and Faye Rodolfos Collection.*

260

Dolls from mold number 260 are also shown in the last Kestsner catalog already mentioned. The face is quite similar to the 257 except through the mouth, and actually appears to be an older version of the same child. The mouth of the 260 is broader, but not so wide open. There are four upper teeth, and the tongue appears as more of a molded ridge than a separate piece. Like the 257, this is also a later mold and apparently one of the last produced by Kestner. This head is found on a variety of body types: typical ball-jointed composition, toddler ball-jointed composition, five-piece toddler, chubby five-piece toddler with starfish hands, and slim ball-jointed girl type. Again, sometimes classifying these molds is difficult, and one wonders if perhaps this mold should have been placed under the character children instead of character babies.

Illustration 338. One of the nicest examples of this mold is this 24in (61.0cm) toddler girl. She has her original blonde mohair wig which, according to dolls pictured in the Kestner catalog, may have been quite puffed out with curls and topped by a large hair bow. She has sleep eyes with nice long lashes. The lovely molding around her mouth emphasizes her upper and lower lips. Altogether she has a wistful face and is a thoroughly huggable doll. Her blue wool dress appears made just for her and is trimmed with decorative stitching. Her blue socks are held up by blue ribbons threaded through the tops. Her body is the most desirable type for a character doll -- a jointed toddler style with chubby legs and slant hip joints like the one shown in *Illustrations 299 and 300.* To top off all of this, she is still in her original box with blue cloth-covered packing and dust ruffle which was arranged as a bed. *H & J Foulke.*

Illustrations 339 and 340. This pert little example of the 260 mold is just 12in (30.5cm) tall. She has a brown wig and brown sleep eyes with real lashes. Like other 260 mold dolls, she also has long and dark painted upper and lower lashes. She is on a jointed composition body. Dolls of this type came either fully dressed or wearing only a chemise, not footwear as was common earlier. The owner feels that perhaps her fancy lace dress was original to her, probably made for her after she was brought home from the store. *Sheila Needle Collection. Photograph by Morton Needle.*

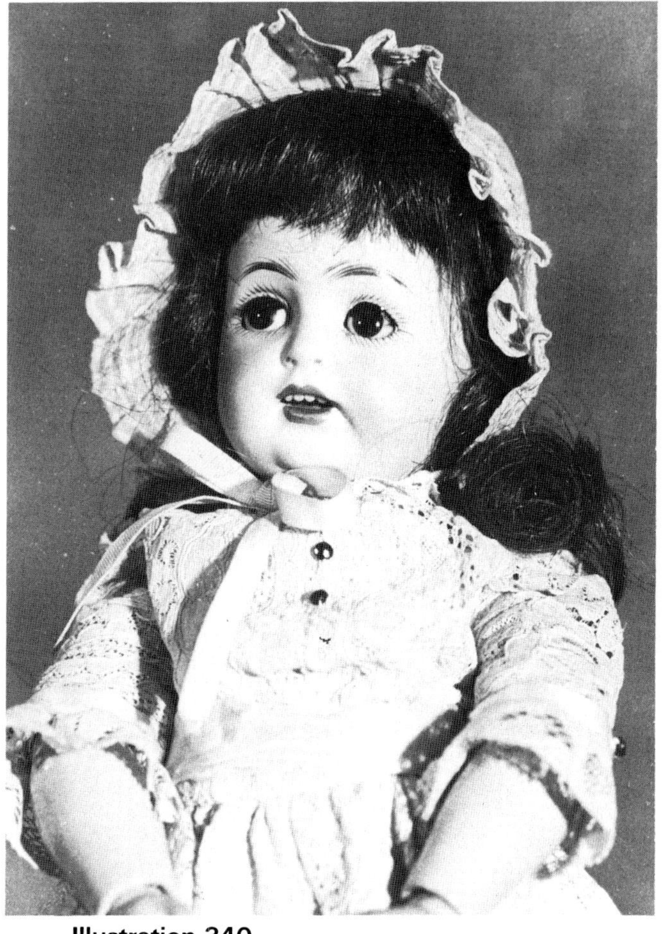

Illustration 340.

Illustration 341. Even in this tiny 8½in (21.6cm) size, the 260 mold has four upper teeth. His mouth is so tiny, it is a wonder that Kestner could fit them all in. Again he has brown eyes. I am not sure whether more really did have brown eyes in these 260 and 257 molds, but certainly most of the examples photographed do. He has a curly blonde wig which is probably original as one of the dolls in the Kestner catalog has one like it. He is wearing his original white linen-type dress with blue buttons, collar, and low slung belt. *Jane Alton Collection.*

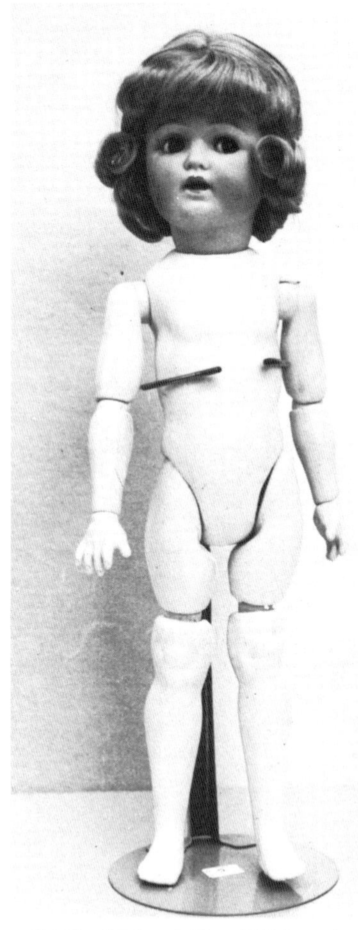

Illustration 342. This 14in (35.6cm) girl has two features which are rare on Kestner dolls. They are flirty eyes and the slim jointed teen-aged body. Both of these features are shown in the 1930s Kestner catalog, however, so there is no doubt that she is right. She has highly colored cheeks, another indication that she is one of the later dolls. Her wig, unfortunately, is a recent replacement. In the Kestner catalog these dolls are shown with short bobbed hair either wavy or straight. Obviously she was meant to be a very stylish girl. The doll could be purchased wearing a fancy lace-trimmed combination or one of ten different dresses in the latest fashion, all dolls wearing bead necklaces and very large hair bows.

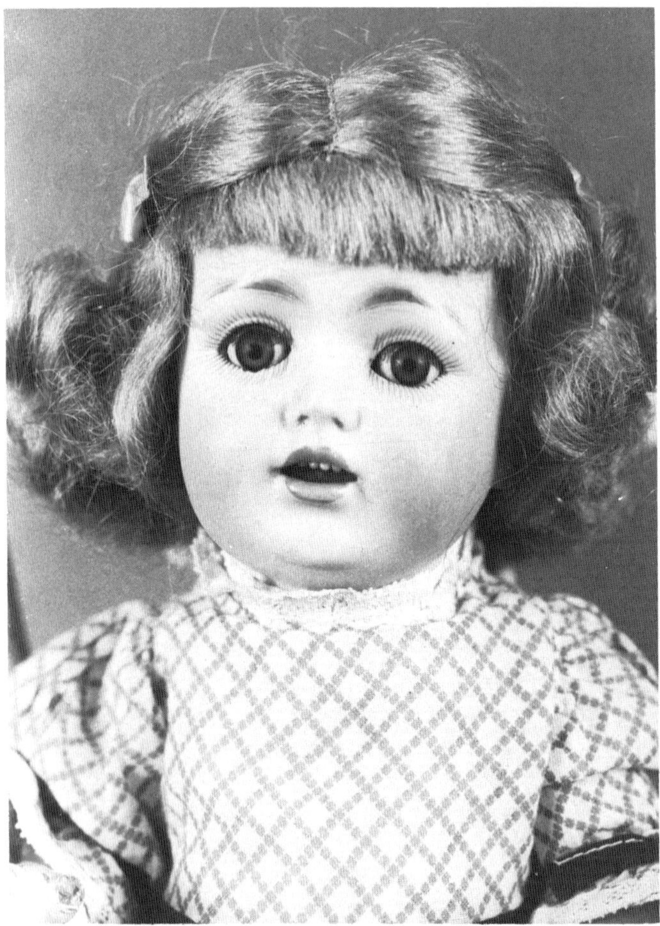

Illustration 343. This small 12½in (31.8cm) 260 mold girl is on a baby body which does not appear to be Kestner-made. Her mouth, which is fairly wide open shows four upper teeth and a separate tongue. Her blue sleep eyes have both real and painted lashes. Her eyebrows are blonde and glossy with many individual strokes. *Ruth Noden Collection.*

Illustrations 344 and 345. This large 28in (71.1cm) 260 girl is on a jointed composition body. She has the usual four upper teeth, but she has a molded rather than inset tongue. She is quite an appealing child with a pleasant expression. You just know she would be great to have around! This little girl has brown sleep eyes with real and painted eyelashes, nice rosy cheeks, and a dimple in her first chin. She is wearing a lovely white cotton dress with a fancy yoke, a gathered waist, and many rows of tiny tucks above the hem. *Mary Lou Rubright Collection.*

Illustration 344.

Illustration 345.

Illustration 346. This 20in (50.8cm) baby is incised with the large and fancy Kestner & Co. factory mark as well as the mold number 262 over size number 44. She has very large blue sleep eyes with painted lashes and light brown feathered eyebrows. Her parted lips, which have accent shading on both the top and bottom, show two upper inset teeth. Very deep molding characterizes her nostrils, her philtrum, and her darling cheek dimples. She wears her original blonde curly mohair wig. Her bent-limb baby body is of excellent quality. *Dr. Carole Stoessel Zvonar Collection.*

Illustration 347. Here is another 262 baby incised with the Kestner & Co. porcelain factory mark. The replaced wig with bangs and long hair makes her look like a slightly older child. This is certainly a cute appealing face with large blue eyes and open mouth posed as though she is just ready to speak. She has a composition baby body with big toes which curl up. She has a cavity for a voice box, but the works are missing. She is wearing her original cotton baby dress with white on white embroidered lace yoke. *Irene Smith Collection. Photograph by Irene Smith.*

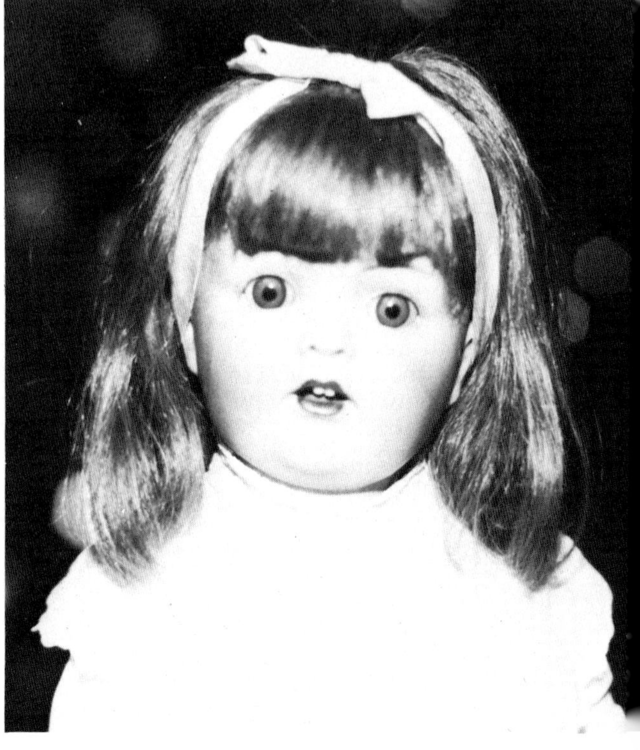

263

Illustration 348. This 25in (63.5cm) baby is incised "263//C.P.," the initials of Catterfelder Puppenfabrik for whom the doll was made. It is well established that Kestner made heads for that company which was located only a few miles from the Kestner factory in Waltershausen, and it is very likely that Kestner made this doll as well, since it also fits in with the Kestner numbering system. A girl doll with mold number 264 carries the marks of both Kestner and Catterfelder Puppenfabrik. This is another typical baby face of the 1920s with wide blue eyes, stroked eyebrows, and painted lashes. Her mouth is quite wide open with two upper teeth and a retractable tongue. Deep dimples add to her pleasant, smiling look. She is wearing her original blonde mohair wig. *Dr. Carole Stoessel Zvonar Collection.*

272

This mold is a wonderful newborn baby incised "Siegfried," sometimes with the mold number 272. The face is very sweet with sleeping eyes, closed mouth, wide nose and pronounced philtrum. It comes as a flange neck to be attached to a cloth body. A few "Siegfried" socket heads have been found on composition baby bodies.

680

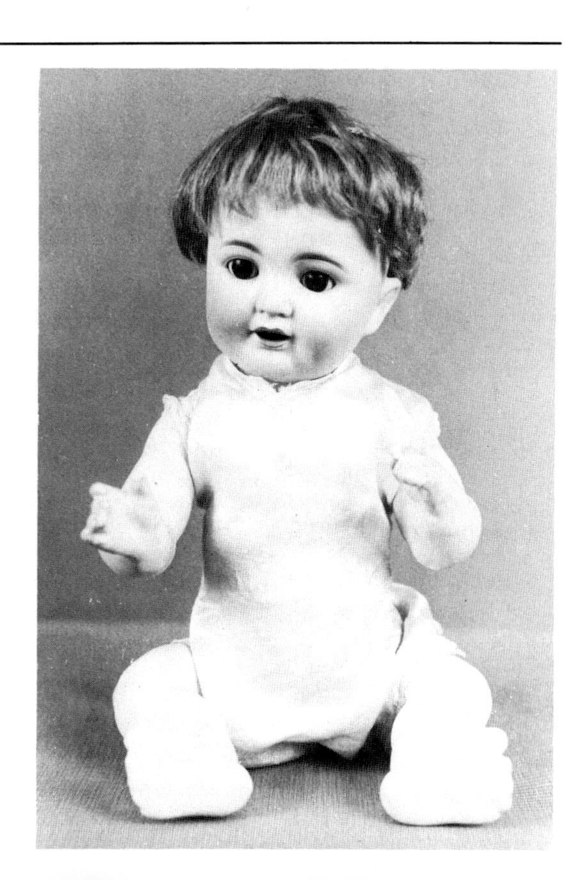

Illustration 350. I was excited to find this doll, because it helps prove my contention that there was probably more cooperation among the large factories in Ohrdruf than we know about. The factories were Kestner & Co., Kley & Hahn, Kling & Co., and Bähr & Pröschild. Anyway, this baby shows some relationship between Kestner and Kley & Hahn since the back of his head contains both marks. He has a very sweet face with brown sleep eyes, brown eyebrows with many individual inner strokes fanning up, and a brown mohair wig. His mouth is open with two upper teeth and a separate tongue. His baby body is not in the Kestner style and has a red circle stamp on it. His mark is:

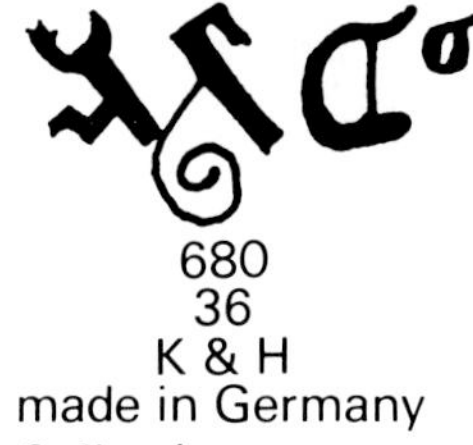

680
36
K & H
made in Germany

Jimmy and Faye Rodolfos Collection.

Bald Head Babies
(No Mold Numbers)

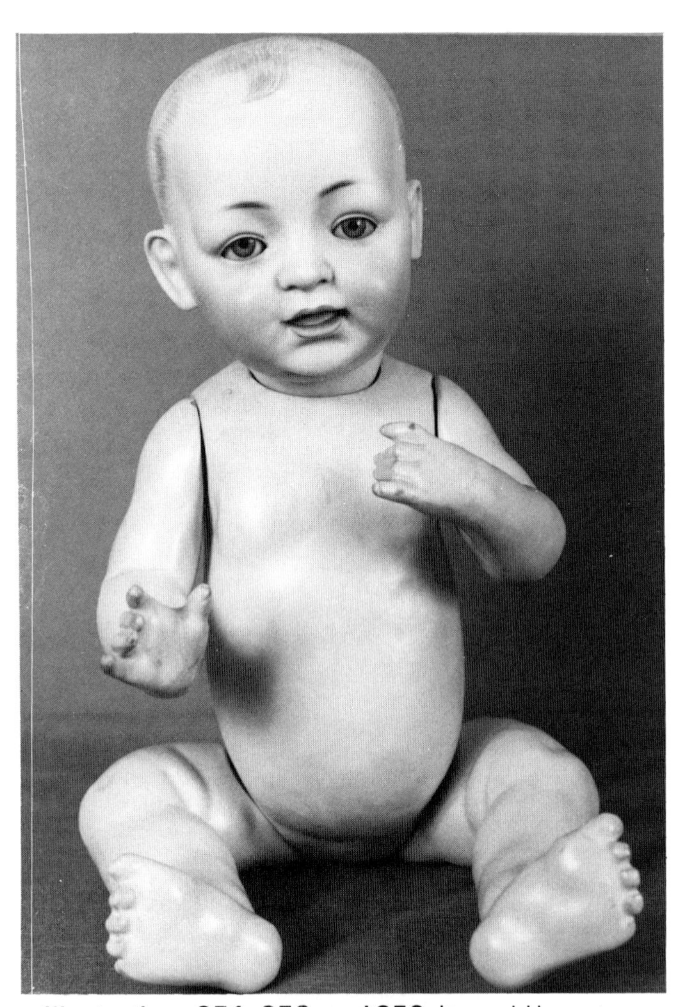

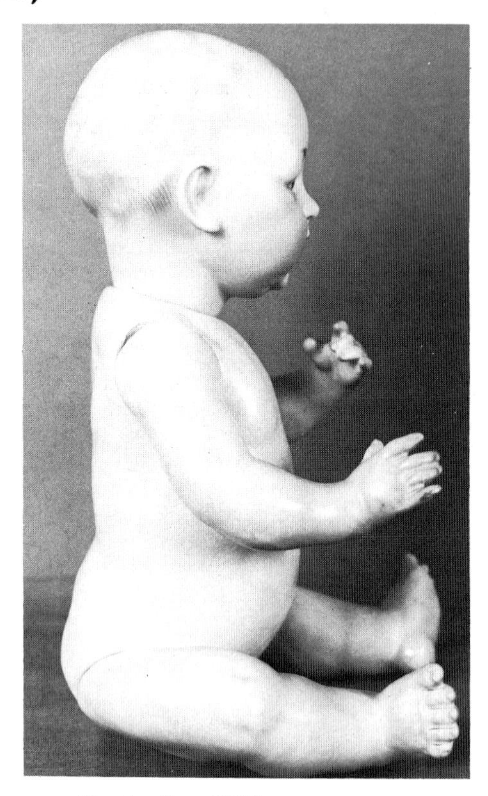

Illustration 352.

Illustrations 351, 352, and 353. It would have been a lot simpler for collectors who like to classify their dolls if Kestner had used mold numbers on these bald head molds, but he did not. Many are confused by the numbers which accompany the J.D.K. initials. These are only size numbers, and as far as I have been able to determine, fit into Kestner's regular sizing system of letter/number without the letter. (Remember when measuring the length of babies, there can be up to 2in (5.1cm) variance on the larger ones depending on whether the person doing the measuring stops at the top of the toe or the bottom of the heel.) A typical mark is "J.D.K.//Made in 7 Germany."

The bald heads are nearly always found on baby bodies. Occasionally one is found on a jointed body which makes up into a marvelous little boy doll. The Kestner baby bodies are of very good quality composition with a shiny finish and coat of top paint in a rosy tone which makes for a very healthy and natural looking complexion. The torsos have distended bellies in the front and shoulder blades molded in the back. Upper arms have chubby rolls and molded elbows; fingers are outlined and nails are shown. It is in the modeling of the legs, however, that Kestner excels. They are masses of chubby creases and crevices even at the ankles. The toes are little works of art with outlining of nails and joints. Even the bottoms of the feet show natural details! *Esther Schwartz Collection.*

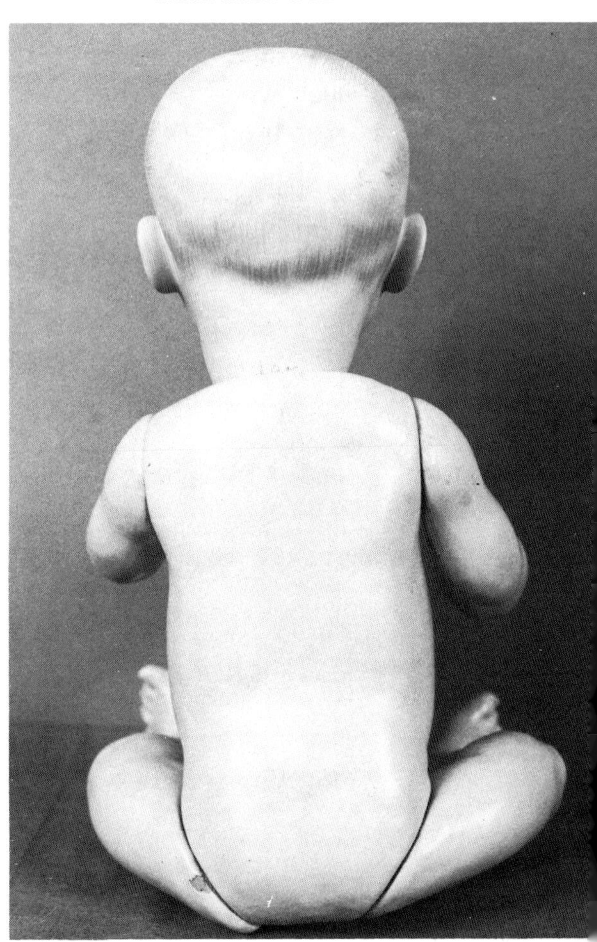

Illustration 353.

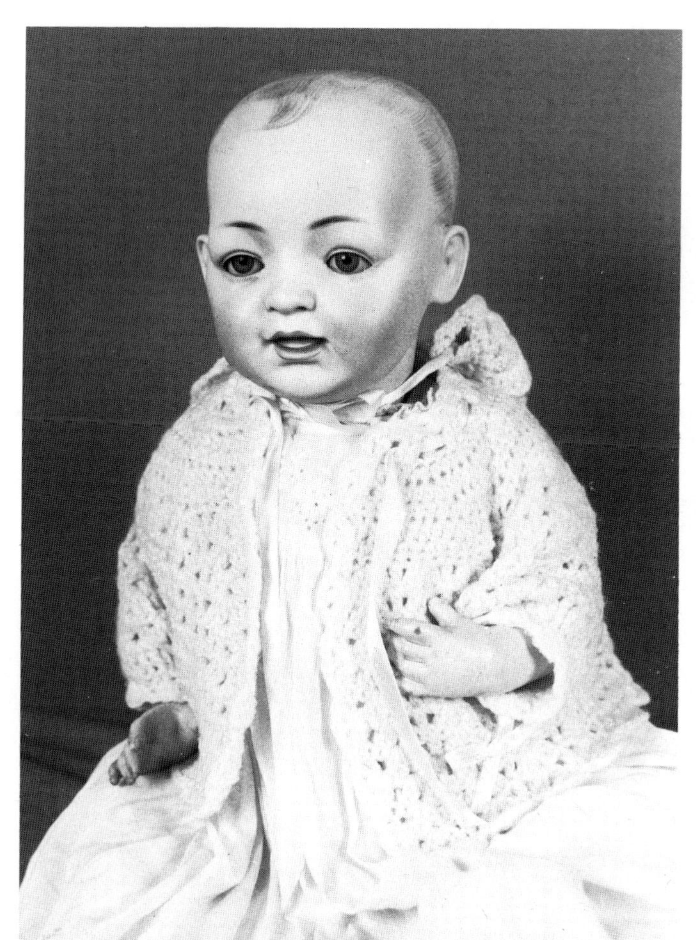

Illustration 354. This 18in (45.7cm) little fellow is incised "J.D.K. 14," and is the same doll which is shown naked in *Illustration 351, 352, and 353.* He appears to be a bald head version of the 211 mold. This is particularly noticeable in the modeling around his mouth, which is like the open/closed mouth version of the 211. "Bald" is really appropriate to describe this baby, who has a molded curl only on the top front of his head with the rest of his head smooth. His blonde hair is painted on in light strokes following a natural hair pattern. His sleep eyes are blue. *Esther Schwartz Collection.*

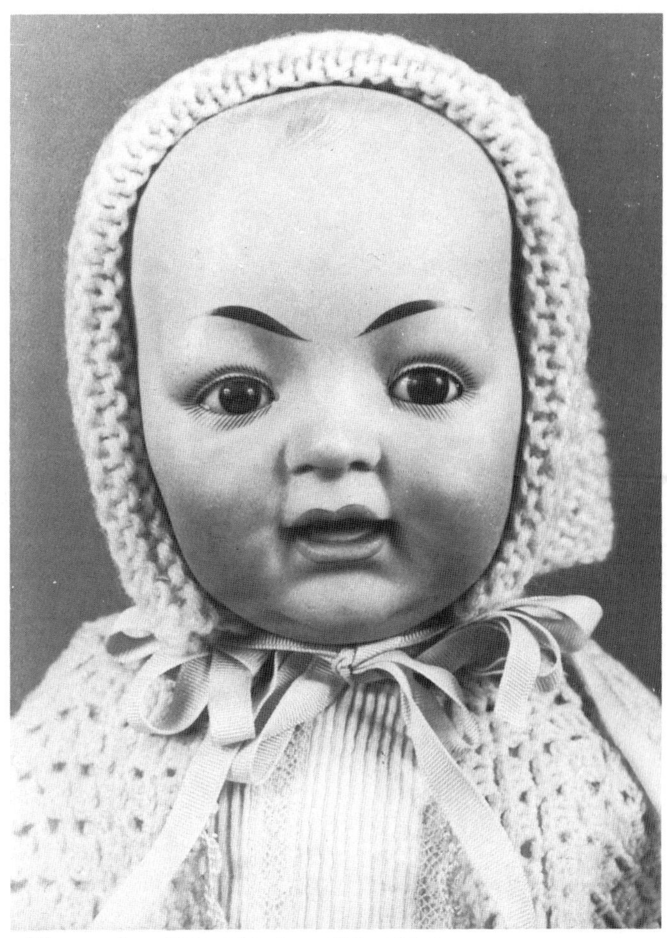

Illustration 355. This is a 17in (43.2cm) version of the same baby, shown in *Illustration 354*, also incised with the size number "14." He is striking, however, because of those curious wavy one-stroke eyebrows which one does not expect on such a large doll. He is on a bent-limb baby body in the Kestner style. *Ruth Noden Collection*.

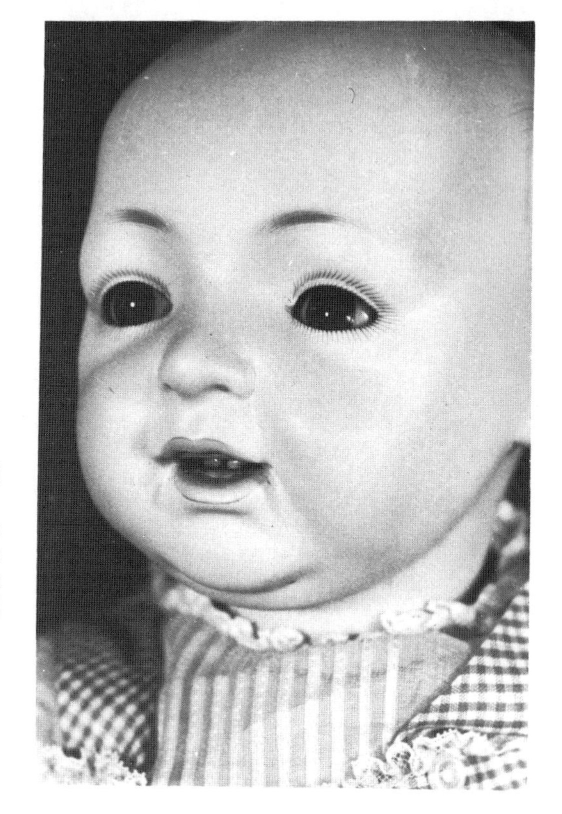

Illustration 356. This 18in (45.7cm) baby, incised "J.D.K. 14," is a variation on the doll in *Illustration 354*. His face is basically the same, but he has a fully open mouth with molded tongue and two lower teeth. He has brown sleep eyes with painted lashes and very nice feathery eyebrows. This close-up view shows his good molding and facial contours. *H & J Foulke.*

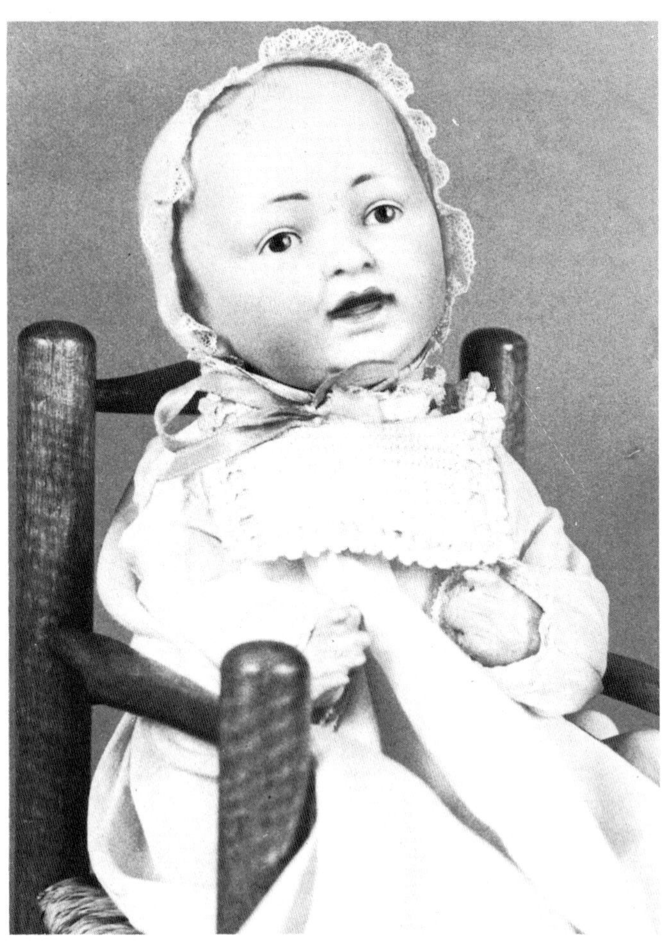

Illustration 357. This 11in (27.9cm) baby, incised "J.D.K. 7," has brown painted eyes. He has a molded upper eyelid with a black eye line and a higher red eye line and a white dot highlight in each eye. He has an open/closed mouth with a molded tongue, prominent ears, and a molded cleft in his chin. His Kestner baby body has good modeling for such a small baby and is in unplayed-with condition. His clothes are completely original. *Esther Schwartz Collection.*

Illustration 358. Here is another example of a Kestner number 7 baby, but this one has no JDK; however, he is exactly the same mold as the baby in *Illustration 359*. He has gray painted eyes, an unusual color, and the incised eye liner at his upper eyelid. His eyebrows also differ in that they are made with one wavy stroke. He is on the usual Kestner baby body, and is 11in (27.9cm) tall. It is possible that these bald heads which often appear without the J.D.K. initials were also sold to producers who put them on their own bodies. C. M. Bergmann's *Baby Belle* was advertised in the January 1914 *Playthings*, yet it is known that Bergmann did not make bisque heads; therefore, these dolls were purchased from another maker. The doll pictured looks like this Kestner bald head model, and it is very possible that Kestner sold character heads to Bergmann as their factories were both in Waltershausen. *Jimmy and Faye Rodolfos Collection.*

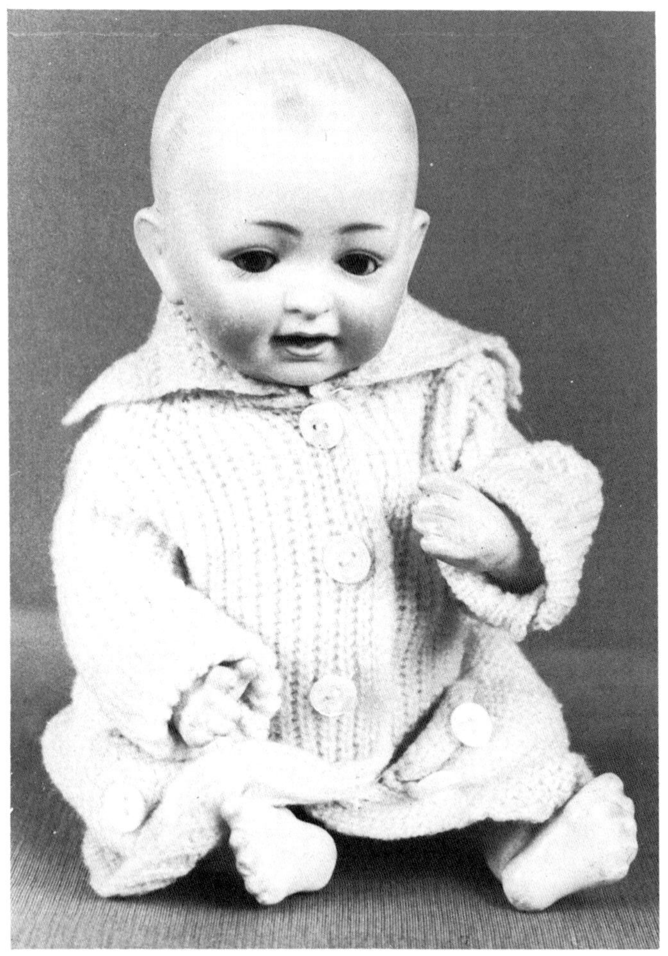

Illustration 359. Definitely a twin to the baby in *Illustration 358*, this one is exactly the same size, but has the incised "J.D.K. 7" marking. He also varies in that he has brown sleep eyes and two lower teeth. *Jimmy and Faye Rodolfos Collection.*

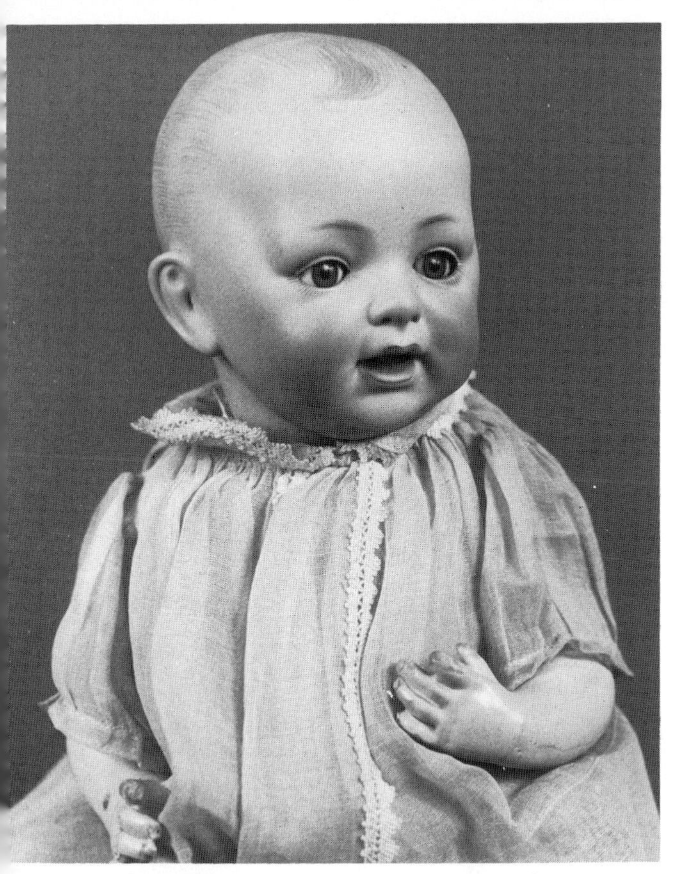

Illustration 360. This is a lovely example of a marked J.D.K. bald-headed baby on which the bisque and coloring are outstanding. He really is a bald model since there is not even a molded forelock, simply a painted swirl on top of his head. He is incised size "12," which makes him 17in (43.2cm) long. His blue eyes are a little larger than on most of these babies and this gives him a more alert look. He has the open/closed mouth with two lower teeth. He is wearing his original dress showing that not all of these babies need to have long white gowns. *Maxine Look Collection.*

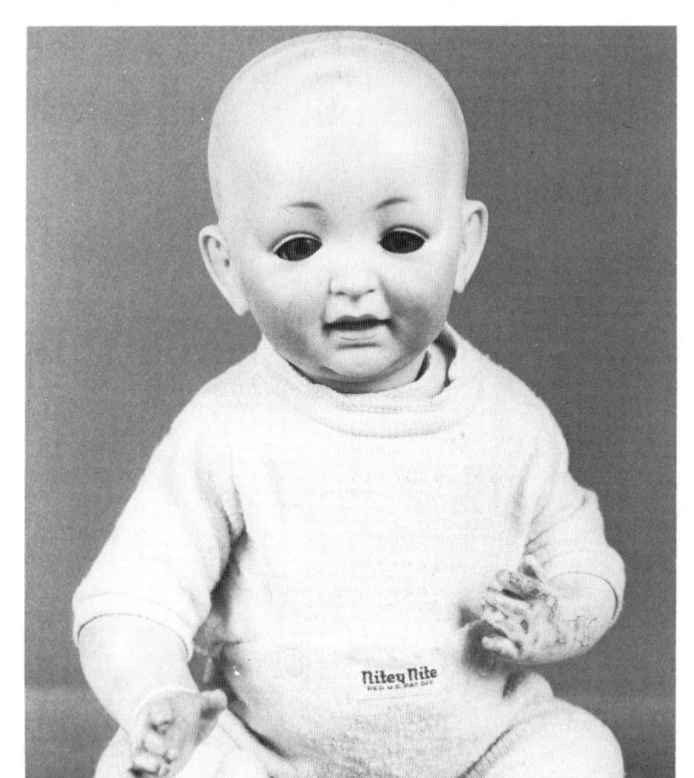

Illustration 361. Just one size smaller than the baby in *Illustration 360*, this one is incised "J.D.K. 11." His ears are particularly interesting. Not only are they large and protrudant, but there is very little modeling to the inner part. This is strange since the outer fold is so nicely done. Again, the doll has the smaller blue sleep eyes, and very tiny and lightly painted lashes. Below his toothless open/closed mouth is a fairly pointed chin with a dimple. *Jimmy and Faye Rodolfos Collection.*

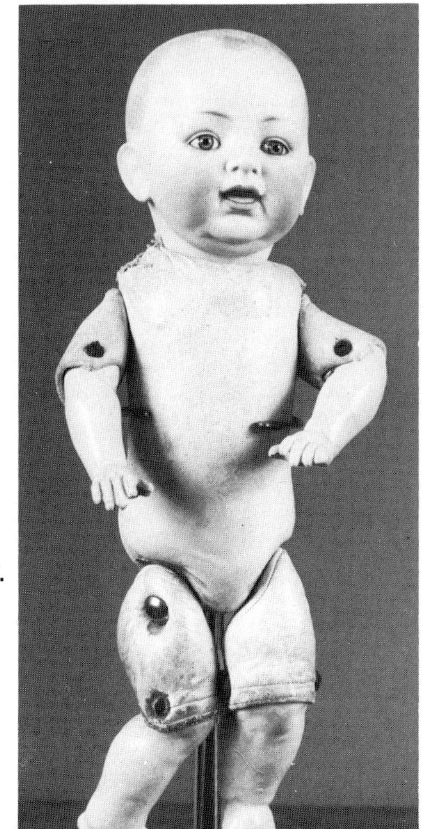

Illustration 362.

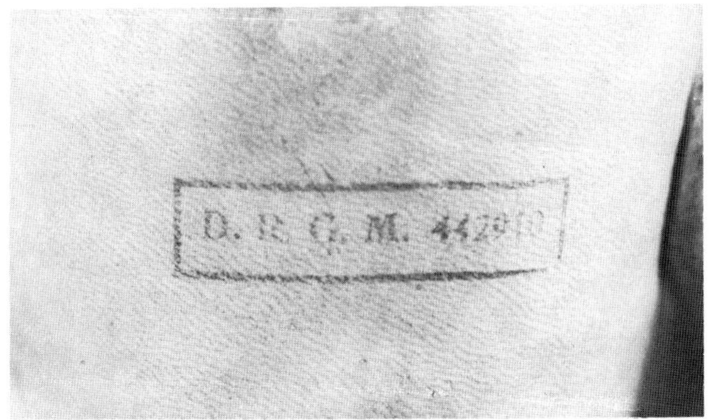

Illustrations 362 and 363. While the majority of the Kestner character babies were on composition bodies, a few models did appear on the rivet-jointed kid bodies. This 16in (40.6cm) baby is a good example of a marked J.D.K. solid head in size 12 on such a body with the rivet-jointed arms and legs; the lower limbs are composition; his shoulder and hip rivets have been replaced. Although the rivets allowed the joints excellent flexibility, the wires have a tendency to rust through and break with age. The body has a red rectangular stamp on the back "D.R.G.M. 442910." According to the Ciesliks, this body was registered by Rudolf Walch in 1910. It is not known whether this head is a replacement or whether Walch used Kestner heads. Another Walch body just 12in (30.5cm) tall has been found with a Hertel, Schwab & Co. head. The head has blue-gray sleep eyes with stroked eyebrows and painted lashes. He has the open/closed mouth and surprisingly, no chin. His hair is painted in naturally lying strokes with a molded forelock and a molded fontanel as well. *Dr. Carole Stoessel Zvonar Collection.*

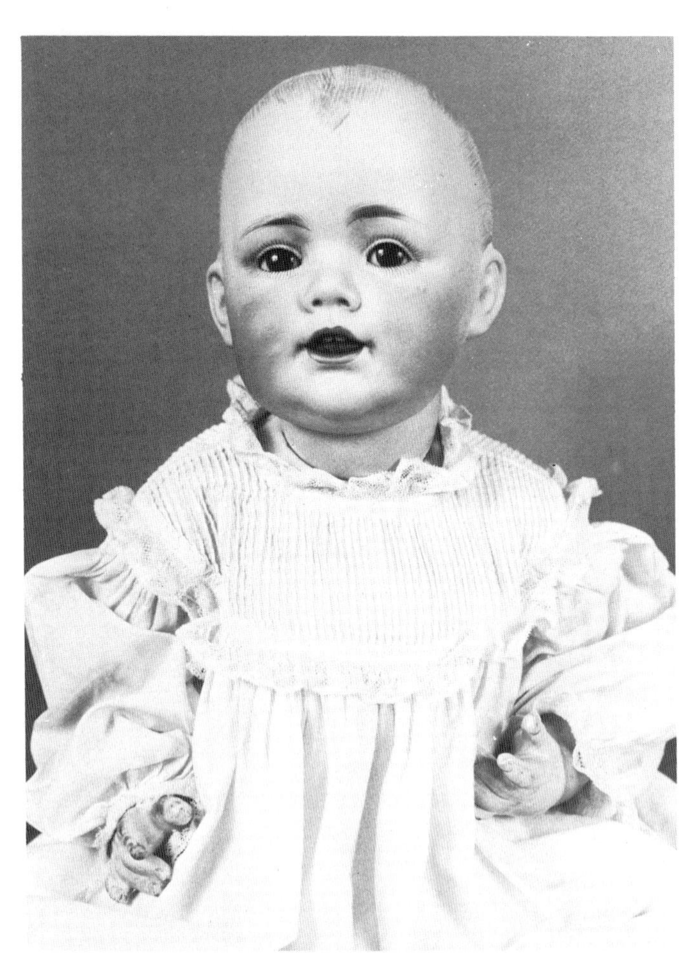

Illustration 364. This nice large 20in (50.8cm) baby is one of the harder-to-find Kestner characters. His face is much wider and quite a bit longer with a very substantial double chin. His chubby cheeks have large dimples on either side of his mouth. He has brown sleep eyes with quite long painted lashes. Both his thick eyebrows and his painted brush-stroked hair are blonde. The shape of his mouth is particularly nice because the upper lip is attractively bowed with two inset upper teeth and a free floating tongue. His ears are not so prominent as those on other bald head babies. He is a very attractive and sought-after baby. In fact, I think perhaps that this doll should be a *she. Esther Schwartz Collection.*

Illustration 366.

Illustrations 366 and 367. This 24in (61.0cm) boy is still another example of the same mold. He is incised only "J.D.K." His eyebrows are very thick and so glossy that the camera light shines off his left one. The close-up view also shows the myriad of individual strokes. His painted lashes are very dark, but not as closely drawn as those on some of the 211 characters. His open mouth is similar to *Hilda's*, but wider; he also has an inset tongue and two upper teeth. His clothes are replacements, but show how cute a baby can be dressed as a boy doll. *Mary Lou Rubright.*

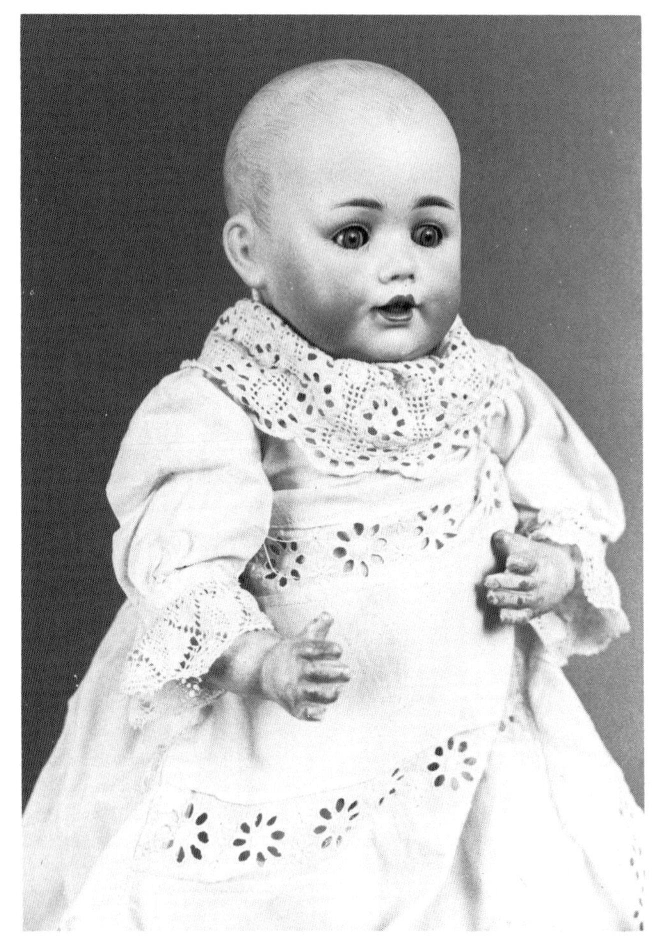

Illustration 365. This small 13in (33.0cm) baby is a little sister to the doll in *Illustration 364* and is incised "J.D.K. 8." She also has the bald head with painted blonde hair and molded front curl. The molding detail is very good for this small size and she still retains her chubby cheeks and cute dimples. The mouth treatment is the same open type with two upper teeth and separate tongue. Her bent-limb baby body is marked "Made in Germany," which is quite unusual as few Kestner baby bodies are marked at all, whereas nearly all of the open-mouth girl bodies are marked. *Jimmy and Faye Rodolfos Collection.*

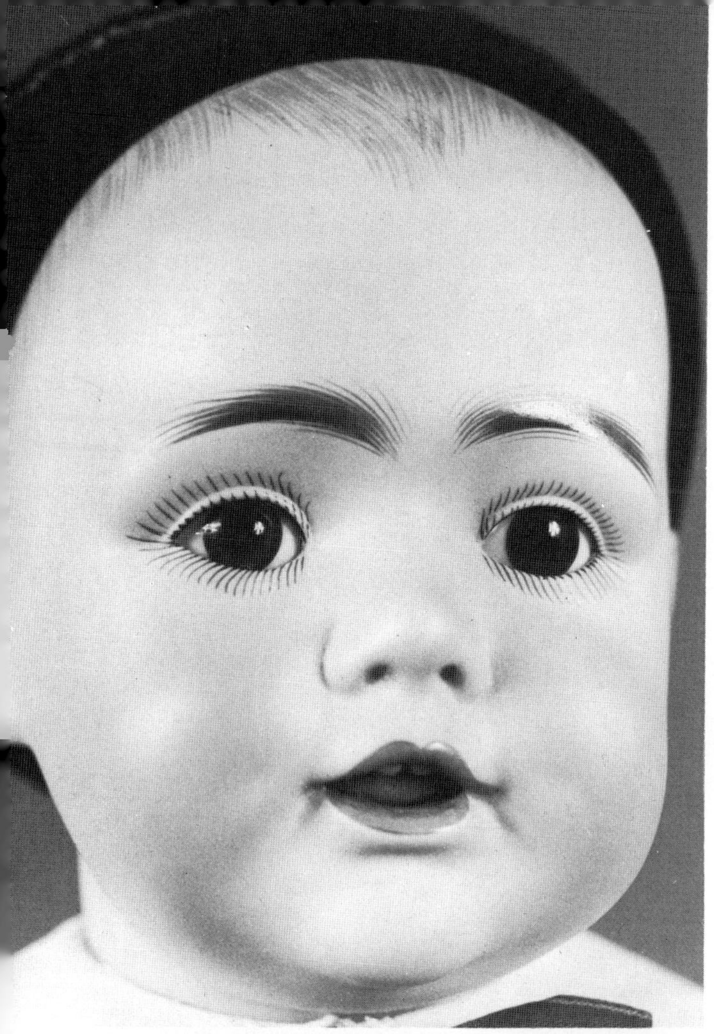

Illustration 367.

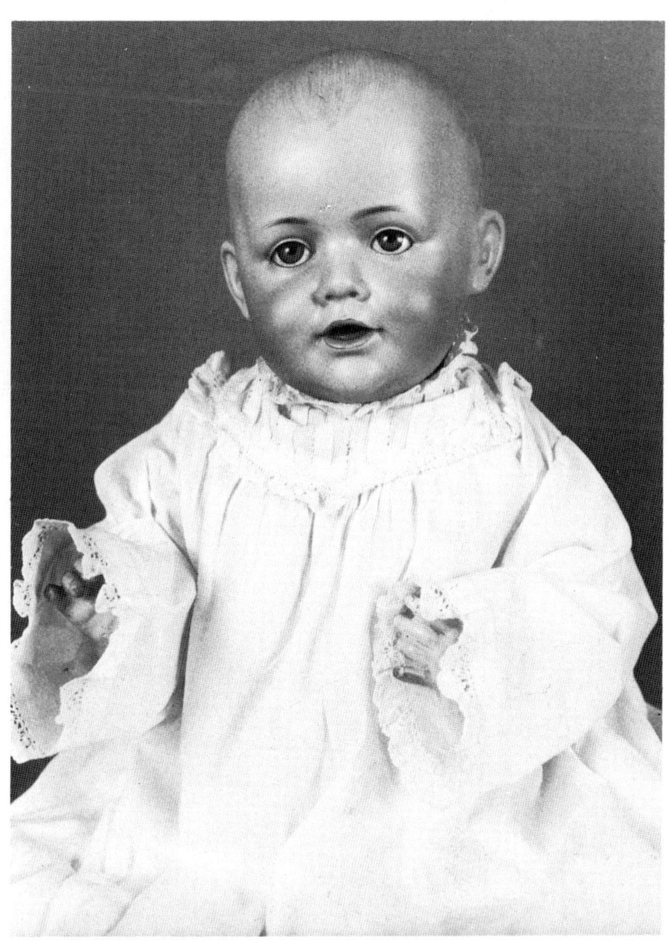

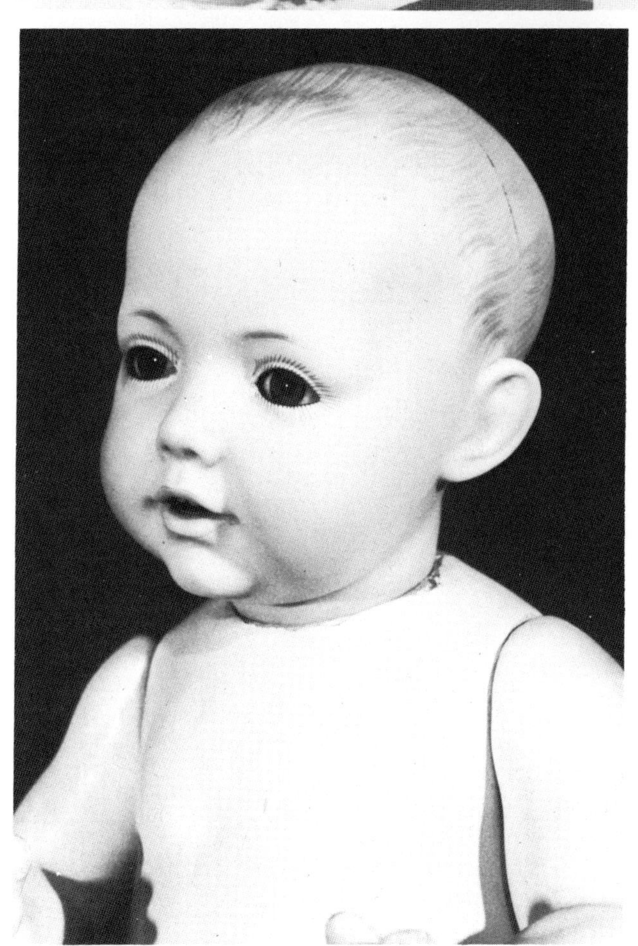

Illustration 368. This is my favorite of the bald-headed Kestner characters, and a very hard one to find. Incised "J.D.K. 14," she appears to be a bald head version of the 247 mold. The giveaway is the unusual shape of the mouth with a bowed upper lip which curls up on the ends, and two upper teeth which are set back fairly far so that they are difficult to see in the illustration. She has a molded curl at the front of her hair and blonde eyebrows which match her hair color. Her finish is shiny, the type collectors refer to as greasy or oily bisque. On a Kestner bent-limb baby body, she is 18in (45.7cm) tall. *Esther Schwartz Collection.*

Illustration 369. Although several bald-headed *Hildas* are shown in another section, I could not resist showing another one here, since she fits in so well in this part also. This 16in (40.6cm) sweetheart is incised as shown on page 149. This is a particularly good view for showing her cheek and forehead profile as well as her nicely molded chin and ear. She has a sweetly expressive irrestible face! *H & J Foulke.*

This J.D.K. baby with molded cap is a wonderful as well as a rare doll. It appears to be a special model of the very expressive 247 face with the addition of the molded blue cap and blonde hair. The cap has a molded decoration over the crown as well as a ruffle framing the face. A great amount of detail is shown in the molded blonde hair including deep curls and comb marks. Her lightly painted eyebrows are formed with many short individual brush strokes. The modeling of the lower face is quite realistic, including deep dimples on either side of her upper lip. The mouth contains two upper teeth and lips are accented on the top peaks and lower edge. She is 17in (43.2cm) tall on a jointed baby body. *Courtesy of Christie's, South Kensington.*

13in (33cm) J.D.K. 211 character baby with original blonde mohair wig. For more information on this mold see page 137. *H & J Foulke, Inc.*

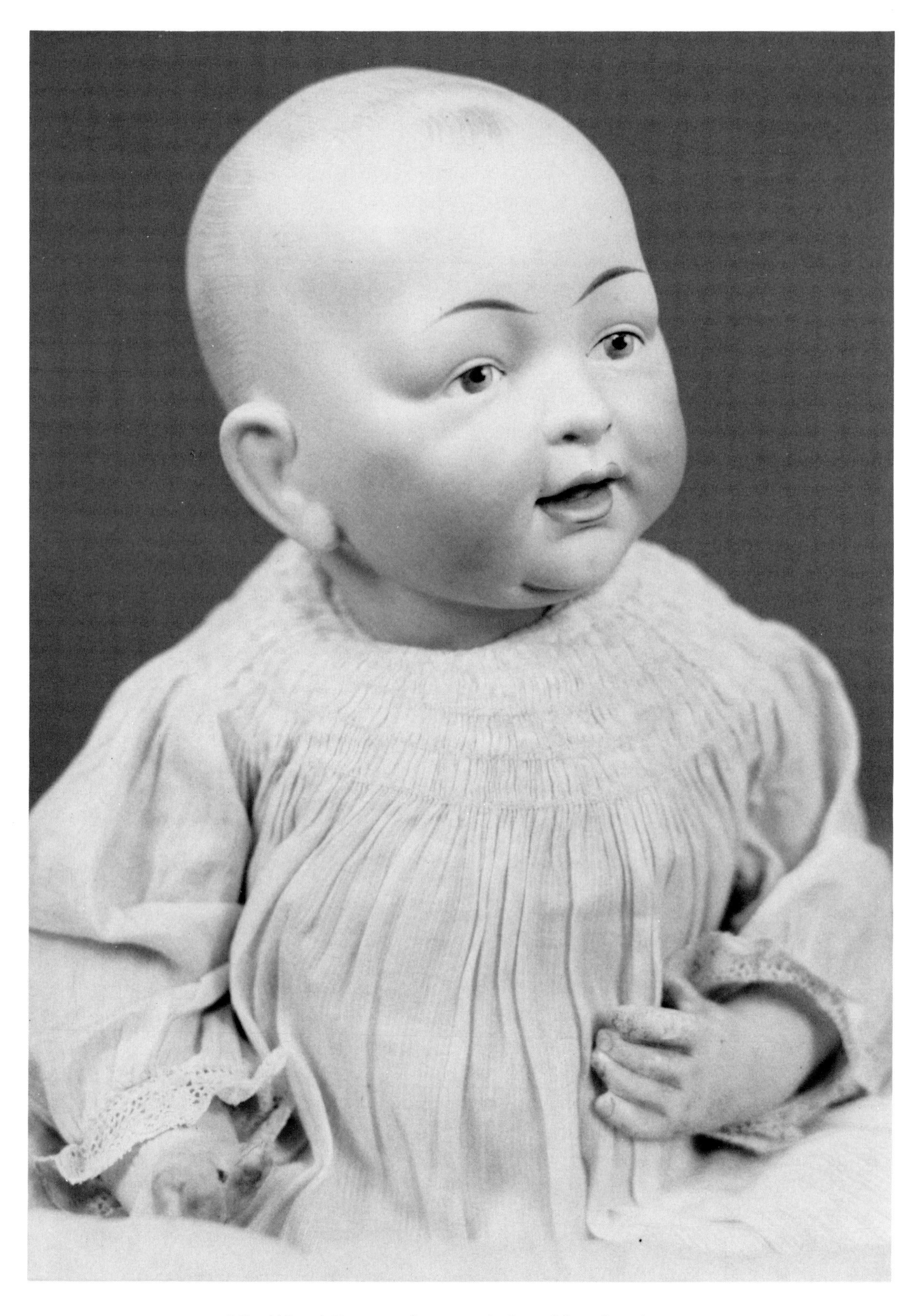

16in (41cm) Kestner character baby with painted eyes. For more information on this mold, see page 164. *H & J Foulke, Inc.*

16in (41cm) Kestner 162 lady doll. For more information on this mold see page 102. *Dolly Valk Collection.*

Illustration 384.

172

Illustrations 384, 385, and 386. The *Gibson Girl* in bisque by J. D. Kestner. The close side view shows how well the feeling has translated into the bisque doll. She has an oval face with delicate features; the upward tilt of her chin gives her a haughty look. She has the same enigmatic smile as the lady on the postcard. Her profile view shows finely chiseled features. The doll is a shoulder head so there is no way to adjust the angle of her head. She is on a kid body which hardly has any shape, but does have rivet joints and nicely modeled bisque lower hands. Her bisque is smooth and the tinting is delicate. Her eyebrows are painted in the usual Kestner manner with many individual brush strokes. Her blue eyes sleep, and she has long lower painted lashes. She is supposed to have real upper lashes, hence no painted ones, although she does have a molded upper eyelid. She has very tiny ears, and a full, but not fat face. Her closed mouth is interesting. The upper lip is full in the center, yet the edges are long and tapering; the bottom lip is fuller all the way across. A darker red line separates the lips. She has a light brown mohair wig done in the typical Gibson Girl style, and she has been appropriately redressed. She is incised "172" and is 21in (53.3cm) tall. *Mary Lou Rubright Collection*.

Lady Dolls

While lady dolls were popular through most of the 19th century, they were overshadowed in the 1880s and 1890s by the advent of the French and German child doll, who looked like a beautiful and perfect little girl. Although lady dolls did not completely disappear during these years, the emphasis was on the child dolls. By about 1910, there was renewed interest in the lady dolls again, and many German firms, such as Simon & Halbig, Dressel, Armand Marseille, and Kestner were making them, but they never again regained their former popularity.

One of the most popular and sought-after of the German lady dolls is Kestner's *Gibson Girl*, which dates from 1910. Some of the dolls have the name "Gibson Girl" stamped in blue on their torsos, and some do not. Some are marked with mold number "172," and some are not, but the look of the real *Gibson Girl* is unmistakable. She was modeled after the ladies made famous by United States artist and illustrator, Charles Dana Gibson. Mr. Gibson started by drawing small humorous sketches for *Life* magazine; these were followed by more serious ones showing the American Girl at various occupations. These caught on, and soon Gibson's Girls were famous. From 1890 until 1910 they set the fashion in America for hair and clothing styles. The Gibson Girl was portrayed as beautiful and dignified.

Illustration 383. A real Gibson Girl drawn by Charles Dana Gibson.

Illustration 385.

Illustration 386.

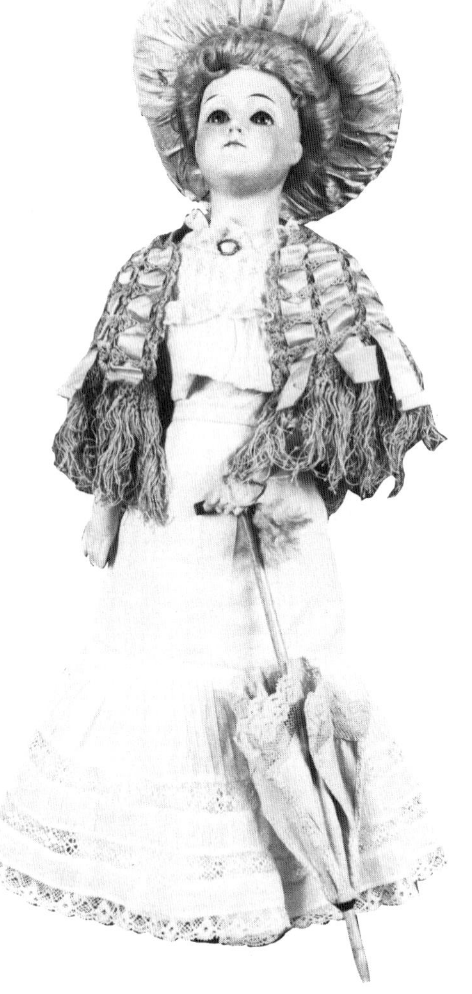

Illustration 387.

Illustrations 387 and 388. This 21in (53.3cm) *Gibson Girl* shows her very tiny ears and her long slender neck. Her eye treatment shows a different hand in that the long eyebrows are painted on in just one stroke, and she has long upper as well as lower painted lashes. Her blue sleep eyes are set into molded eye sockets with a black painted eye liner and molded upper eyelids. Her aristocratic nose has a definite uptilt. She is also a shoulder head (this face does not come in a socket model) on a hinge-jointed kid body with a button shoulder joint. Her bisque lower arms and hands show shading on the knuckles and have a tucked-in thumb. She wears her original blonde mohair wig done in the typical Gibson Girl style. She is particularly amazing because she is all original in her white lawn dress with lace and tucked bodice. The bottom of her gown has a flounce with lace insertion. Her blue crocheted shawl has satin ribbon interwoven through it. Her matching hat has an upturned brim, and she carries a white parasol with lace trim. *Maurine Popp Collection*.

Illustration 389. Here is the same lady again in another of her original outfits, this one of rust taffeta with the very fashionable high lace collar and yoke and the exaggerated leg-of-mutton sleeves. Her gored skirt ends in a ruffle which is feather-stitched around the hem. Her hat is of natural straw and rust satin with a plume *Maurine Popp Collection*.

Illustration 391. This 10in (25.4cm) 172 *Gibson Girl* was the smallest size made. She is a bisque shoulder head on a cloth body with bisque lower arms and hands and bisque lower legs. She usually has white molded hose and black heeled boots with three straps. The small lady has all of the features of the larger doll: long neck, sharp nose, uplifted chin, and closed mouth. She has sleep eyes with upper and lower painted lashes, and one-stroke eyebrows. She has been redressed in an appropriate taffeta gown. *Ruth Noden Collection*.

Illustration 390. Here are two more of the lady's original outfits. The gown on the left is of aqua silk with pleated sleeves and bodice which also has tassel trim. Her gored skirt has a short train with a ruffled bottom and an additional gathered ruffle for trim. The gown on the right is of aqua and black taffeta with a matching jacket, which is also trimmed with black beads.

Because of the crowded conditions of the makeshift studio and lack of appropriate accouterments (we were working in a motel room), we could not properly photograph all of the lady's clothes and accessories, but in addition to the four outfits shown here, she also has a blue ruffled organdy dress, a white lawn morning dress, a peignoir, a black velvet cape, hats, muff, assorted underwear, hose, and many small accessories. It is truly an exception to find a doll with such a marvelous wardrobe of original items. *Maurine Popp Collection*.

Illustration 392. Kestner also made a socket-head lady on a jointed composition body, but her face, though attractive, does not have the chic bearing of the *Gibson Girl*. This 20in (50.8cm) lady has blue sleep eyes with a molded upper eyelid and painted lower lashes only. She was intended to have real upper lashes. Her eyebrows are done in the typical full Kestner style with many individual brush strokes. Her face is pleasant, slimmer than that of the girl dolls, but not as adult looking as the *Gibson Girl*. She has a brown mohair wig over her plaster dome. Her mouth has the Kestner shape and style with shading at the top of the upper lip and the bottom of the lower one and four upper inset teeth. Her jointed composition body has longer arms and legs than those on the child dolls; it has a molded bosom and nipped-in waist with rounded stomach. The body is marked with the red boxed "Germany" used by Kestner. She has been redressed in a maroon and navy silk striped dress with a black velvet cape and hat. *Emma Wedmore Collection.*

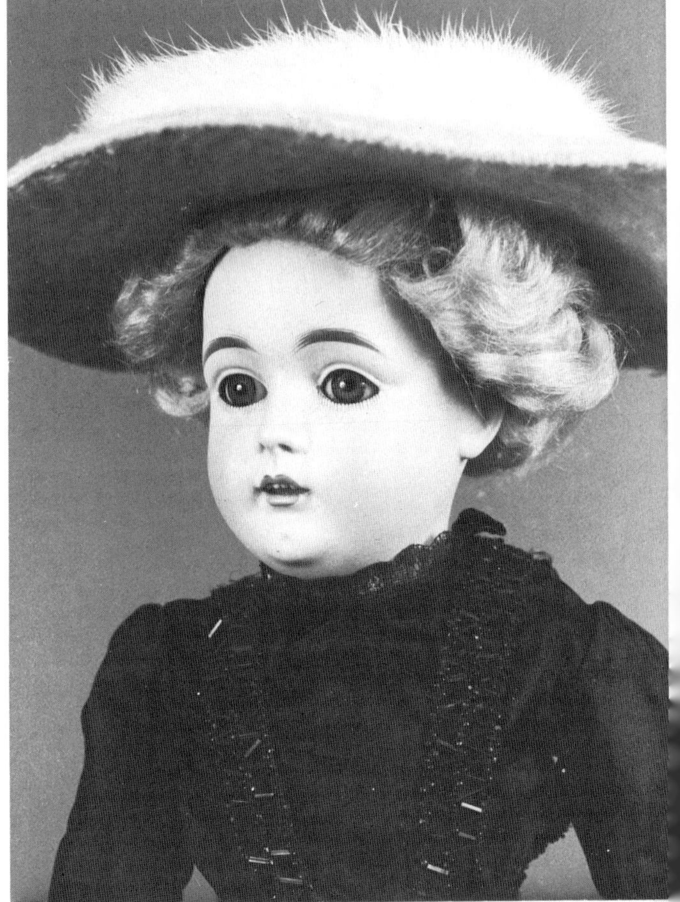

Illustration 393. Here is another 162 lady, this time in the 20in (50.8cm) size. She has glossy blonde eyebrows done in the Kestner style. Like the lady in *Illustration 392* she was intended to have real upper lashes and painted lower lashes only. The lack of the upper lashes always raises the question of what to do about the missing ones. And they are usually missing as they often fell out of the wax or just plain disintegrated or were eaten by insects. However, it takes a practiced and talented hand to properly restore real eyelashes. Most often an amateur makes a very bad job of it, and in most cases it is usually better to do without the real lashes than have those which look so artificial. This lady has a very prominent chin dimple and quite a pointy nose. She has the same mouth with four upper teeth. She is on her proper jointed composition body with the nipped-in waist and molded bosom and the slender arms and legs. She has her original blonde mohair wig and is wearing her original black silk dress with jet bead trim. *Ruth Noden Collection.*

Googly-Eyed Dolls

These whimsical dolls with roguish expressions are great favorites with doll collectors. Just how popular the googlies really are becomes a specific reality in the sales or auction rooms where bisque head ones over 10in (25.4cm) tall are bringing well into four figures. These round-eyed charmers with side-glancing eyes began to appear in 1911. Certainly they were influenced by the work of Rose O'Neill, creator of the *Kewpies,* and Grace Drayton, creator of the *Campbell Kids, Dolly Dingle,* and other similar dolls. The style caught on right away, and German producers, such as Gebrüder Heubach, Kämmer & Reinhardt, Armand Marseille, and J. D. Kestner, were including googlies in their lines. This style doll was popular for about 20 years. Googly-eyed dolls with numbers 163, 165, 172 and 173 had long been attributed by collectors to Kestner, but the Ciesliks have determined these heads to have been made by Hertel, Schwab & Co. founded in 1910 in Ohrdruf, the same town where the Kestner & Co. porcelain factory was located. Given two factories in one small town, it is not surprising that both made heads of similar style and quality. It is likely there was considerable movement of personnel between the two companies.

221

Illustration 394. The 221 googly is one of the most desired of Kestner's dolls. Seldom does it fail to bring a smile to the face of the viewer. Its expression is arresting; it cannot be ignored. This particular boy seems to be questioning in a surprised manner: Am I really going to be punished for such a small misdeed? He is hoping that his winning expression will get him off the hook. Just 13½in (34.3cm) tall, he is on a jointed composition toddler body, which is chubbier and has shorter limbs than a regular doll body. The arms and legs also have little fat rolls and creases. The hip joint is slanted like that on a baby body, and on the smaller dolls the wrists are unjointed. His round brown sleep eyes, which look to the right, have tiny painted upper eyelashes only. His eyebrows consist of one heavy brown stroke, thick in the center and tapered at each end with smaller and shorter lighter brown strokes for accent and shading. Over his plaster dome is a blonde mohair wig. His watermelon mouth is molded with a protruding but curled under upper lip and a narrow lower lip. He has a broad nose and chubby cheeks. Of course, the emphasis on the modeling is on his round eyes and watermelon mouth. He is wearing a brown corduroy coat and hat and cute brown leather shoes. He is incised:

F. made in 10.
Germany
J.D.K.
221.
ges. gesch.

Mary Lou Rubright Collection.

The J.D.K. 221 googly was still being offered as late as the catalog from about 1930 which is in *Die Deutsche Puppenindustrie 1815-1940* by Anka and Gauder. Four models were offered. A little girl with a bobbed mohair wig wore a checked bloomer with very full pant legs gathered above her knee; she is number 3562D. A boy and girl pair in lovely detailed Tyrolean outfits are catalog number 3562/40a and "Tiroler" for the boy and "Dirndl" for the girl. Another is offered with a bobbed brown mohair wig and very large hair bow, but wearing only a chemise. The doll must have still been a good seller for four different models to have been prepared.

Illustration 395. This 14in (35.6cm) example of mold 221 is especially interesting because she is wearing her original outfit. She is all dressed up as a nurse, a popular World War I style for dolls. Her dress is light blue with a white apron which has a tiny cross on the bib. *Ruth Noden Collection.*

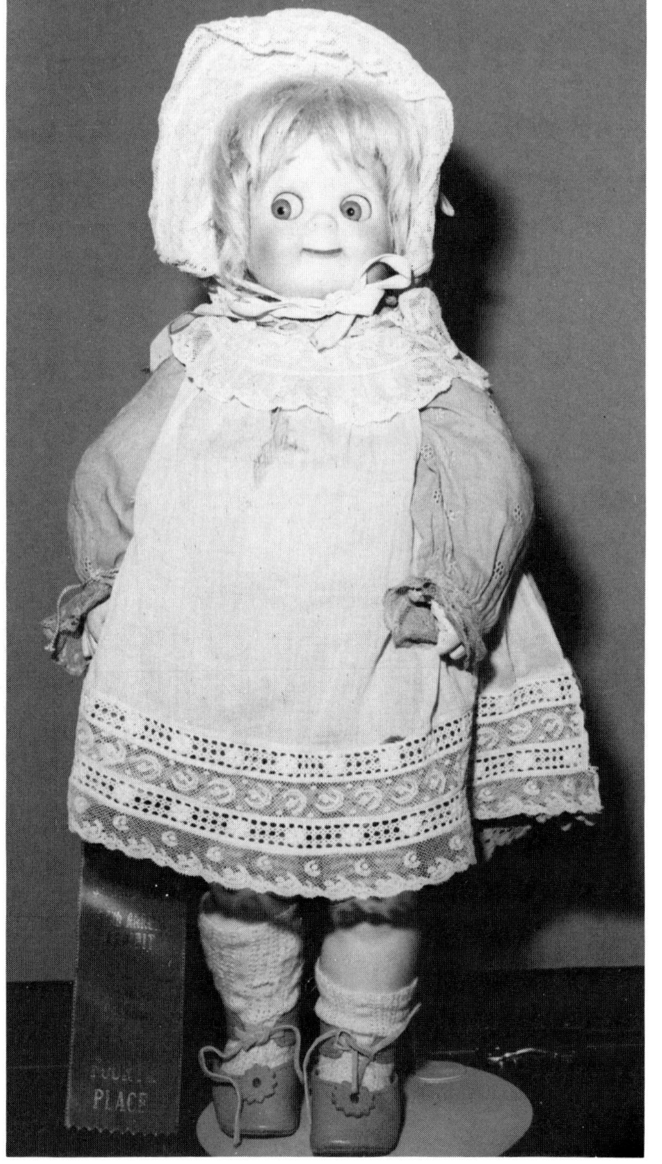

Illustration 396. This 18in (45.7cm) example of mold 221 is the largest size we have seen. Her face is just so mischievous and appealing, one has to immediately grant whatever she is asking for as she dances on tiptoe in anticipation. Her little pug nose is charming; her chubby cheeks are pinchable. She is on a fully-jointed composition body. *Jackie's Dolls of the Valley.*

Illustration 397. Kestner all-bisque googlies with jointed elbows and knees. For more information on these dolls, see pages 183 and 184. *Jan Foulke Collection.*

All-Bisques
189

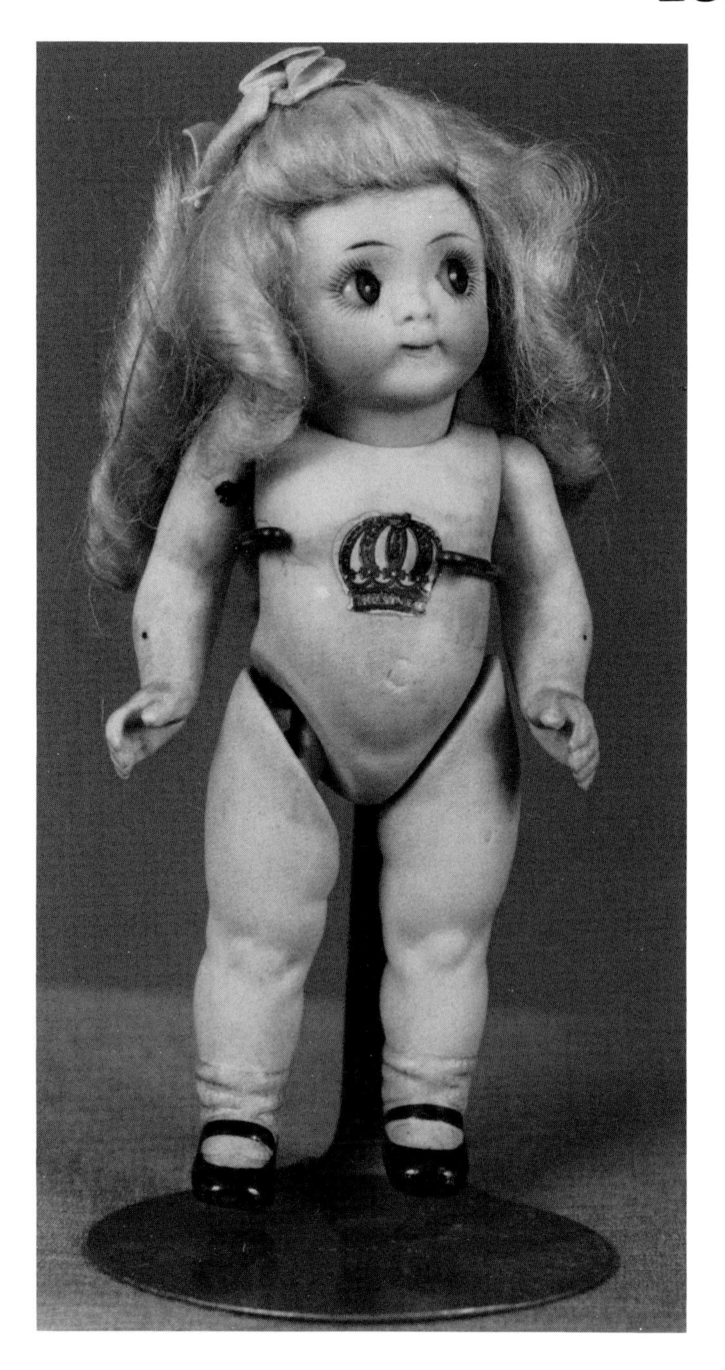

Illustration 402. Just one glance at the face of the 6½in (16.5cm) googly shows why these dolls are such favorites with doll collectors. Their faces are so roguish and full of life. This little girl is incised "189//2/0" on her head and torso; "179 2/0" on her legs; and "2/0" on her arms. 179 is the mold number of a painted eye version of this doll, so Kestner, being a good business-man, would use the limbs interchangeably. And most wonderful of all, she still has the Kestner crown label on her torso. Her face has short one-stroke eyebrows, large round blue sleep eyes looking to the side with closely spaced painted lashes. Her nose is wide and round; she has the cute watermelon-shaped mouth with protruding upper lip so desired by collectors. (So named because it curves around and up on the ends like a piece of watermelon.) Also, she has the sought-after swivel neck. Her body has the chubby healthy look; her torso shows some molding detail; her arms are rather mundane, but in her legs Kestner again shows his strong point. The thighs are chunky with fat creases and the knees are quite dimpled. Her footwear shows that she is of later design since she has low molded blue socks which show wrinkles as though they have slid down and plain black one-strap Mary Janes. Her original wig is of blonde mohair on a cloth cap over a plaster dome. These googlies came with both stiff or swivel necks and with both painted or glass eyes. One found in the original box had two labels, one with the J.D.K. crown and streamers trademark and the words "Biscuit Babies//Made in Germany" and the other "Kestner's Tiny Tots." *Richard Wright Collection.*

Illustration 403. These 5in (12.7cm) twins are excellent examples of the 189 googly with swivel neck. This pair of boy and girl twins follows the description of the doll in *Illustration 402* except that their eyelashes are painted much further apart and their dark brown eyes look to the left. Which way to make the eyes glance was apparently an arbitrary decision. Their molded socks are white, also a variation. They are dressed in completely original provincial costumes, a style that has been popular for at least 100 years continuing to the present. *Jane Alton Collection*.

Illustration 404. This 6½in (16.5cm) googly is incised "189." Again she is dressed in her original provincial costume. She is of fine quality bisque with rosy cheeks. Her watermelon mouth has a protruding upper lip; her grey sleep eyes, set to the side, are surrounded by long painted lashes. Her brown one-stroke brows are glossy. She is wearing her original blonde mohair wig in bobbed style with bangs. Her body is chubby with good knee dimples. Her footwear consists of white molded low socks and plain black Mary Janes. *Richard Wright Collection*.

This pair of 5in (12.7cm) googlies with swivel necks are from mold 292, which seems a likely Kestner product. They appear to be related to the doll in *Illustration 406,* although the girl's mouth is not quite as long. Her rounded side-glancing glass eyes are stationary, with no evidence of ever having slept. Her arms are loop-strung with the elastic passing though holes in either side of her neck. Her legs are wire-strung through a hole on each thigh. At first glance one would think she was a put-together because of the two different stringing methods but she is not. Her short ribbed stockings are blue; her one-strap shoes are black. She retains her original bobbed mohair wig over a cardboard pate. She is wearing her original Swedish outfit. Her brother also retains his original clothing. A charming pair! *H & J Foulke, Inc.*

This 6½in (16.5cm) fellow is unmarked, but he certainly appears to be a painted-eye version of the 189 mold. His eyeball is molded; blue eyes are enameled with a white dot highlight; brown eyebrows are one long stroke. Pug nose and watermelon mouth complete his pixie face. *Private Collection.*

Illustration 405. The *1914 Marshall Field & Company Doll Catalog* of their Kringle Society Dolls shows quite a few all-bisque dolls which are likely Kestners. One is this googly which appears to be a stiff-necked version of mold 189. The catalog uses the term "Peek-a-boo baby" for all of these side-glancing dolls. This particular doll is described as having "sideways glancing, opening and closing eyes; fine mohair wig; rubber cord jointed, straight limbs; 4½ inches; each in box, Dozen $4.00...7 inches $8.00." Since the confidential trade discount was 40 to 50 percent, these would be fairly close to retail prices for the dolls. She is also shown with bent baby legs and arms.

Illustration 406. Quite a few of these small googlies have been examined, but it is difficult to attribute some of them since other companies beside Kestner, such as Goebel, did make small dolls. This 4in (10.2cm) swivel-necked girl incised "201," seems a likely Kestner, however. She has good molding and decoration, and she is wearing what is probably her original chemise. Her blue sleep eyes are emphasized by long one-stroke eyebrows and very long painted lashes. Her pug nose is a bump on her face, and she has the typical watermelon mouth. Her blonde mohair wig has a circlet of braids. Her footwear consists of low blue stockings and black Mary Janes. *Richard Wright Collection.*

Illustration 407. Here is another 4in (10.2cm) googly which could possibly be a Kestner. She is incised "293.8." She has blue sleep eyes glancing to the side with many painted lashes on the right eye, and fewer on the left one. Her mouth is the open/closed type with full painted lips. She has the same blue low socks and black Mary Janes as the doll in *Illustration 406*. Her plaid dress is possibly original. *Richard Wright Collection*.

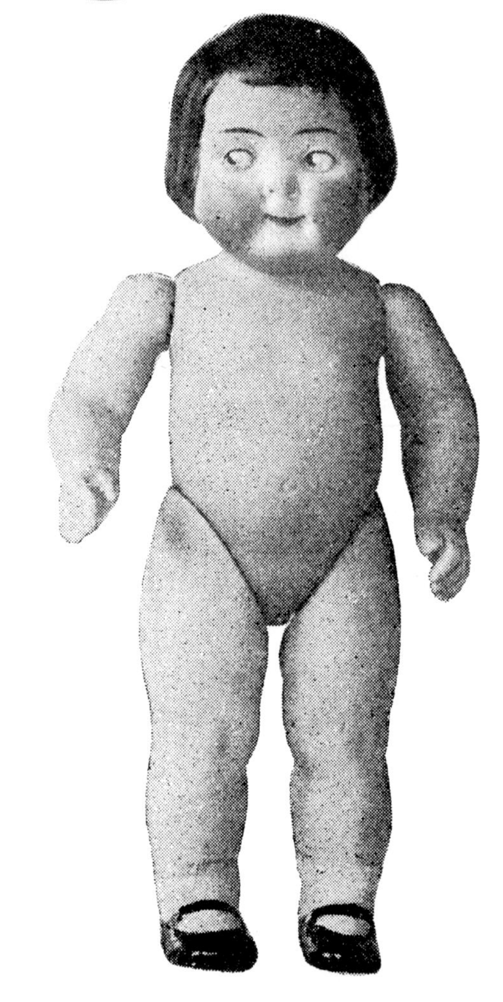

Illustration 408. This *Campbell Kid*-type googly from the *1914 Marshall Field & Company Doll Catalog* is another type of Kestner googly, this one with both painted eyes and molded and painted hair. Again,she is called a "Peek-a-boo baby." She has red-orange molded hair, although the color varies somewhat from doll to doll. Her eyes are usually blue, always side-glancing with a black upper eyelid. Her cheeks are rosy and she has the desirable watermelon mouth. Although stiff-necked, she has rubber cord jointed arms and legs. Apparently the Marshall Field Company felt this was important, as it is always mentioned in the doll descriptions. Her arms and legs are the 179 type with the low molded socks in either blue or white and the plain black Mary Janes. She was offered in three sizes: 4½in (11.5cm) at $1.80 per dozen; 5in (12.7cm) at $2.50 per dozen; and 6½in (16.5cm) at $4.20 per dozen. This doll is also shown in the Kestner catalog from the 1930s which is reprinted in *Die Deutsche Puppenindustrie*, so there is no doubt about her being a Kestner product. Her catalog number was 179, 182.

111 and 112

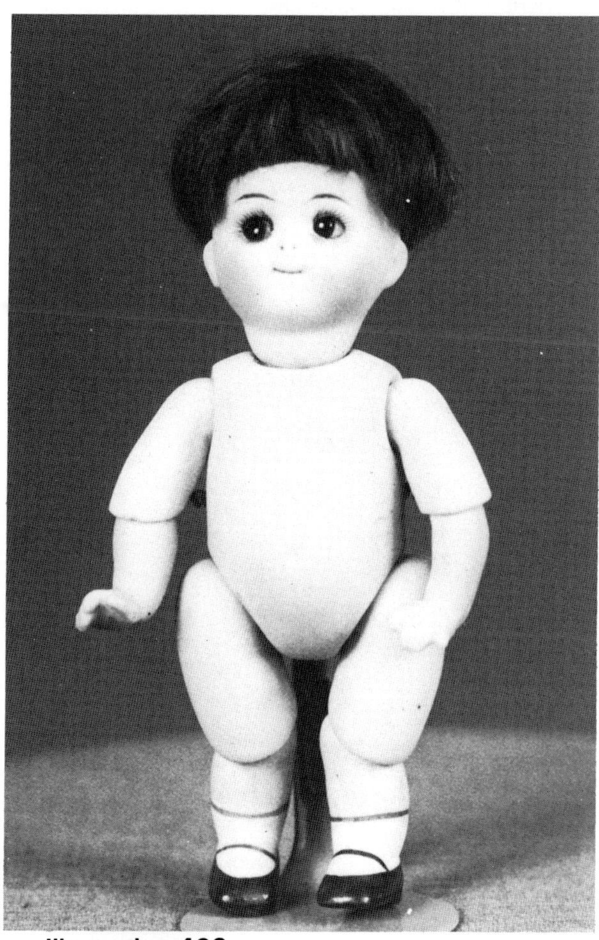

Illustration 409.

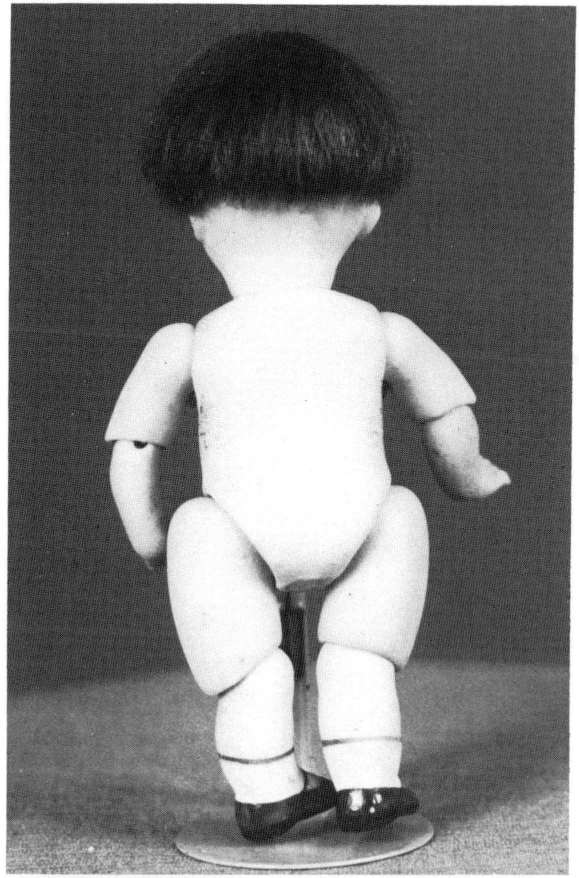

Illustration 410.

Illustrations 409 and 410. This darling googly with a swivel neck as well as jointed elbows and knees is certainly one of the most popular and desirable of all-bisque dolls. Just looking at him causes one to smile: he is just so appealing and roguish. Most people just cannot resist him. His face is round with nice rosy cheeks. His brown side-glancing sleep eyes are surrounded by many closely-painted long eyelashes. His eyebrows are made with one glossy brown stroke. His tiny mouth has faintly molded upper and lower lips which are separated by the one-stroke curving mouth. His chin barely juts out, just a little. His original brown bobbed mohair wig is on a brown cloth cap. His torso molding consists of small breasts, rounded stomach, delineated navel, and dimpled buttocks. His hands are nicely done with molded fingers of different lengths with a free thumb and little finger. His thighs are chubby. Footwear consists of low painted white socks with blue bands and brown molded one-strap shoes with low black heels and black outlined soles. The soles and heels are brown also. Just 5in (12.7cm) tall, he is incised "112 1" on the head and torso; "111 1" on the upper legs. *Jan Foulke Collection*.

Illustration 411. Here is the same little boy in an old light and dark blue crocheted suit with matching cap. *Jan Foulke Collection*.

Illustration 412. This 4 3/4in (12.2cm) little girl is incised "111 0" on her head and legs. She does not have a swivel neck, so apparently this accounts for the different mold number, although her face is exactly the same as the 112 doll. She has the blue-gray Kestner eyes looking to the side. Her original blonde mohair wig is on a paper cap and has a red silk bow at the back. She is dressed in turquoise cotton with white polka dots, the type of dress which little girls made by the dozens for dolls of this type. *Jan Foulke Collection*.

Illustration 413. Here is another of the 5in (12.7cm) ones with swivel neck, incised "112 1." She has her original stringing which is so loose that she has to sit in a chair for her photograph. She has dark brown eyes looking to the side and her original dark brown mohair wig with a red silk bow on a paper cap. None of these googlies has pates. Her neck and leg sockets are kid-lined; her head is strung on a metal hook with elastic running to her knees. She is wearing her original outfit, either made at home or made for her by a store as she would certainly have been a luxury item available only in the larger department or better toy stores. Her dress is white silk trimmed with lace and blue smocking and topstitching. Her white cotton petticoat has the same lace trim. Her hat has a lace brim and two light blue silk rosettes on the sides. She also has an undershirt, a diaper, a nightgown with blue topstitching and French knot trim and a flannel half petticoat. Also the blue crocheted suit on the first 112 boy came with her clothes. She is still stored in the candy box in which her original owner kept her. *Jan Foulke Collection*.

Illustration 414. These gogglies are so charming that I just could not resist showing one more with the jointed elbows and knees. She also has brown eyes and a brown mohair wig. She has been redressed in a pink silk and ecru lace dress with a matching straw-brimmed bonnet. *Ruth Noden Collection*.

Kewpies

Rose O'Neill's first *Kewpies*, those adorable whimsical chubby creatures with elfin features, a cherubic face, tiny blue wings, and wispy topknot first appeared in 1909 as illustrations in magazines to accompany monthly adventures about them. In Janet Johl's *The Fascinating Story of Dolls*, Rose O'Neill tells how the idea of the *Kewpies* developed from her observations of her baby brother. The smile, the topknot, the bright kind eye, the round tummy -- all were features of his babyism. She simply added a pair of wings and turned "the temporary baby into the immortal elf." It was in a dream that she conceived of the *Kewpies* as little "fairy elves ... bent on doing good deeds in a funny way." They named themselves *Kewpies*, short for Cupid.

By 1912 *Kewpie* had become a doll. Joseph Kallus helped with the *Kewpie* design. George Borgfeldt produced and distributed the dolls. *Playthings* magazine of January 1913 tells about Rose O'Neill going to the Kestner factory to help with the production of her new doll. The first *Kewpies* were all bisque, but later some had bisque heads on composition or cloth bodies. (*Kewpies* were also made of celluloid, cloth, and rubber, but this will only touch on the bisque dolls made by Kestner.) The *Kewpies* were so popular that within a year 20 other European factories were making them. Of course, since the factories did not put their mark on the small Kewpies, it is virtually impossible to tell which ones are from what factory.

Illustration 415.

Illustration 416

Illustrations 415, 416, 417, 418, 419, and 420. Just one look at this adorable elf tells it all: no further explanation of the nearly universal appeal of the *Kewpie* is necessary. *Kewpie* fever spread throughout the United States, and manufacturers provided an almost limitless number of all types of *Kewpie* items: china dishes, linens, paper dolls, glassware, postcards, and jewelry among others. The *Kewpie* made Rose O'Neill a millionaire, and is still a popular little cherub today over 70 years later. This irresistible *Kewpie* has a bisque head nearly bald expect for several clumps of molded hair at the sides and back of his head and, of course, his precarious topknot, all of which are painted a dark blonde. He has large round brown eyes looking to the side. His eye sockets are outlined with black, and he has very tiny, but closely painted eyelashes. His eyebrows are two short and thick dabs of brown paint. His nose is hardly there; his chin is almost lost in his wide smile, but his mouth really is just a molded and painted single watermelon line terminating at his pudgy cheeks. He has an excellent jointed composition body which is uniquely his with his long torso, rounded tummy, and his extended arms with the starfish hands. His hip is slant jointed and he even has bending knees. Just 12in (30.5cm) tall, he is signed on the back of his neck "Ges. gesch.//O'Neill J.D.K." so there is no doubt as to his maker. Because it is quite rare, this style *Kewpie* brings a very high price in the sales or auction room. This *Kewpie* is especially desirable because he has a tagged pink flower print cotton dress and bonnet with lace trim. *Esther Schwartz Collection.*

Illustration 417.

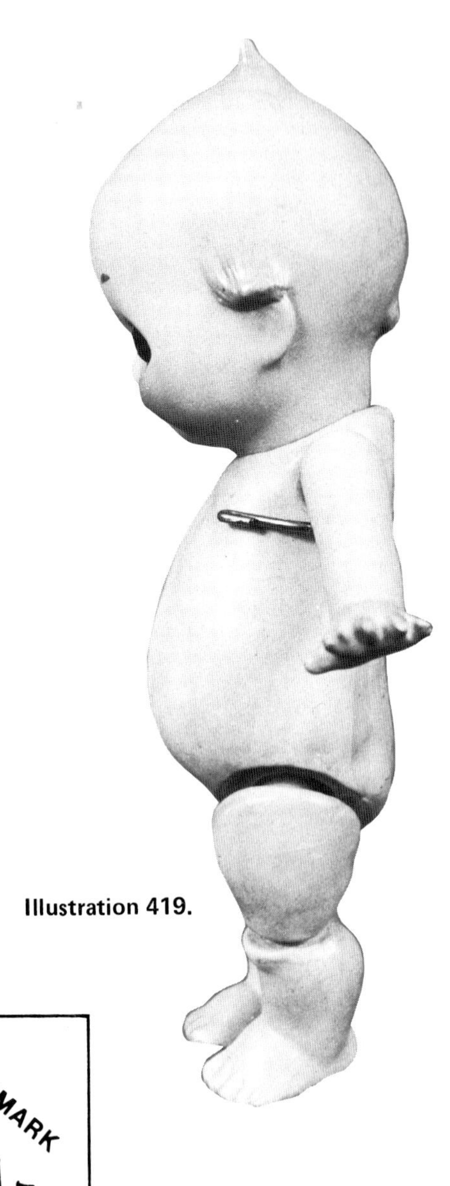

Illustration 419.

TRADEMARK
KEWPIE
REG. U.S. PAT. OFF.
COPYRIGHT
BY ROSE O'NEILL
PAT. MAR. 4 1913

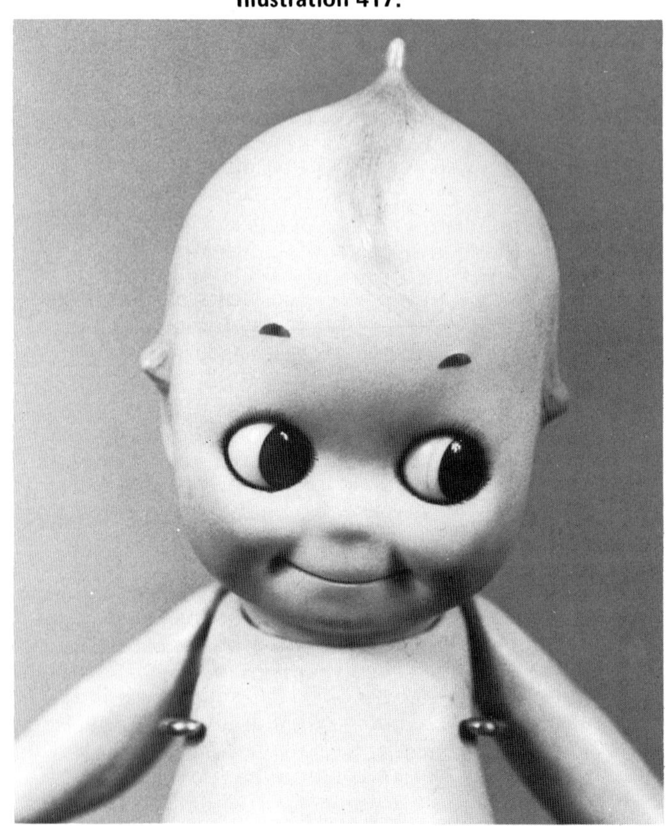

Illustration 418.

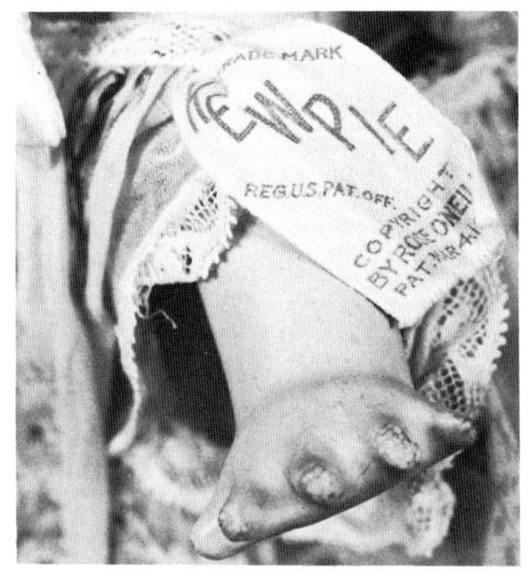

Illustration 420.

Illustration 421. Here is another example of the bisque-head *Kewpie* on the composition jointed body. This one is 11in (27.9cm) tall on a five-piece body; there is no joint at the knee. Again, his large brown eyes are a magnificent feature. He is also marked on the head "O'Neill" and "J.D.K." *Richard Wright Collection.*

Illustration 422. Here is a 7½in (19.1cm) example of the usual standing *Kewpie* which is the one most commonly found today, but this one is a fairly large size. The majority of all-bisque *Kewpies* are in the 4in (10.2cm) to 5in (12.7cm) size. He has a molded-in-one-piece head, torso, and pedestal legs. His wide eyes are painted to the side with a white dot highlight in a molded eye socket. He has a lot of tiny upper painted lashes. His eyebrows are two dabs of paint; his nose is barely there; his watermelon mouth is simply one painted and molded line. It is his eyes and smile which make him such an endearing creature. He has molded tufts of hair at the sides and back of his head; his topknot ends in a slight painted curl on his forehead. His arms are jointed at the shoulders and end in the starfish style with all five fingers extended. His bisque is rather good, but the bisque varies from doll to doll because of the millions of these made and the numbers of factories involved. Many times there are molding imperfections and debris or black specks in the bisque. This *Kewpie* is nicely blushed on his navel, cute dimpled knees, toes, cheeks, elbows, and tops of his hands. He still retains his original red heart sticker and is incised with the O'Neill signature on his foot "O'Neill" Of course, it is impossible to say that Kestner made this particular doll, but if not this one, then millions like him! *H & J Foulke.*

Illustration 423. Here is a 6in (15.2cm) *Kewpie* in his original box. He has the red shield label which says "Kewpie//Germany." He is also signed "O'Neill" on his foot; most *Kewpies* are signed on the foot, but a good many are not. *H & J Foulke.*

Illustration 424.

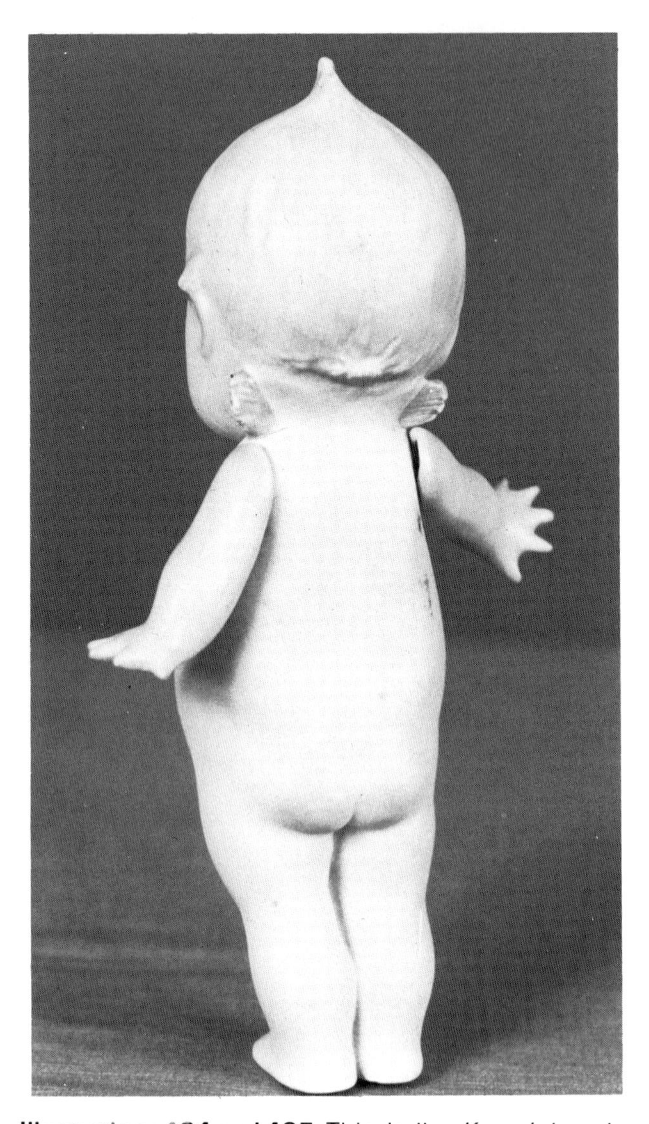

Illustrations 424 and 425. This darling *Kewpie* is a nice big 9in (22.9cm) tall. He is just plain cute to look at. The back view shows his tiny molded blue wings, an important *Kewpie* characteristic. This *Kewpie* is unusual because of his cute molded hair. Most *Kewpies* are bald except for their four molded tufts of hair, but his is slightly molded all over his head! *H & J Foulke.*

Illustration 426. This 6in (15.2cm) *Kewpie* is glancing in the opposite direction from the others, but this is an arbitrary decision, and *Kewpies* can look in either direction. What is really rare about him, however, is the fact that he has jointed hips, a feature which seldom occurs in *Kewpies*. *H & J Foulke.*

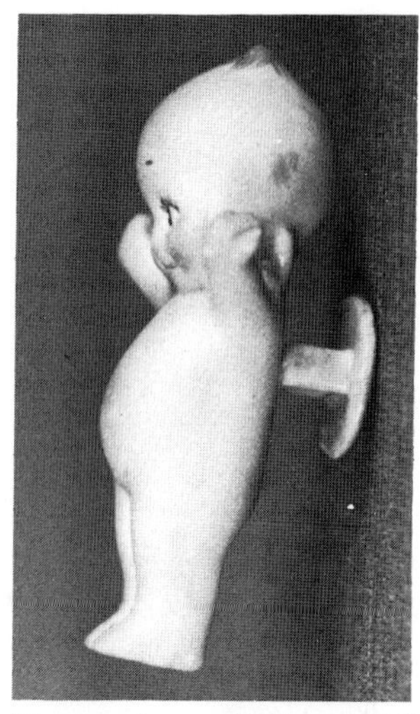

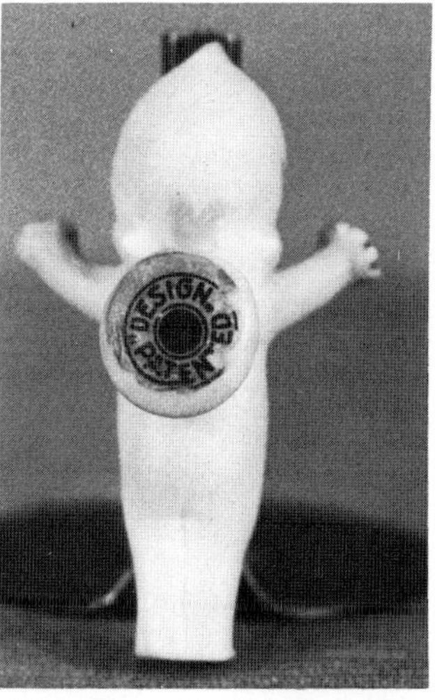

Illustrations 427 and 428. This little *Kewpie* novelty was designed to be worn in the buttonhole of a lapel, hence its name "Buttonhole *Kewpie*," but with the high price of these little items today, few collectors dare to wear them in their buttonholes! Just 2in (5.1cm) tall, he has his original label on his back. Many *Kewpies* of all kinds still retain labels like this which sometimes read "Copyright Rose O'Neill" in a circle instead of "Design Patented" as this one does. *H & J Foulke.*

Illustration 429. Another *Kewpie* which is fairly difficult to find is the black one which came on the market in 1914; however, it was not until the following year that these were named *Hottentots*. This one is 5in (12.7cm) tall. I have seen more of these in celluloid than in bisque. *H & J Foulke.*

Action Kewpies

Collectors refer to *Kewpies* other than the plain standing up ones as "Action *Kewpies*." These were made in hundreds of varieties, and many collectors specialize in finding as many different ones as possible. These *Kewpies* were made in all sorts of positions and situations; they were also made to represent many different occupations. In one partial page from a Borgfeldt catalog shown in *The Antique Trader Weekly's Book of Collectible Dolls*, over 40 different models are shown. To present all of the various action *Kewpies* would take a book in itself, so we have chosen a few, just as a sampling. Again, it is known that Kestner made all-bisque *Kewpies*, but other companies did also, and it is not possible to know whether the Kestner factory made these specific ones.

Illustration 430. This 4½in (11.5cm) *Kewpie* is very unusual because of her painted footwear. She is wearing short white socks with blue bands and black slippers with straps and painted bows. She is eagerly sought after by *Kewpie* collectors. *Richard Wright Collection.*

Illustration 431. This 4½in (11.5 cm) policeman *Kewpie* was very possibly a souvenir from a policeman's convention or some patriotic event since he has a flag decal on his tummy. There are quite a few *Kewpies* with hats representing various military and civilian service groups. *Richard Wright Collection*.

Illustration 432. This 4¼in (11.5cm) fellow with the molded yellow hat is referred to as "The Farmer." He has a hole molded through his fist so that he can hold some sort of implement. *Richard Wright Collection*.

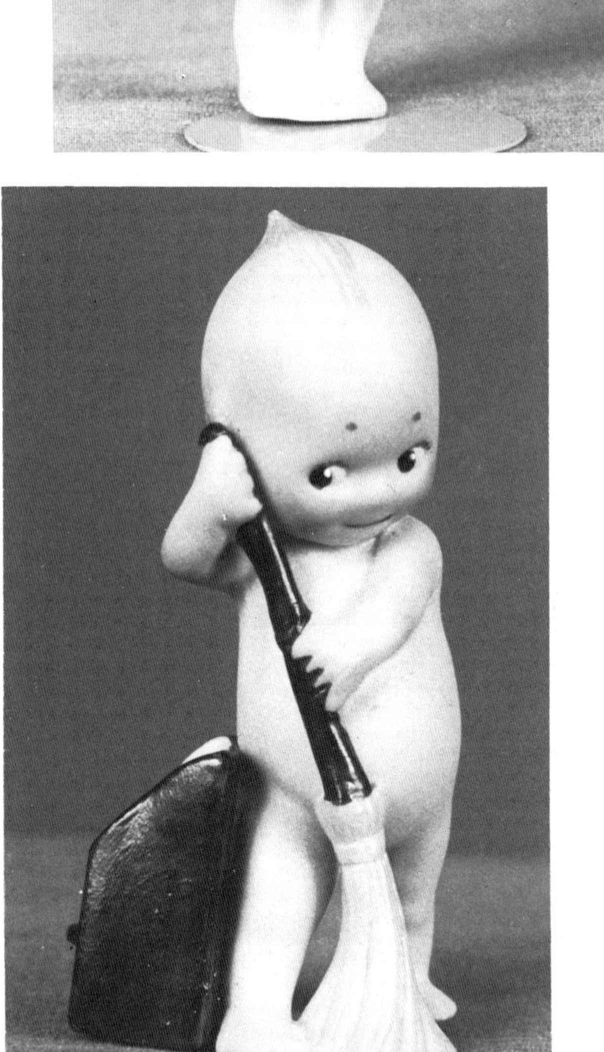

Illustration 433. This 4¼in (11.5cm) *Kewpie* is called "The Sweeper" for obvious reasons. The dustbin beside him is open at the top, so it could have been used for pins, trinkets, matches, or other items. *Richard Wright Collection*.

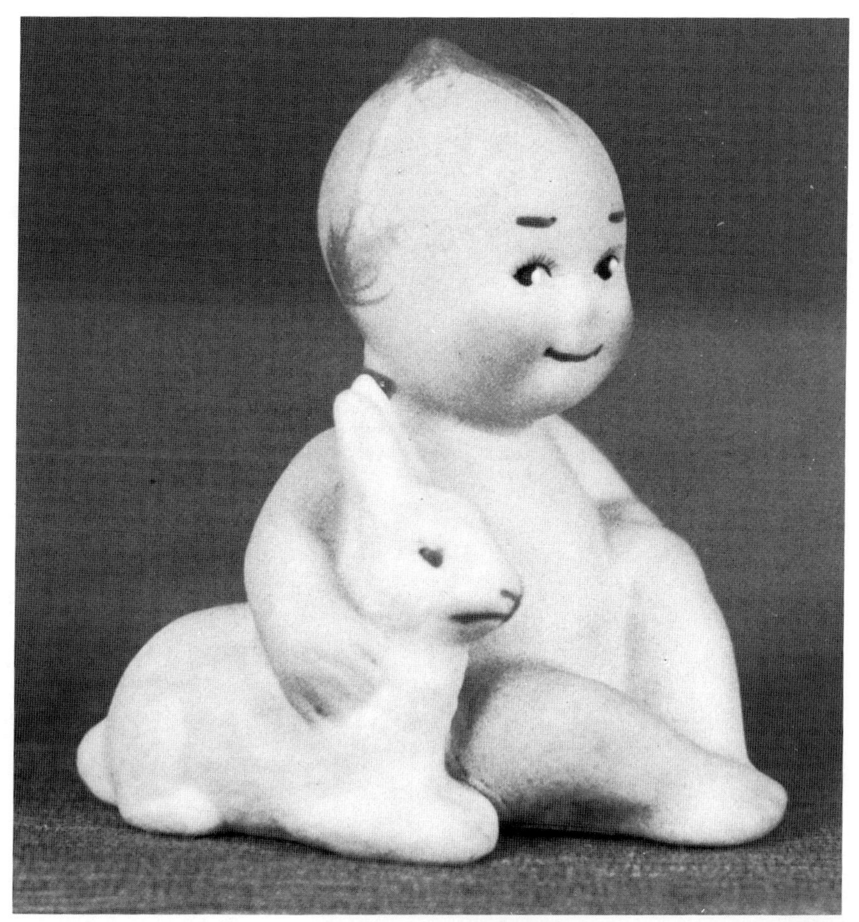

Illustration 434. Here is a delightful 2in (5.1cm) sitting *Kewpie* with his pet white bunny rabbit. The detail on the bunny is excellent for such a small item; even his coat has an uneven texture to simulate fur. *Richard Wright Collection*.

Illustration 435. Any of the action *Kewpies* could be attached to a container with an open top to make a decorative vase, trinket box, toothpick holder, or catch box of any sort. This one is "The Guitar Player." *Richard Wright Collection*.

Illustration 436. This 4½in (11.5cm) Kewpie holding a molded woven basket is another item which would have been used as a catch box for any type of small item. *Richard Wright Collection*.

Illustration 437. Certainly one of the rarer *Kewpie* items is this molded basket with the *Kewpie* head and the molded flowers on one side. It is a delightful ornament, just 3in (7.6cm) tall. *Richard Wright Collection.*

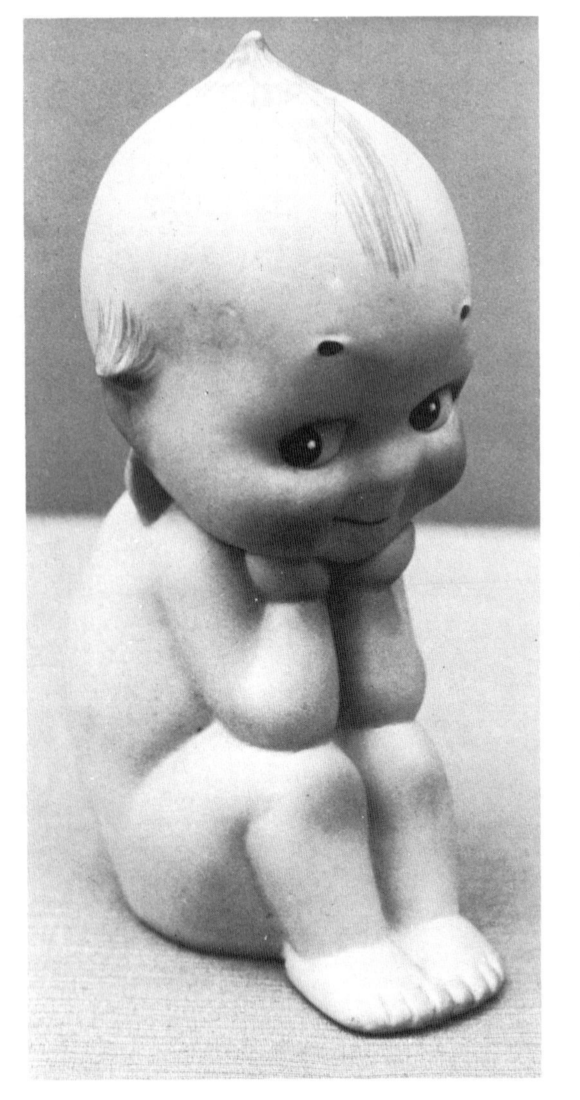

Illustration 438. Known as "The Thinker" this 6½in (16.5cm) *Kewpie* is a favorite with collectors. He is not rare, but he is especially desirable in this large size. He just makes such an appealing figure with his knees drawn up, his chin in his hands, and that perfectly marvelous face, a combination of cherub and imp. He has molded as well as painted eyebrows. *H & J Foulke.*

Century Dolls

In 1925 the Century Doll Company of New York, New York advertised that it had a new series of bisque-headed dolls made by Kestner. With the enormous popularity of the *Bye-Lo Baby*, quite innovative because it was a newborn baby, all of the other doll companies were striving to get infant dolls on the market.

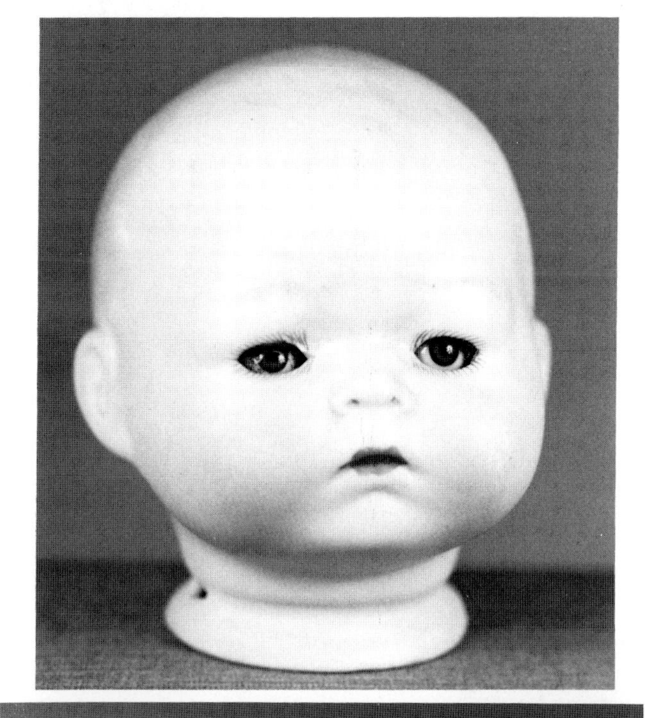

Illustration 439. This frowning baby has an 8½in (21.6cm) head circumference. It has a molded fontanel and slightly molded hair overall with an even painting, perhaps sprayed, but not done with the brush strokes of the character babies. He has tiny blue sleep eyes with both upper and lower painted lashes; his eyebrows are very light, more washed than stroked on. This is a flange neck which was meant to be sewn on to a cloth body with composition hands. Marked:

K Germany
CENTURY DOLL Co.
O

Emma Wedmore Collection.

Illustration 440. This Century baby is 12in (30.5cm) long with the same face as the doll in *Illustration 439*. His complexion is nice and pink; his sleep eyes are brown. He is on a cloth body with composition limbs. The bodies of the Century babies vary from doll to doll. *Jimmy and Faye Rodolfos Collection.*

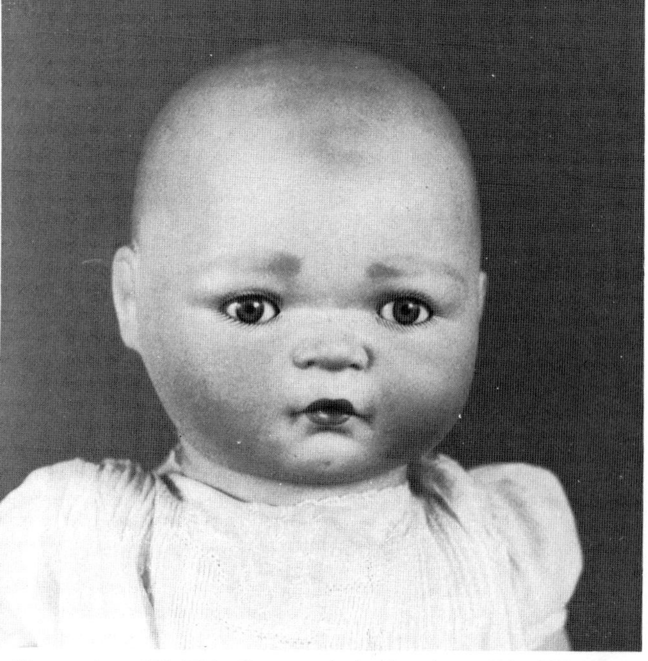

Illustration 441. This Century baby is a large 18in (45.7cm) long. He is on a hard cloth body with swivel hips and full composition arms. His small blue sleep eyes are quite wide set with long painted lashes. His eyebrows are very lightly painted but quite heavy for a baby. His nose is short, yet broad with flaring nostrils. The molding around his eyes and across the bridge of his nose, as well as his downturned mouth contribute to his frowning look. His hair has very good molding, which is difficult to tell from the illustration. *Jimmy and Faye Rodolfos Collection.*

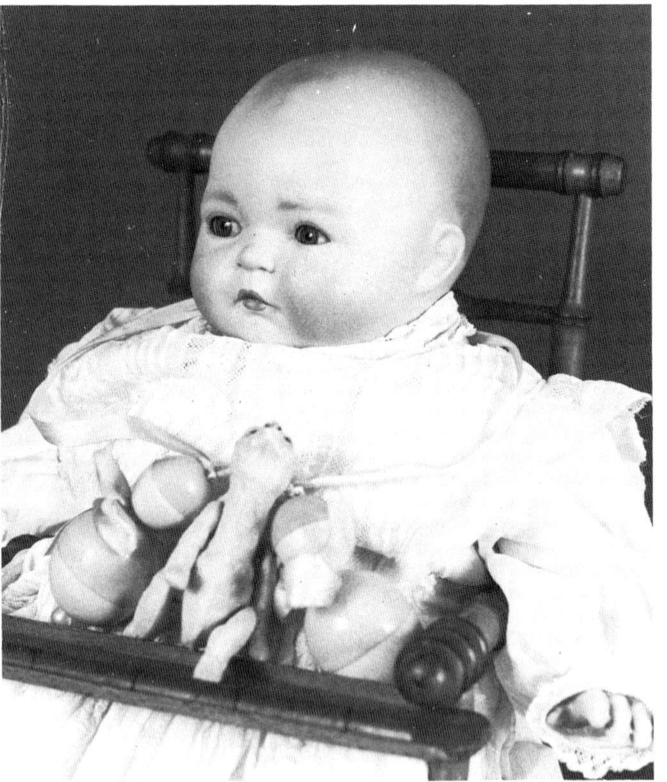

Illustration 442. This Century baby has a head circumference of 13in (33.0cm). This is a particularly good angle for seeing the face molding, the small eyes, the broad nose, the fat cheeks, and the mouth in typical baby shape with a protruding upper lip and a molded philtrum as well as a cleft under his lower lip. He also has a molded fontanel and a curly forelock. Some of these babies have more detail in the hair molding than others. *Richard Wright Collection.*

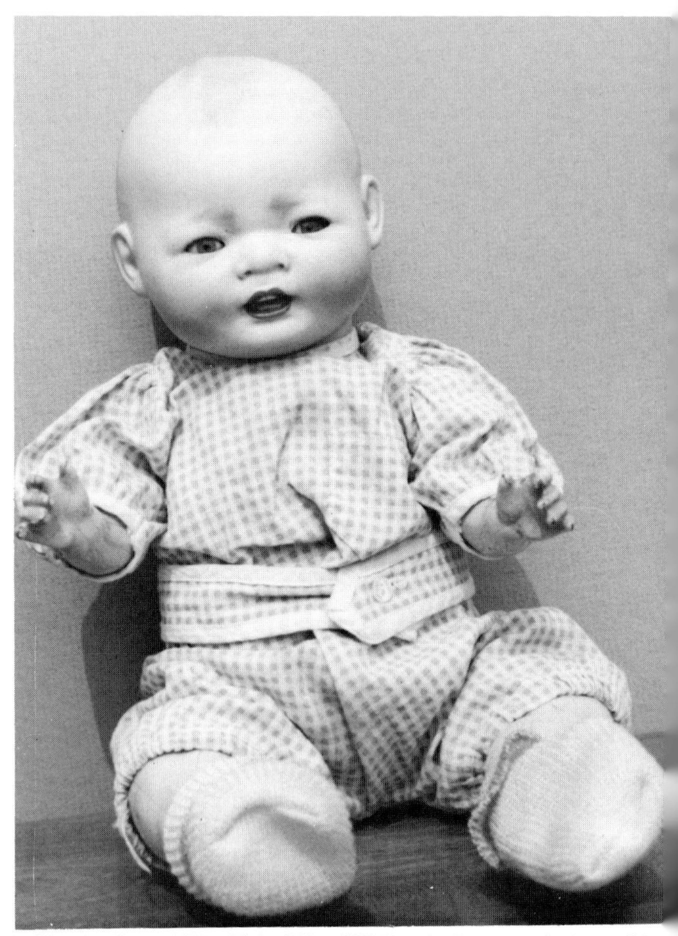

Illustration 443. This 14in (35.6cm) Century baby with an 11in (27.9cm) head circumference is a very unusual one with a smiling face. He has tiny blue sleep eyes with laugh creases at the sides. His lashes are painted, and he has wide, lightly washed on eyebrows. His smiling mouth is in the open/closed style with full lips, a molded tongue and two molded upper teeth. Obviously he is not intended to be a newborn. His solid dome head has a molded peak of hair right on the top and center front. He is on a hard cloth body with composition arms and legs. An interesting version of this face is shown in the United Federation of Doll Clubs 1981 Convention Book. Actually the head has two faces: this cute smiling one with glass eyes on one side, and a quite frowning face with painted eyes on the other side. The doll shown here is incised: 'Century Doll Co. // Kestner Germany.' *H & J Foulke.*

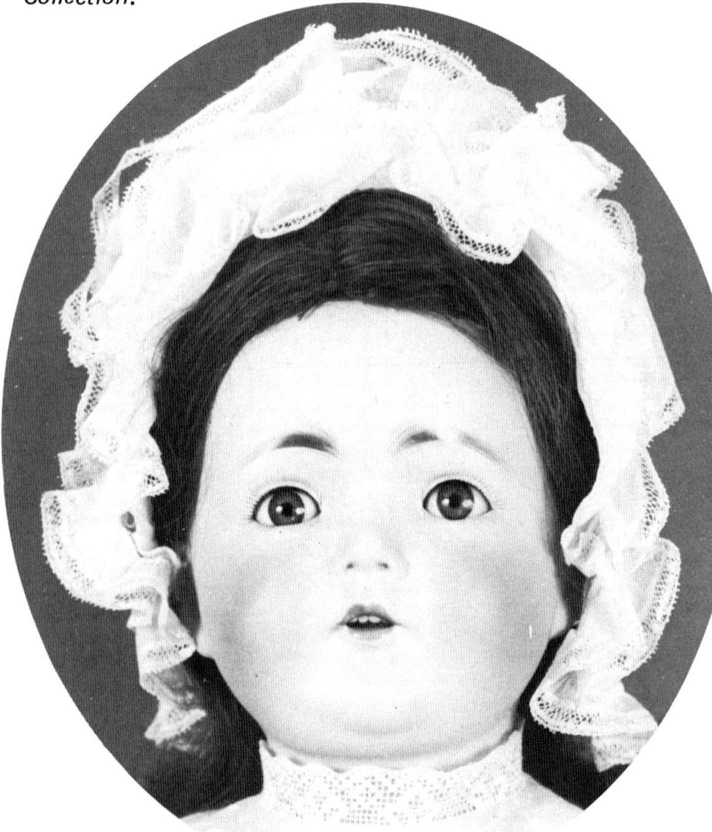

Illustration 444. This rare and interesting 21in (53.3cm) mama doll was made for the Century Doll Company by Kestner. She is mold number 281, a shoulder head on a cloth body with a voice box and composition limbs. Her face is quite wide with chubby cheeks and a substantial chin. Her bisque is lovely with nicely blushed cheeks. Her eyebrows are lovely--long and glossy with many individual brush strokes above, below, and at the ends. She has thick painted eyelashes with small blue sleep eyes. Her small open mouth has a bow shaped upper lip with upturned ends in the Kestner style and four upper teeth. Her brown human hair wig appears to be original. She is marked: "CENTURY DOLL Co.
Kestner Germany
281/5."

This face has also been found on a 24in (61cm) doll w
socket-type head on a composition shoulder plate mar
"Century Doll Co.//Pat. App'd 1924." *Jane Alton Collect*

Bye-Lo Babies

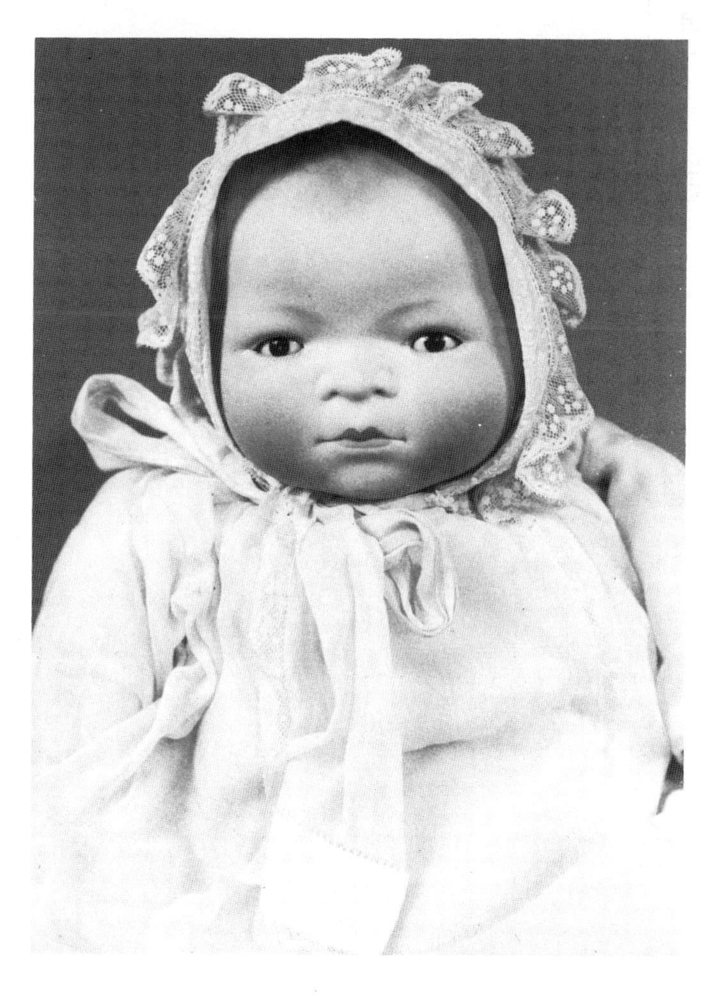

Grace Storey Putnam's *Bye-Lo Baby* came on the market in 1924. After a slow start it became a huge success. In the *Facinating Story of Dolls* by Janet Johl, Mrs. Putnam tells the story of making her *Bye-Lo Baby.* She said that the first Christmas George Borgfeldt, the distributor, could not get enough of the babies and people stood in line to buy the doll, thus it received the nickname 'Million Dollar Baby.' *Bye-Lo* heads were made by Kestner as well as at least five other porcelain factories. The first *Bye-Los* were on composition bodies, but soon it was decided to use a soft one which Mrs. Putnam designed with bent lower legs and feet which curled in. However, before Mrs. Putnam's design was ready, some dolls did appear on cloth bodies with straight swinging legs. By 1925, *Bye-Los* were offered in seven sizes, ranging from 9in (22.9cm) to 20in (50.8cm), selling for $3.00 to $14.95. Heads could be purchased separately for replacing broken ones or for use on homemade bodies.

Illustration 445. A *Bye-Lo Baby* with a 15in (38.1cm) head. She has a solid dome head with a molded fontanel and deeply molded hair painted with a spray or wash, not individual strokes. The amount of molding in the hair varies from doll to doll, some having more and deeper curls than others. Her tiny brown sleep eyes are supposed to be just half open and have short painted eyelashes and soft looking but curved eyebrows brushed on. Her cute little face is fat like a new baby's, and there is a wrinkle in her nose which is quite broad. Her full lips are closed; she has a molded philtrum and a cleft in her chin. She is on a marked cloth *Bye-Lo* body with celluloid hands. She is wearing her original dress of white batiste with lace trim down the front. Of course, it is not certain that Kestner made this particular baby, but he certainly did produce thousands just like it. *H & J Foulke.*

Illustration 446. The all-bisque *Bye-Los* were definitely made by the Kestner porcelain factory; they were advertised in Butler Brothers 1925 wholesale catalog in December of that year. The model shown was the 4in (10.2cm) baby with painted features and hair, jointed arms and legs; the wholesale price was $2.50 per dozen. Because their features are so true to the larger dolls, these tiny babies are very popular with collectors. They range in size from 4in (10.2cm) to about 8in (20.3cm). The doll illustrated is 4½in (11.5cm), incised "20-11", and wearing original clothes. Many of these tiny babies are found with elaborate original clothes, some store made and some homemade. Many have a complete set of underclothes and booties as well as gown, sweater, and cap. Sizes start with number 10 and increase ½in (1.3cm) with each additional number, hence size 11 is 4½in (11.5cm). Stock and size numbers are also incised on the arms and legs, and while the stock numbers may vary on the same doll, the size numbers should match. *Richard Wright Collection.*

Illustration 447. Many variations of the *Bye-Lo* were made in all-bisque. This one is a fairly hard-to-find model with the glass sleep eyes and mohair wig. Since collectors are particularly drawn to this wonderful model, it brings a particularly high price for such a tiny doll. This doll is 5in (12.7cm) tall and is incised "16-12." The stock number 16 refers to the head style of glass eyes and wig. She is wearing her original lacy cap, but has lost her dress. *Richard Wright Collection.*

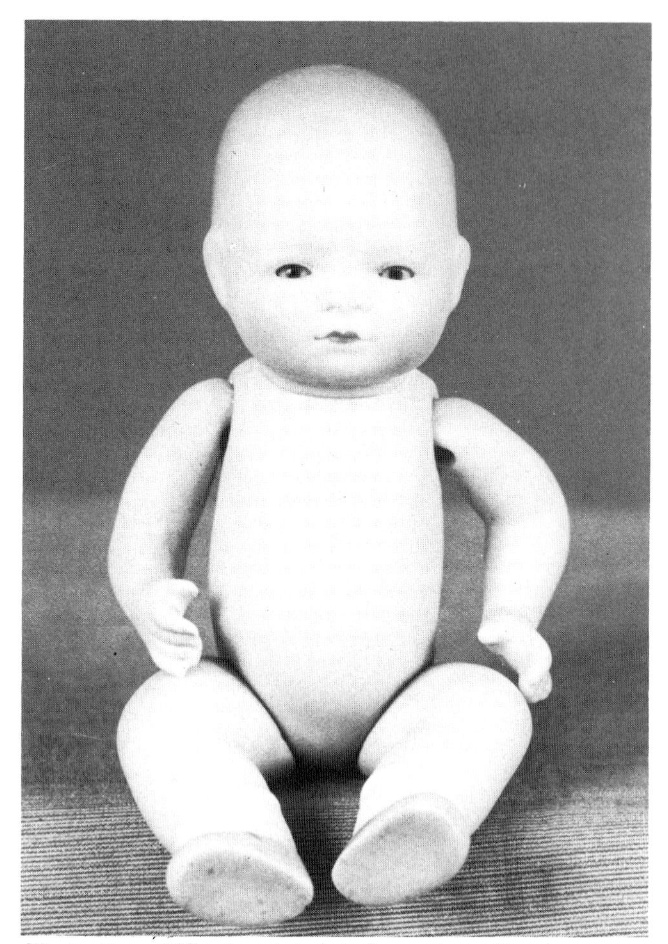

Illustration 448. As stated before, the *Bye-Los* came in many different versions. This tiny 5¾in (14.7cm) model is incised "6-13½." The 6 apparently refers to the swivel neck; the legs with the molded socks and pink or blue shoes are usually stock number 94. This doll is also a favorite with collectors because of the tiny glass eyes and swivel neck. Unlike the wigged doll, this one has a solid dome. *Richard Wright Collection.*

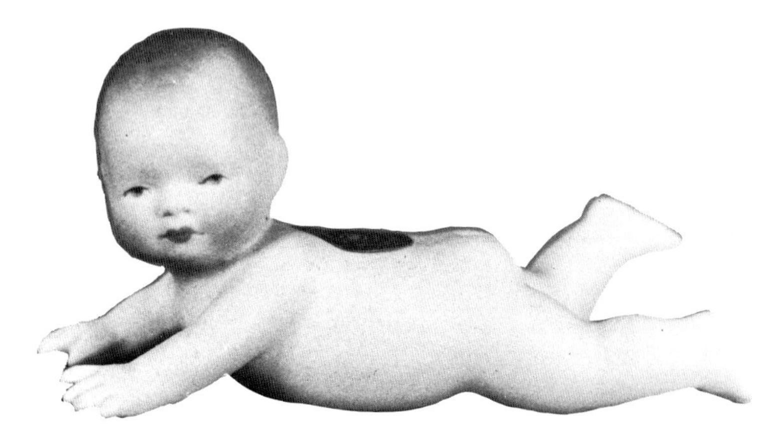

Illustration 449. This 3¼in (8.3cm) long *Bye-Lo* is molded all in one piece. These frozen *Bye-Los* were made in at least three positions. This baby is incised with stock number "4999," and still retains her original green *Bye-Lo* sticker illustrated here. *Richard Wright Collection.*

All-Bisque Dolls
Introduction

The dolls that are made all of bisque are favorites with many collectors. Tiny dolls have also been favorites of little girls for over 100 years. Little girls have always liked pretty little or cute little playthings, small objects just large enough to hold in their hands, or slip in their pockets to take riding in the car, to church, school, or anywhere else they possibly could go.

Dolls all of bisque go back to the 1860s. In 1876 Strasburger, Pfeiffer & Co. advertised "bisque babies or bathing dolls" which had molded blonde hair and bare feet. Their sizes were from 1in (2.5cm) to 6½in (16.5cm) and prices were from 9¢ to $2.50 per dozen. These could be obtained with fixed or movable arms. Although they were called "babies" they were probably not the bent-limb babies which we think of nowadays, but more likely the term "baby" here was used to refer to a small doll. A "bathing" doll was one which could be immersed in water. These would be more of the *Frozen Charlotte* type with a bisque finish, perhaps similar to the 4in (10.2cm) doll shown in *Illustration 450.* We do not know how early Kestner made all-bisque dolls, but he certainly had the porcelain factory at this time and we do know that he made small objects, such as tea sets.

In *Harper's Bazar* of 1881, the doll column mentioned a "tiny doll entirely of bisque with long natural blonde hair, eyes that open and close, and jointed limbs" which was a favorite of little girls "who do not think size everything." The cost was from 65¢ upward. In 1886, *Ridley's Fashion Magazine* advertised small all-bisque dolls with moving eyes, turning heads, jointed limbs, and long flowing hair. Prices were $1.00 and up. Without moving eyes and in smaller sizes, prices were 25¢ to 75¢. So thus it seems that the tiny dolls were firmly into the doll market at a fairly early date.

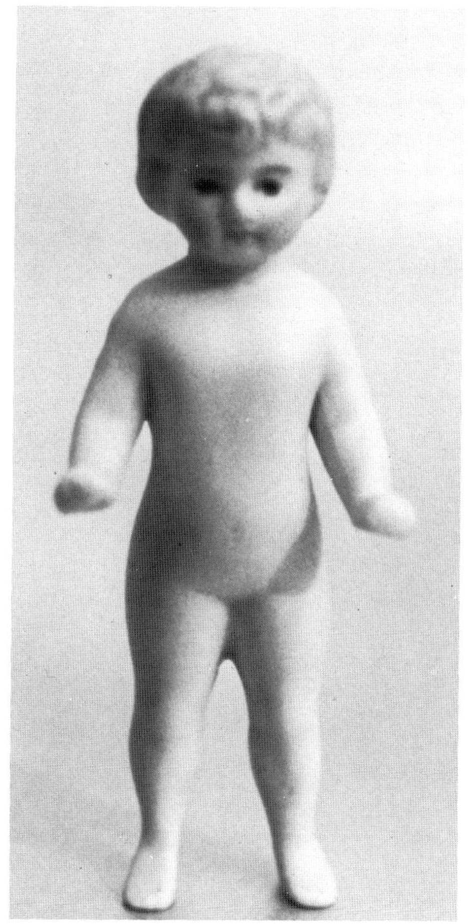

Illustration 450.

Candy Store Dolls

Collectors refer to the small all-bisque dolls as "Candy Store Dolls" because in the early decades of this century, they were sold in the small corner stores which also sold groceries and other household supplies. These little dolls were kept in a special box in the same cases which held the candy as well as a few other inexpensive toys, such as tops, jacks,and balls. Everything to entice a child to spend his precious pennies all in one place! The little all-bisque dolls were priced from 10¢ up, depending upon the size, whether the hair was "real" or molded, whether or not the neck turned, and just how nice the doll was. In her book, *All-Bisque and Half Bisque Dolls*, Genevieve Angione recounts her memories of being a child collector during this period. She tells how the little girls studied the dolls in all of the stores they could get to for miles around before making their purchases to make sure they were getting the biggest and best dolls for their money and how they even compared dolls with children from another town to be sure that their storekeepers were giving them a good buy for their money!

Judging from the numbers available today, these small all-bisque dolls were extremely popular and were imported into this country by the millions from Germany. Kestner, being such an enterprising firm, was not to be left out of this very lucrative market. After examining many of these small products, one is amazed at the fine detail and work which went into such a small and inexpensive item. They were still shown in the Kestner catalog of the 1930s.

Because of their common characteristics collectors have long attributed the dolls using mold numbers 130, 620, 310, 208, 160, 184, and other models with yellow boots to Kestner. The 150 dolls are definite Kestner products.

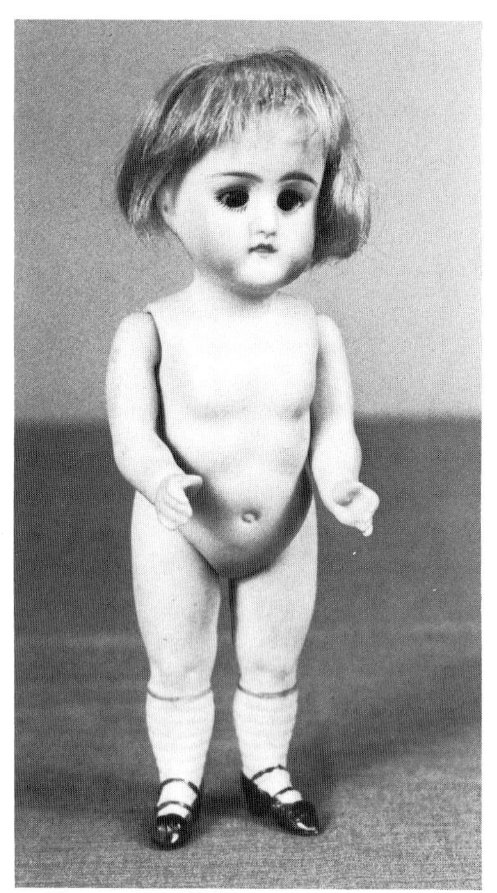
Illustration 451.

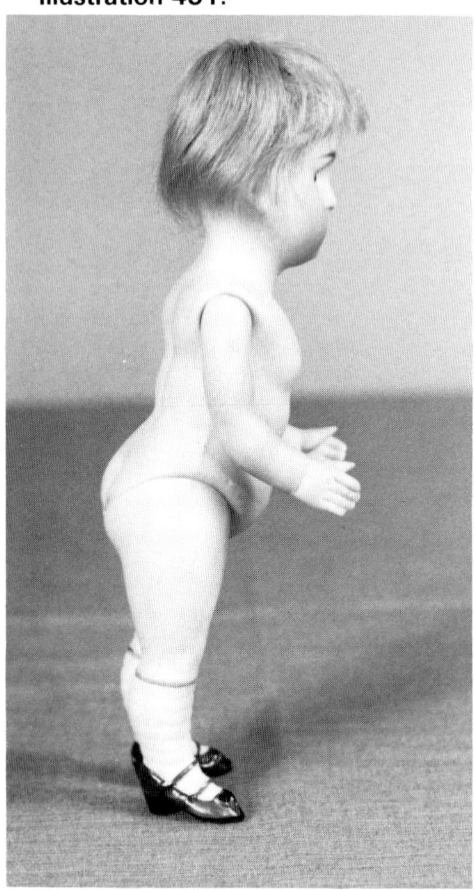
Illustration 452.

Illustrations 451, 452, and 453. This 7½in (19.1cm) example of mold 130 shows how well modeled these small dolls were. She has blue sleep eyes accented by painted multi-stroked lashes and light brown eyebrows, also painted with many strokes. Her face is long with a pointed chin and a round nose. The wooden bar inside her head is a Kestner characteristic which provides a stopper for the weighted sleep eyes. Her mouth is done in the open/closed style with a white space between her lips perhaps to indicate teeth. The molding on her body is perhaps her greatest asset. Her torso shows a large navel, slight breasts, a rounded rib cage and stomach, shoulder blades, waist indentation and buttocks with two dimples. Her arms show both fat and muscle detail with dimples at the slightly bent elbows and a fat roll at the wrists. Her hands are well-done with dimples and finger creases and free thumbs. Her legs are also well-done with chubby thighs and dimpled knees. The molded shoes and socks on these dolls are nearly a special study in themselves. Her shirred white stockings have a molded ridge just under her knee with a painted blue band just under the ridge. The stockings have a shirred pattern with heavy horizontal and lighter vertical lines as shown in *Illustration 454*. Her very good shoes are brown with molded black bows and two straps which button. There is black trim around the soles and on the heels; the soles themselves are brown. This doll is completely marked "130" for her mold or stock number and "7" for her size number on her head, arms, legs; the limbs are loop strung. The view of her head shows the wooden eye-bar plastered inside her head and also illustrates a fault with many of these tiny dolls: the heads warped during the drying process leaving some with deformed heads, but since this defect could be covered with the wig, Kestner did not discard the dolls. *H & J Foulke.*

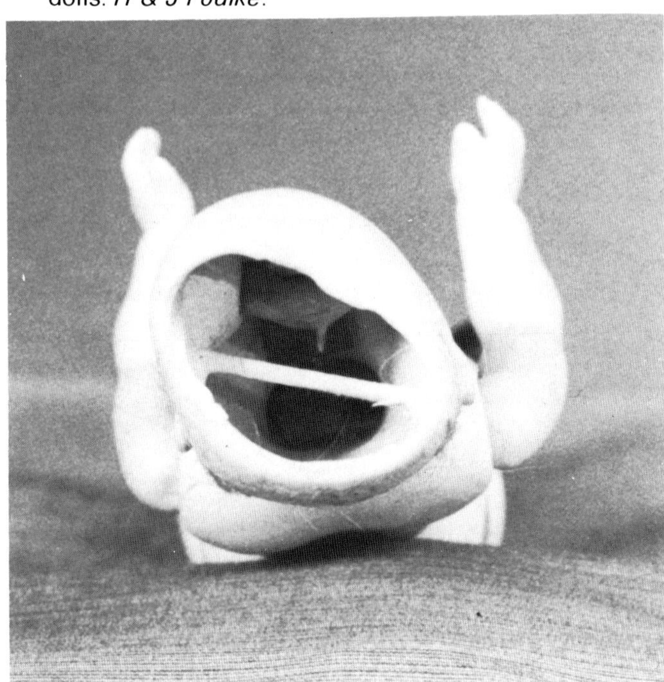
Illustration 453.

Illustration 454.

Illustration 455. Here is an example of an all-bisque in her original clothes with the same face as the doll in *Illustration 451*. Since these tiny dolls were usually purchased naked, little girls often spent hours and hours making not one dress but a wardrobe of clothes for them. It is exciting for a collector to find a doll with her original child-made clothes. Often also the dolls are found with clothes made by more experienced hands of mother or grandmother. The outfit is simple yet attractive with a matching lacy bonnet. *Esther Schwartz Collection.*

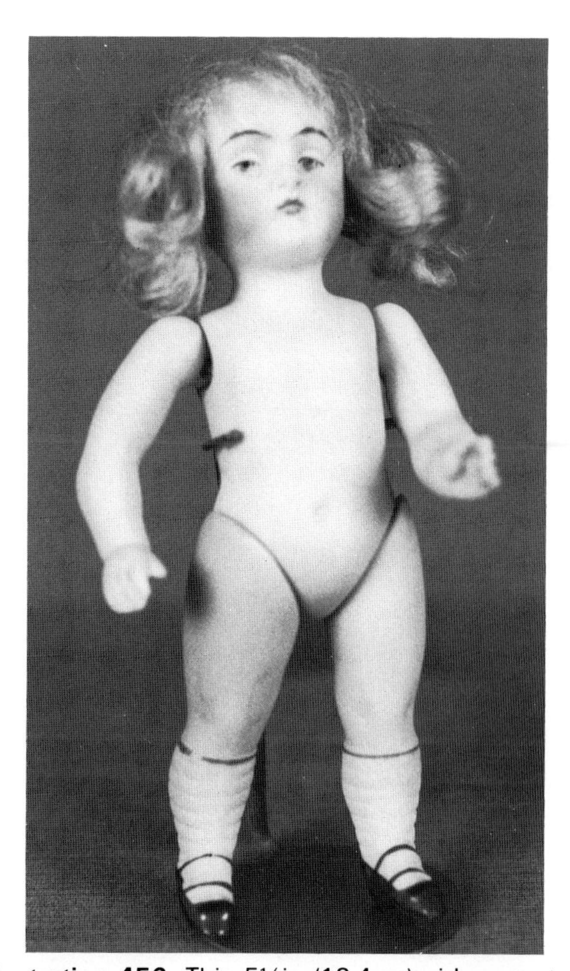

Illustration 456. This 5¼in (13.4cm) girl presents a variation of mold 130 which would have been a less expensive doll because she has painted eyes. Her modeling is very good, however, with molded eyelids and sockets. Her eyebrows are put on with one stroke each; her mouth is of the open/closed type with a hint of molded teeth. Her bisque is of excellent quality with a pale grayish cast to the arms and legs. She has the excellent body modeling of the larger 130 doll with very hefty arms and well-developed thighs. She has the same 130 legs; all of the legs with this number have the same type of footwear. She is incised "130//3½" on her head and legs and "3½" on her arms. Her wig is original of blonde mohair in the Rembrandt style as seen on so many of these Kestner all-bisques. Dolls which look like this model with painted eyes and long curly mohair wigs in this style were offered as late as 1914 in the *1914 Marshall Field & Company Doll Catalog*. Dolls 4½in (11.5cm) tall sold for 80¢ a dozen wholesale. Apparently these dolls were on the market for about 20 years. *H & J Foulke*.

Illustration 457. This 5in (12.7cm) example of mold 130 with the painted eyes is dressed in her original crocheted outfit. This method of dressing the small dolls was extremely popular, and since many of them were sewn onto the doll, they have survived intact. *H & J Foulke*.

190

Illustration 459. This girl with lovely bisque is from mold 190 and has a swivel neck. She has a very cute face with a Kestner look and she is certainly of exceptional Kestner quality. Her eyebrows are beautifully painted with many tiny strokes, sleeping eyes are brown, and she has nicely painted lashes. She still retains her original mohair wig in Rembrandt style. The shoulders and hips are pegged, generally regarded as an earlier type of jointing. Her molded stockings are yellow with a darker yellow garter; her black one-strap shoes have molded boots. *H & J Foulke, Inc.*

620

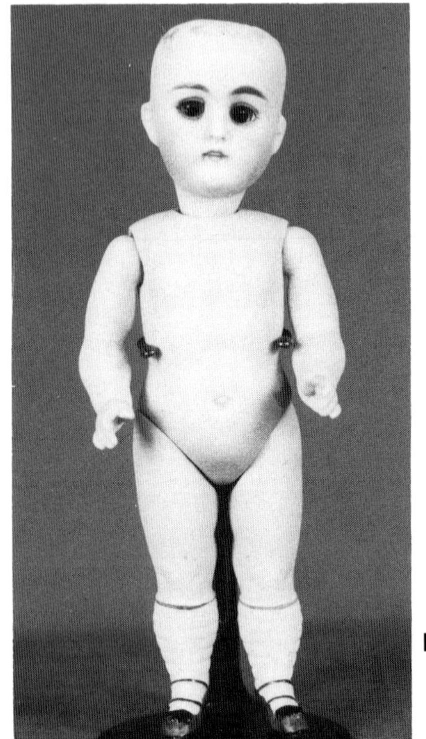

Illustration 460.

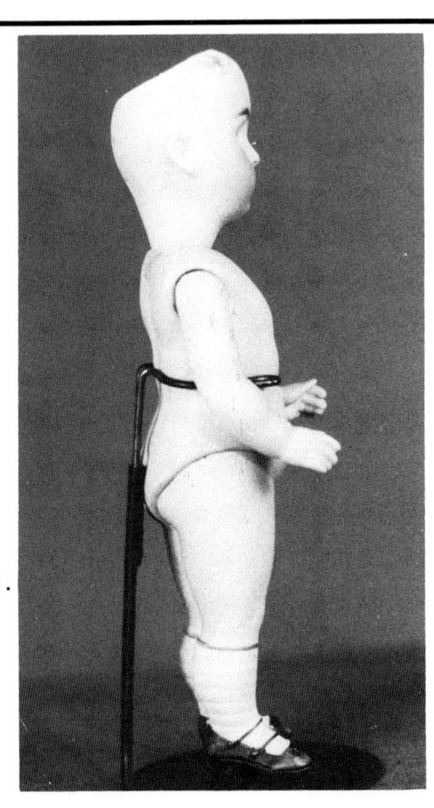

Illustration 461.

Illustrations 460, 461 and 462. This 8½in (21.6cm) doll is incised "620//9//+" on her head and torso, "130.9" on her arms and "130 9" on her legs. Actually, she is simply a swivel-head version of the 130 doll with the exact same face. However, in this large size she has molded eyebrows, which are multi-stroked ones. She has painted lashes and brown sleep eyes. Her head has the molded string loop, a feature which does not allow as much flexibility as the neck plug. She has a substantial nose with red nostril dots and a prominent cleft above her upper lip as well as small ears. Although her torso is incised "620," it is exactly like that one on the 130 doll with exceptionally fine detail in the breasts, rib cage, stomach, shoulder blades, waist, and buttocks. Her legs are kid-lined with chubby molded thighs and dimpled knees. It is not to be supposed that her arms and legs are wrong because they are 130 stock numbers. In fact, they are correct and simply illustrate the flexibility of the Kestner factory in using limbs already in production to complete this doll. One often finds variance in stock numbers between torsos and limbs on Kestner dolls, and it is the size number which should be matched to insure proper parts, allowing for the fact that the limbs should be of the same period as the torso. Even so, it must be admitted that parts may sometimes vary on size also. *H & J Foulke*.

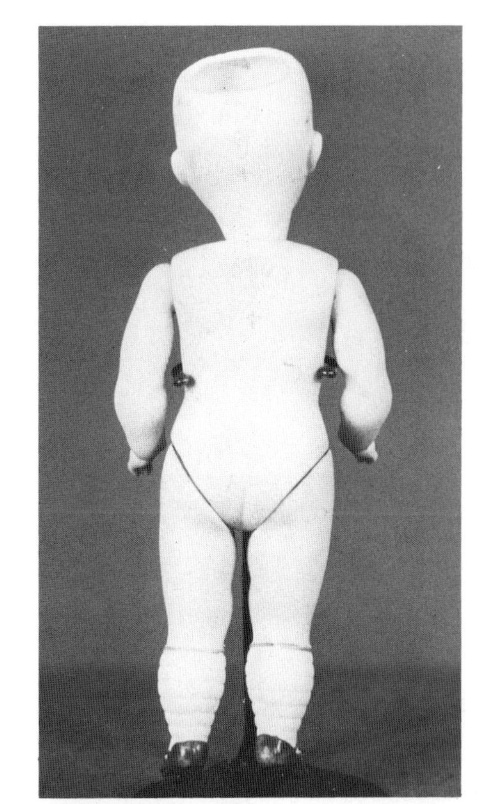

Illustration 462.

Illustration 463. This tiny 4in (10.2cm) example of mold 620/130 combination has her original blonde mohair wig in Rembrandt style, blue set eyes, one-stroke blonde eyebrows, long painted eyelashes, and the same open/closed mouth with a tiny white space to indicate teeth. All of these dolls with the 130 legs have the same type of footwear with shirred white hose and brown two-strap slippers with black bows. The 620 swivel-neck doll also comes in a version with painted eyes, just like the 130 mold. She is wearing a plain pink silk dress which probably some small owner made for her; with her moving head, she certainly would have been a desired doll. *H. & J Foulke*.

Illustration 464. This 10in (25.4cm) girl is incised "620//12" on her head and torso with the "130 12" limbs. She is a variation of the preceding 620 dolls in that she has an open mouth with four upper teeth. She has lovely glossy molded eyebrows painted with many strokes and blue sleep eyes. Of course, her modeling is the same as that on the previous dolls and she has lovely hands with very detailed fingers, palms, and free thumbs. She has been beautifully redressed. *Excelsior* line bisque dolls which appear as though they could be these large all-bisque dolls were advertised in the 1899 Butler Brothers wholesale catalog as having "bisque turning heads, natural eyes, painted eyebrows and eyelashes, good features, open mouth, exposing [a] row of teeth, hip and shoulder jointed, painted shoes and stockings, underwear, full costume and bonnet." These were priced per dozen from 8in (20.3cm) for $1.80 to 9½in (24.2cm) for $2.00. Fancy bridal dolls at 9in (22.9cm) were $2.20 per dozen. *Edna Black Collection.*

Illustration 465. Here is a 5¾in (14.7cm) example of the 620/130 girl again with open mouth and four upper teeth. She has brown sleep eyes with lashes and one-stroke eyebrows. She has been dressed in a rosebud print cotton dress. Her size number is "4". *Edna Black Collection.*

310

Illustration 466. This 5in (12.7cm) girl is incised "310//4" on her head. This 310 mold has a sweet face and is a swivel neck with a side stringing loop. All of her body parts are also loop strung. She has a closed mouth and brown sleep eyes. She has molded yellow stockings with an overall crosshatch pattern; her black two-strap slippers have molded pompons. This footwear is not found as often as the 130 leg. *Richard Wright Collection.*

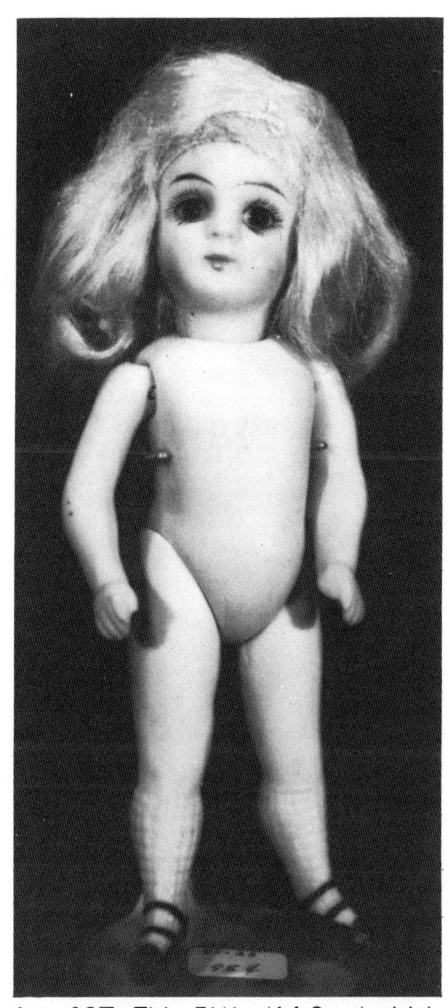

Illustration 467. This 5½in (14.0cm) girl is another example of the 310 face, this time without the clothes. She has her original blonde mohair wig and brown sleep eyes. She has the same yellow stockings and black shoes as the doll in *Illustration 466*. This body does not have quite as much molding detail as the 130 body, although the bisque and finishing is excellent. *H & J Foulke.*

Illustration 468. This sweet little girl is a good example of Kestner combinations. She has the 310 head, the 620 torso, and 130 legs! She has gray set eyes and a closed mouth. She has her original blonde mohair wig on a cloth cap in the Rembrandt style, so often found on these tiny all-bisque dolls. She is 5½in (14.0cm) tall wearing a sweet old dress. *H & J Foulke.*

Illustration 469. This doll does not really fit here, but I am at a loss as to where exactly she fits in, so I will place her here near the dolls with the yellow stockings since hers are basically the same except they are orange. She is quite unusual in that she has two upper and two lower molded teeth. It certainly is rare to find a doll with both. She is 9in (22.9cm) tall and unmarked, but a likely Kestner product. Her body molding is excellent with pegged joints and a swivel neck. Her hands are quite dainty with raised index fingers and free thumbs. Her footwear is also different in that her shirred stockings are orange and she wears black two-strap shoes with molded bows. She has blue sleep eyes with painted lower lashes only which indicates that at one time she had real upper lashes. She has her original blonde mohair wig, and is wearing her original outfit: a bright blue silk dress with lace and flower trimmed hat. *Richard Wright Collection.*

150

Illustration 470. This mold is a great favorite with Kestner collectors who often try to find one of these in each size. They do come large for all-bisques, the tallest one we have seen is size 7 at 12in (30.5cm). This is a substantial doll and certainly would have been a luxury item carried by only a few stores. The Butler Brothers wholesale catalog for 1908 lists Kestner solid bisque dolls, but the largest size given is 9⅜in (23.8cm) at an expensive $8.40 per dozen! This 11in (27.9cm) number 150 is size 6. She is a very sturdy and chubby little girl with chubby cheeks and double chin with a tiny pointed "first" chin. Her blue sleep eyes are small in the Kestner manner. She has painted lower lashes only, no painted ones at the top because she was intended to have real eyelashes, a characteristic that would put her after 1900. She has lovely blonde eyebrows lightly painted with many individual strokes. Her mouth has lips parted and painted in the Kestner style with two upper peaks and upturned ends. Inside are four upper teeth. She has her original blonde mohair wig and wears an outfit crocheted from a contemporary pattern, which could have been obtained from the yarn company. Her molded footwear consists of blue shirred hose with heavy horizontal ridges and lighter vertical ones, but no painted top band. Her low-heeled black shoes have one strap with a molded pompon. She is hip and shoulder jointed with elastic strung through loops. *Edna Black Collection*.

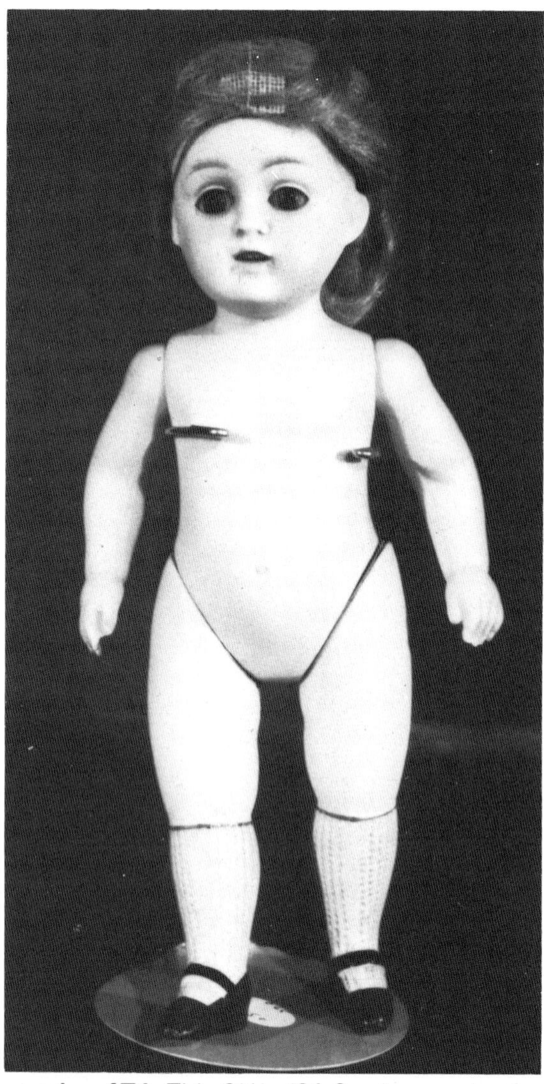

Illustration 471. This 8½in (21.6cm) example of mold 150 is shown undressed so her body modeling can be studied. She is well made, but not nearly so detailed as the 130 doll. Again, we assume that this 150 mold is a later doll; at least the molds were made at a later date. Her arms are slightly bent at the elbows and show muscles, wrist creases, and fingernail and knuckle detail, although her thumbs are not free as they are on the larger doll. Her torso has molded breasts and navel; her thighs are chubby and knees dimpled, but not to the extent of the 130 doll. The footwear on this doll consists of pink stockings in an overall pattern of vertical dots with darker pink tops. Her black one-strap slippers do not have bows. *H & J Foulke*.

Illustration 472. This 9½in (24.2cm) girl again shows the great appeal which these 150 dolls possess. She has a very alert and intelligent visage. She has brown sleep eyes with both upper and lower painted lashes. Again, her eyebrows are extremely well painted. She has a very long straight brown mohair wig. Her footwear consists of blue hose with rows of vertical dots and a dark blue top band. Her black one-strap shoes do not have bows. She is dressed in old white embroidered organdy. Her size number is 4. *Edna Black Collection*.

Illustration 473. This 8in (20.3cm) girl is size number 2 and she is an exact little sister of the doll in *Illustration 472*. *Edna Black Collection*.

Illustration 474. This 7in (17.8cm) example of 150 is size 1. She has her Kestner crown label on the front of her torso like the one shown in *Illustration 402*, so there is no doubt that Kestner made this series. She has the pink stockings with the impressed rows of vertical dots and the plain black one-strap shoes for footwear. Her sleep eyes are brown with lower painted lashes and real upper ones. Her light eyebrows have many high individual strokes at the inner corners. In the Butler Brothers 1910 wholesale catalog "Kestner's Sleepers" are advertised in a special box. A line drawing of the doll similar to the 150 is shown with the notation that she has "moving glass eyes." A stout doll with sewn wig, side curls, two ribbon bows and sleep eyes was advertised in the 7in (17.8cm) size for $3.98 per dozen. Kestner dolls were noted as being "Known the world over." *Edna Black Collection*.

Illustration 475. This doll is puzzling, because although it also is incised "150," it has a very different face. A good guess is that it was an earlier model. She is 4¼in (10.9cm) tall and incised "150//0½" on the head and "160 0½" on her limbs. She has a definite character type face with dimples and an open/closed mouth with two upper molded teeth, a definite relative of *Illustration 476.* She has gray painted eyes in nicely molded sockets with black pupils and black lid lines. Her cute round nose has red nostril dots. Her bisque is excellent and she has very good body molding. Her footwear consists of white stockings with a waffle-weave design, a pattern different from the 130 legs, although she still has the brown two-strap shoes with black bows and heels. Although her limbs are the 160 style, they are correct. Every 150 with this smiling face that I have seen has had 160 limbs. Apparently Kestner used stock numbers interchangeably with torso numbers. Her brown mohair wig is original, as is her cotton Russian style dress which was popular for little girls in about 1910. *H & J Foulke.*

Illustration 476. This 7in (17.8cm) example of mold 150 is the same face and mold number as the painted-eye doll in *Illustration 475.* She has the same smiling mouth with two molded teeth and dimples; she has tiny blue sleeping eyes, painted eyelashes and glossy eyebrows with many individual brush strokes. Her chin is pointed. Her original crocheted white wool dress with fancy matching hat are typical for her period. She is, indeed, a very appealing doll with a character face. *Ruth Noden Collection.*

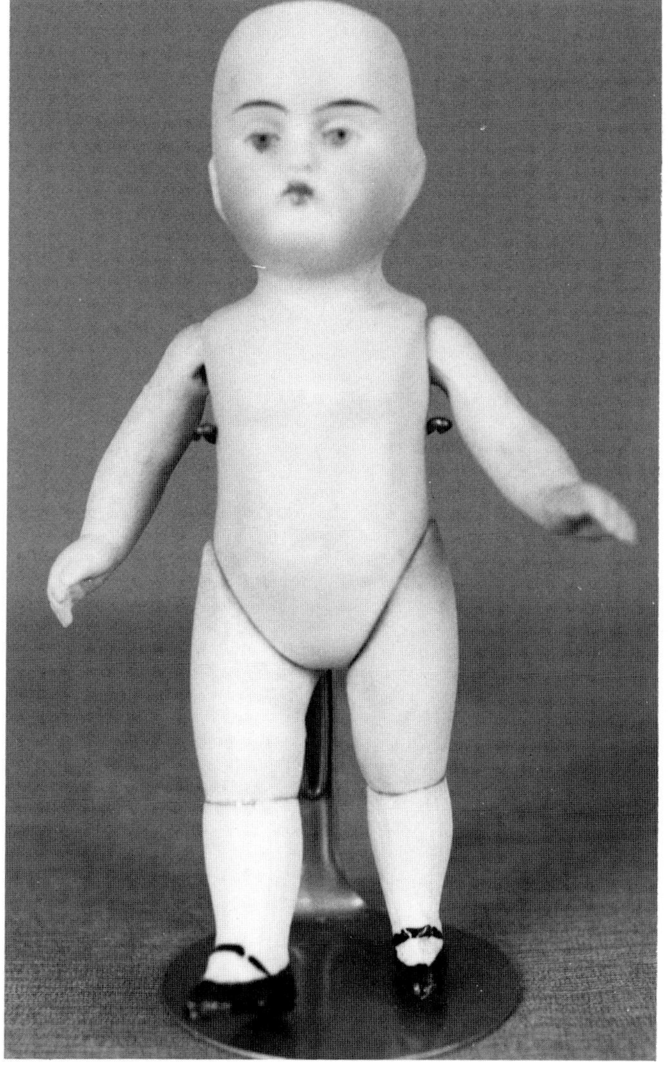

Illustration 477. Here is another doll with the 150 number, this one's size is "4/0" on her arms and "150 4/0" on her head and legs. She has quite a chubby face with rosy cheeks and a small closed mouth with thick lips in a rosebud shape. Her blue painted eyes with black pupils and eyelids are in a molded socket. She has brown one-stroke eyebrows. She has molded breasts and navel, but not so much detail in her torso as the 130 girl; however, there is no back molding detail at all. Her arms and legs are roughly finished. She has molded white hose with vertical rows of dots like the larger 150 dolls have. Her top bands are glossy blue while her shoes are black low heels with one strap. *H & J Foulke.*

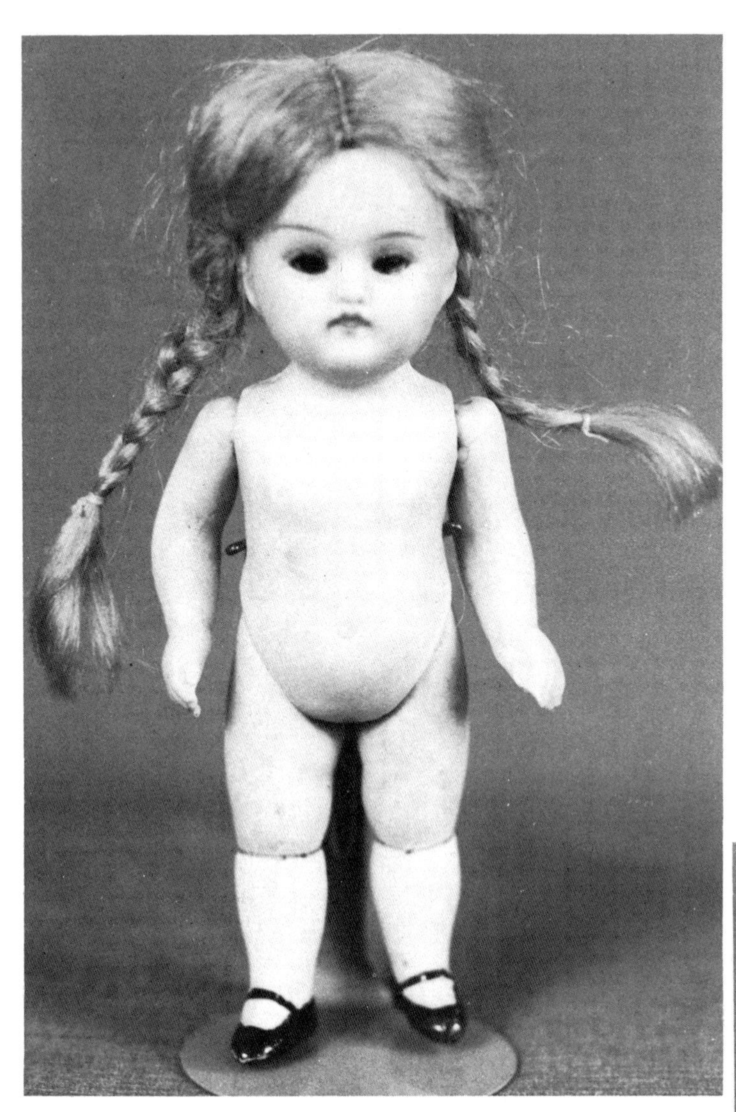

Illustration 478. Here is another 5in (12.7cm) example of mold 150, size 4/0. She has blonde one-stroked brows and tiny very dark sleep eyes with dark painted lashes. Her open/closed mouth has two molded upper teeth. Her blonde mohair braids have a sewn center part on a cloth strip. She has the same footwear as the doll in *Illustration 477* with 150 legs. The soles of her shoes are yellow and the heels black. *H & J Foulke*.

Illustration 479. This 5in (12.7cm) girl is marked indentically to the doll in *Illustration 478*. She still has her original tiny plaster pate and blonde wig with coiled braids on a yellow cloth cap. Her open/closed mouth with two upper molded teeth is a little easier to see. She is wearing a sweet old organdy print dress with lace and ribbon trim to show how easy it is to dress these little ones attractively. In the *1914 Marshall Field & Company Doll Catalog* little dolls similar in looks and size were offered with moving eyes, snail braided mohair wigs with ribbon bows and rubber cord jointed limbs for $4.00 per dozen, although they announced a 40 to 50 percent confidential trade discount. *H & J Foulke*.

Illustration 480. This 5½in (14.0cm) girl is another example of Kestner combinations. Her head is incised "160 3/0;" her limbs are "208 3." Her facial painting is excellent for such a small doll, much better than that of the dolls shown in *Illustrations 477, 478 and 479*. Her eyebrows are glossy and multi-stroked, her eyelashes are long and carefully done. Her closed mouth is pouty with thick lips while her tiny brown eyes open and close. Her body molding is adequate, not as detailed as the 130 model. Her footwear consists of molded white stockings with vertical stripes and short irregular crossing lines as shown in *Illustration 481*. Her black one-strap shoes have low heels and tan soles. Her original brown mohair wig is sewn with a center part and brown cloth cap. Contemporary catalogs make much of *sewn* wigs as opposed to wigs which were just clumps of hair pushed into a hole in a crown cover or glued onto a crown. She has been attractively redressed in blue satin. *H & J Foulke*.

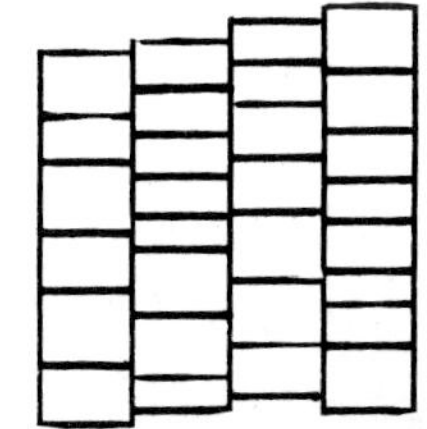

Illustration 481.

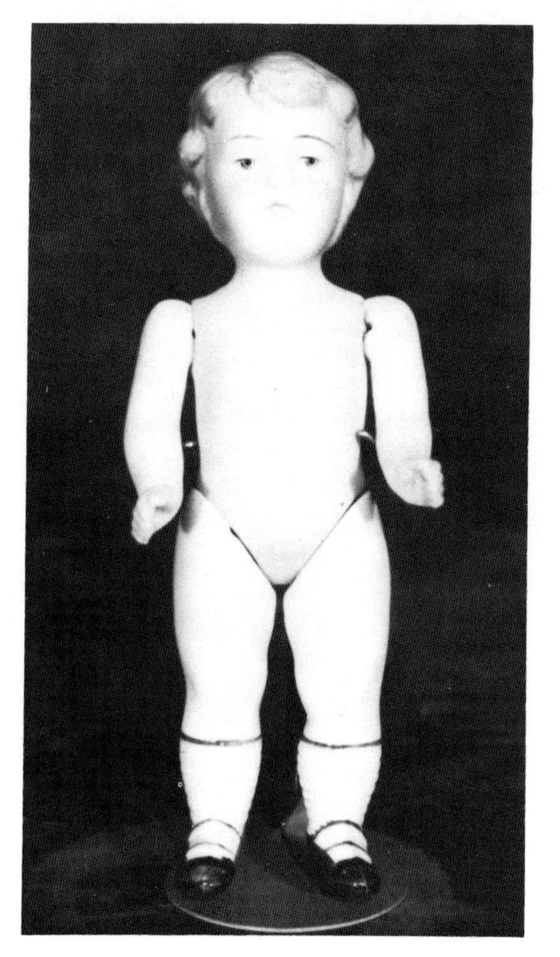

Illustration 482. This 5¾in (14.7cm) boy is incised "160." He has a lovely face showing much detail in the eye and mouth features as usually found on Kestner dolls. He has molded blonde hair with a side part and lovely curls and comb marks. He is definitely related to the molded blonde-haired babies shown in *Illustrations 504, 505, and 506*. He has white shirred hose and the familiar brown two-strap shoes with molded black bows. Although the bisque and decoration of this doll are fine quality, it was designed as a cheaper doll. The painted eyes and molded hair precluded having the added expense of glass eyes and a wig. *H & J Foulke.*

Illustration 483. The dolls with yellow boots are extremely popular with all-bisque collectors. They are very well-done and certainly the most attractive footwear on Kestner dolls or dolls attributed to him. The yellow color varies from fairly pale to quite deep gold, probably depending upon the individual painter. The boots have molded points in the front and back which are outlined in black. The toes, soles, and heels are also black. The white stockings have vertical line ribbing with shiny blue bands painted at the very tops of the stockings. Some of the boots have molded gussets at the sides. Some are incised "184" and a size number; some have only the size number. Some were designed to be used with a pegged joint (generally considered an earlier type), and some with loop stringing. This particular doll is 5in (12.7cm) tall, a fine example of mold 184, a stiff-neck model. She has blue glass eyes with painted lashes and blonde eyebrows painted with just one stroke. Her pouty mouth is closed. She has her original blonde curly mohair wig on a cloth cap with a tiny satin bow. Her original crocheted dress and hat contribute to a most attractive doll. *H & J Foulke*.

This 5in (12.7cm) example of mold 184 has a particularly pert face. It is the appealing face and the yellow boots, decidedly different footwear, which make the doll so desirable to collectors. In addition, this example has a swivel neck. She still retains her original mohair wig and is daintily dressed in old lace. *H & J Foulke, Inc.*

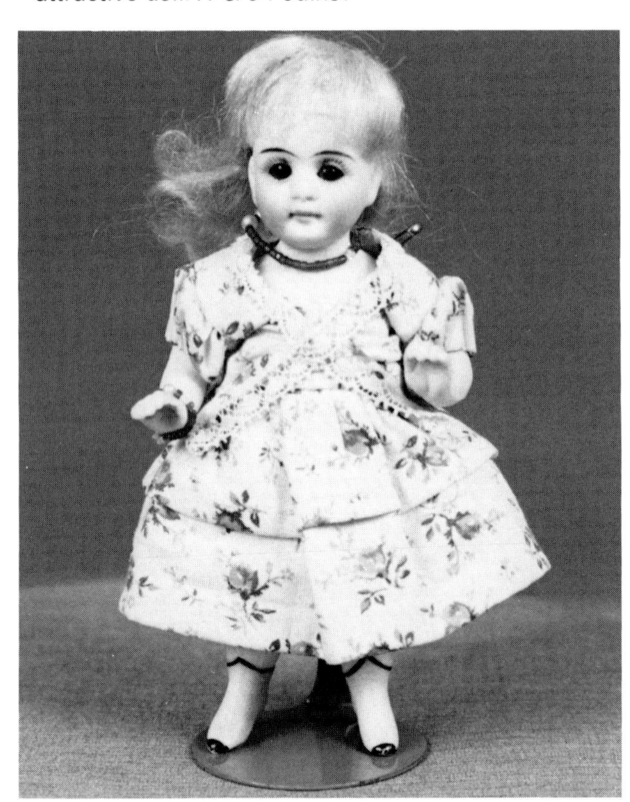

This 5in (12.7cm) little girl is unmarked, but is perhaps a forerunner of the 184 mold. Her quality is excellent with overall smooth and creamy bisque. She has brown sleep eyes and her original mohair wig. She is peg-jointed at shoulders and hips and has the same yellow boots as the 184 model. Her hands are molded in a different position however, as her wrists are raised and her fingers slightly curled. *H & J Foulke, Inc.*

Illustration 484. This 5¾in (14.7cm) girl has head and torso incised "208//5½," legs incised "184 5½," and arms incised "5½." She is another example of the Kestner habit of interchanging body parts. Although she has a swivel neck, she is a less expensive model because she has the painted eyes. Her eye sockets are molded with blue eyes, black pupils and lid lining as well as an extra red accent eye line. Her low eyebrows are made with one brush stroke. Her blonde mohair wig in Rembrandt style with a satin bow is on a paper cap, another economy. She is loop strung. It is the 184 model yellow boot legs which have the inset gussets and horizontal ridging at each side of her boot. This mold is also found with glass eyes and arms incised "184." *H & J Foulke.*

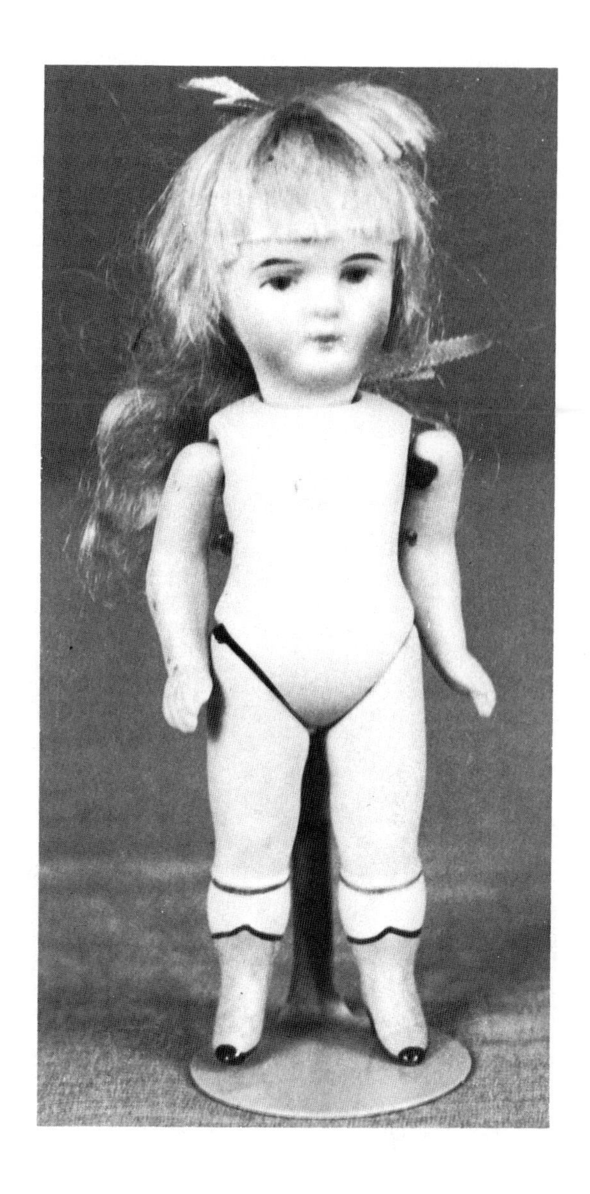

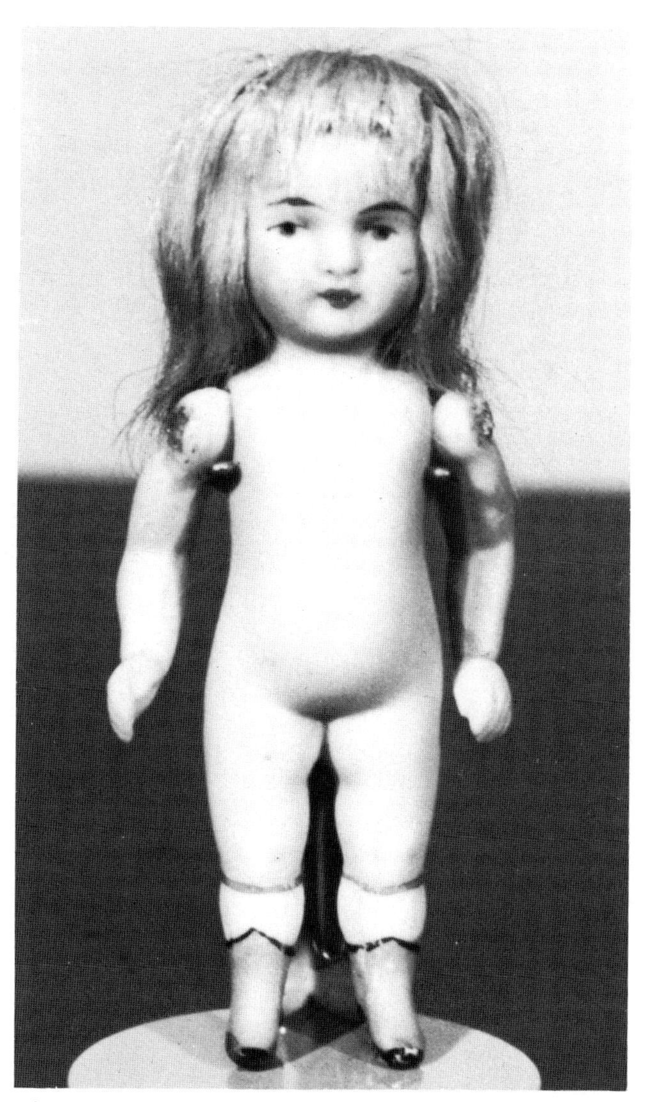

Illustration 485. This 3½in (8.9cm) child does not really fit here as far as face numbers are concerned, but I wanted to place her with the yellow boot models. She is possibly an earlier version because of her peg-jointed arms, and certainly a less expensive one as she does not have the jointed hips. She is incised "103" on her head and arms. Her face is slightly fatter than the 208 doll and she has fuller lips. The modeling of her torso is good with breasts, navel, rounded stomach, ample thighs, and dimpled knees. *H & J Foulke*.

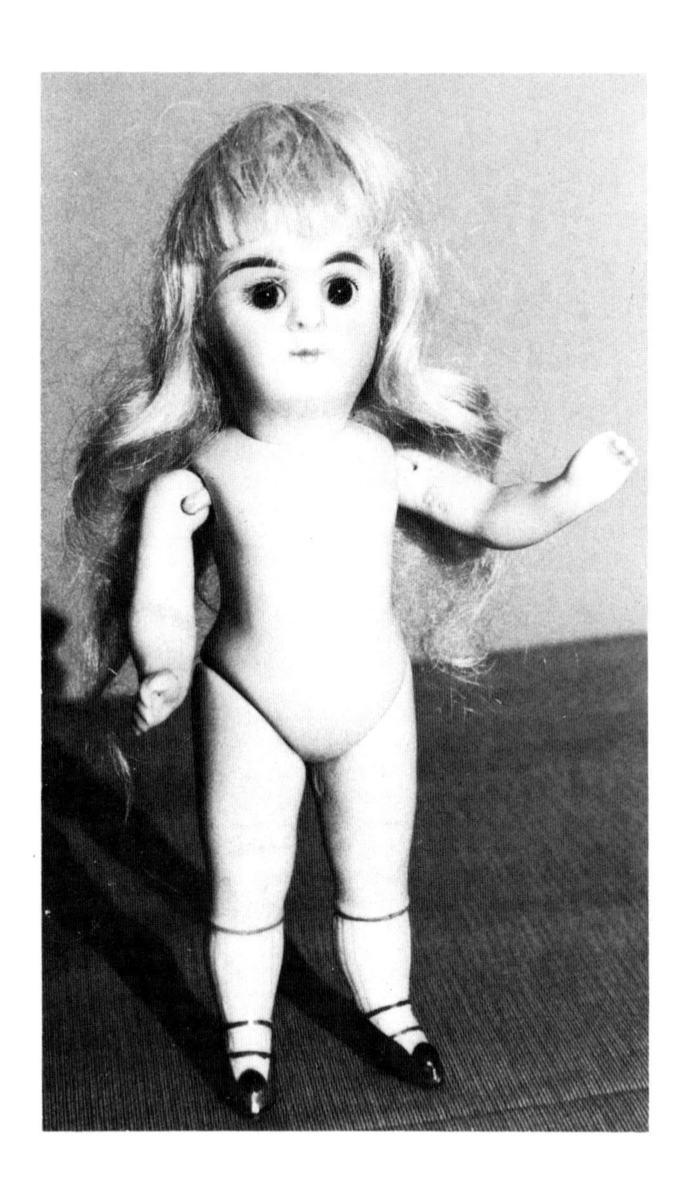

Hertel, Schwab & Co., also of Ohrdruf where the Kestner & Co. porcelain factory was located, is also supposed to have used this 208 mold number for an all-bisque doll. Whether Kestner or Hertel, Schwab & Co. made the dolls shown here is not certain, but they certainly do have Kestner qualities.

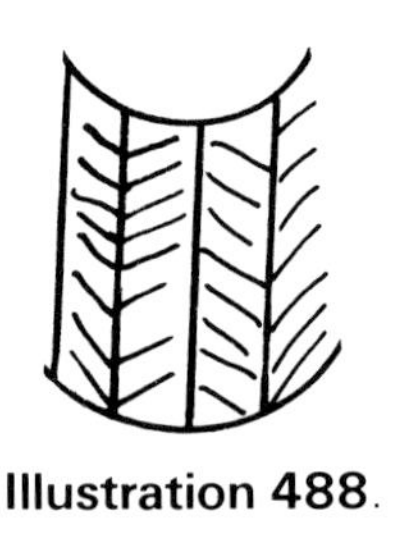

Illustration 488.

Illustrations 486 and 487. This 6½in (16.5cm) girl is incised "208//6" on head and torso, another swivel-neck model with a molded stringing loop in the neck and loop elastic strung arms and legs. She is incised "6" on her arms and legs. Her large brown eyes, which have always been stationary, give her a pert, alert expression. She has brown eyebrows with many brush strokes; her closed mouth has an upturned upper lip. Her bisque is very pale, of excellent quality, with nicely tinted cheeks. She has a chubby torso with molded breasts and rib cage and upper legs with fat rolls and dimpled knees. Her footwear consists of white stockings molded in a herringbone pattern as shown in *Illustration 488* with blue painted bands just short of the top of the molded hose edge. Her brown two-strap slippers have black pompons. She has her original blonde mohair wig in a long curled style with a blue satin bow. *H & J Foulke*.

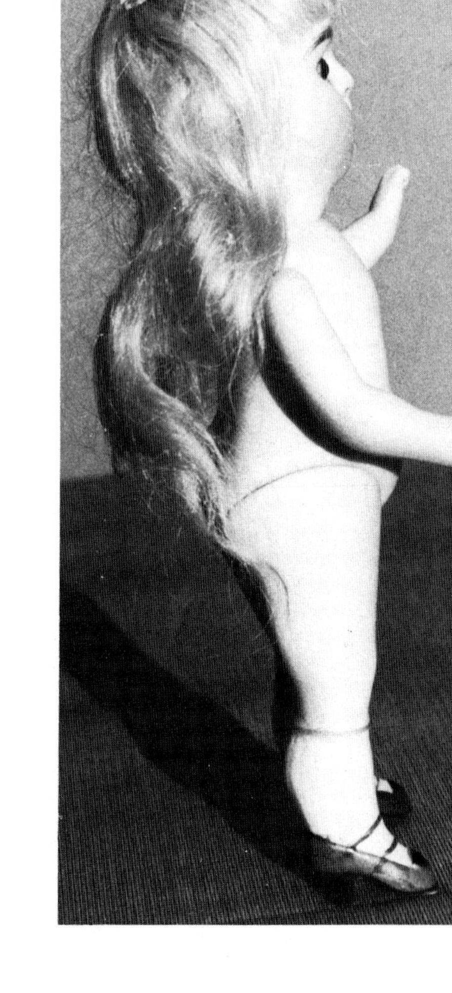

Illustration 489. This 5½in (14.0cm) girl is incised "208//4" on her head and torso; "4" on her legs; arms illegible. She differs just a little from the doll shown in *Illustration 486* in her one-stroke brown eyebrows and blue sleep eyes, which are tiny and close together. She has a dainty closed mouth. Her original blonde mohair wig is on a cloth cap. Her footwear consists of white molded hose in the herringbone design, like the doll in *Illustration 486*, and the same brown slippers. We have never found a stock number on these legs with herringbone design, but they often appear with the 208 heads. She is wearing a nice old blue silk dress. *H & J Foulke*.

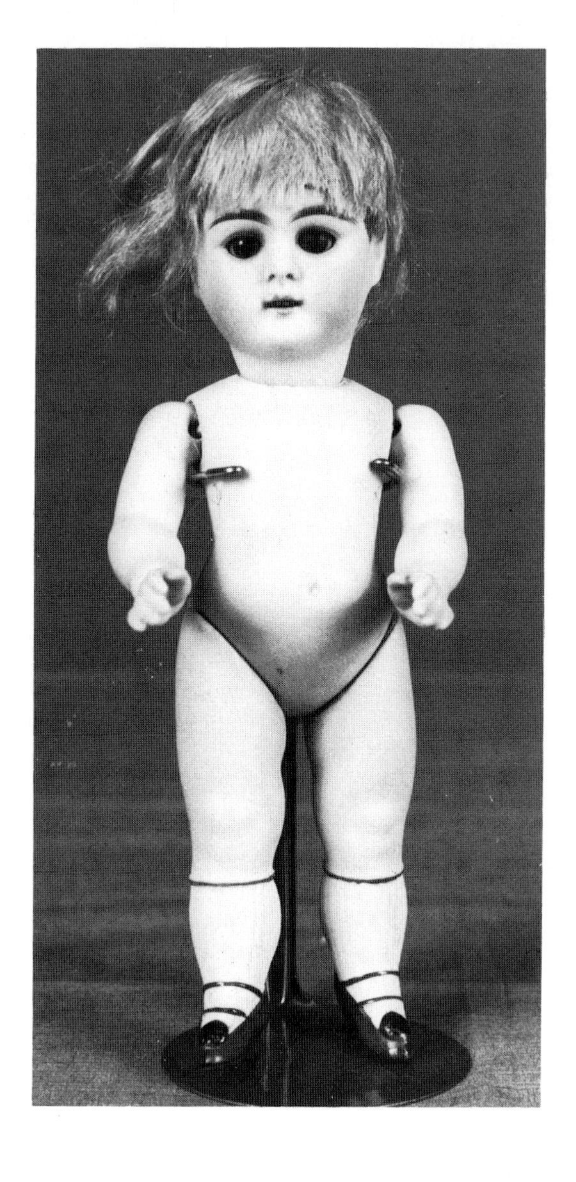

Illustration 490. This lovely 9in (22.9cm) example of mold 208 is size 12 on all of her parts. She has beautifully stroked brown eyebrows painted low to her brown sleep eyes. This version has an open mouth with four upper teeth. The bisque, as on the other 208 dolls, is very pale and of fine quality with red cheeks. She has the same herringbone design stockings. *Emma Wedmore Collection*.

Illustration 491. This lovely 10in (25.4cm) girl, incised "208//9," is a marvelously large size and makes an extremely pretty doll. She has the Kestner gray sleep eyes with eyebrows which have many high individual strokes at the inner corners, and upper and lower painted lashes. Her parted lips are full with the typical Kestner look of the upper lip with two rounded peaks and upturned corners. Her molded footwear consists of grayish white hose in the typical 208 design shown in *Illustration 488* along with the plain black one-strap Mary Janes. She is dressed in a completely original outfit of a low waisted organdy dress with lace trimmed ruffled skirt and large back sash. Her blonde mohair wig is center parted with curls and her hat is a replacement. *Kathleen Moyne Collection.*

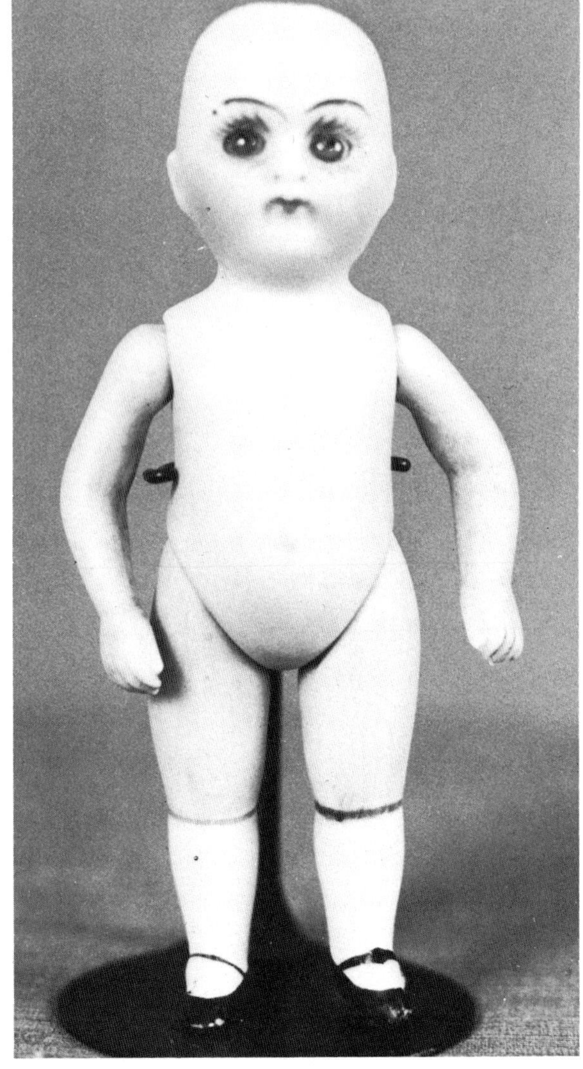

Illustration 492. This 4¼in (10.9cm) example of mold 208 is probably quite a later model. She is incised "208//0½" on her head and legs; "0½" on her arms. She has light blue sleep eyes, gray long and heavy painted lashes, and simple one-stroke eyebrows. Again, this doll has the petite closed mouth and rosy cheeks. Overall her arms and legs are not well finished. Her footwear, which is the typical 208 design, is only faintly molded in the crosshatch design. Her black slippers and straps are of two different widths. She is probably a later and much less expensive model. *H & J Foulke.*

Characters

155

Illustration 493. This 5in (12.7cm) little girl is incised "155//2" on head and arms and "2" on her legs; she is thought to be another Kestner product. She is a darling character with a smiling face and open/closed mouth. She has tiny gray sleep eyes with heavy black painted lashes and one-stroke eyebrows. Her bisque is excellent with rosy cheeks. She has slender arms nearly straight down with dainty hands, slightly flicked up. Her legs are short and chubby, of the toddler variety with dimpled knees. Her footwear is indicative of a later era with the short white molded socks with no design and blue bands. Her brown slippers are one-strap with plain rounded toes. She wears her original dark mohair wig on a paper cap, and a turquoise satin dress. *H & J Foulke*.

156

Illustration 494. This is another 5in (12.7cm) character with just about the same face, but mold number 156, size 2. Her eye sockets are cut just a little larger, but she has the same thinner face with pointed chin. Her arms and legs are molded exactly like the doll in *Illustration 493*. She has her original blonde mohair wig and a cute plain satin dress. *H & J Foulke*.

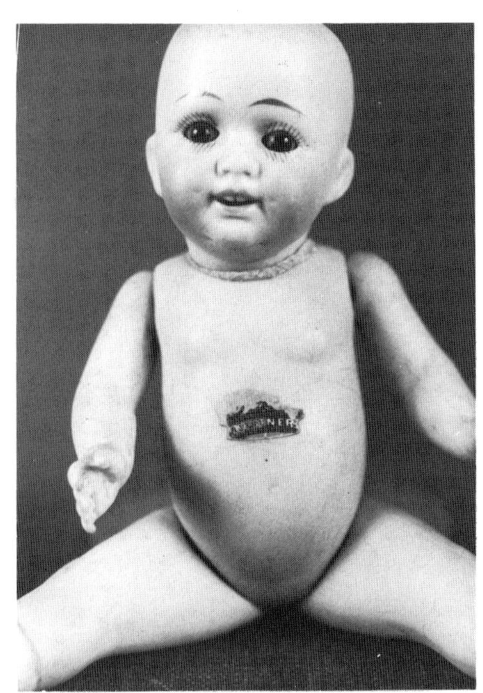

Illustration 495. This 6in (15.2cm) character has the cutest face. His smiling mouth is open to reveal two molded upper teeth. Dimples grace each cheek. He has a swivel neck with a kid lining. He is incised "602" and most importantly he has the remains of his Kestner crown label on his torso. His one original leg is also incised "602;" it has low molded white socks and black one-strap shoes. The legs currently on the doll are modern replacements. *Dr. Carol Stoessel Zvonar Collection.*

Illustration 496. This 5¾in (14.7cm) character toddler is another example of 602. She is wearing her original blonde mohair wig and crocheted dress. Her original box has a printed "Badekinder" label and stamped number "602." *H & J Foulke, Inc.*

Babies

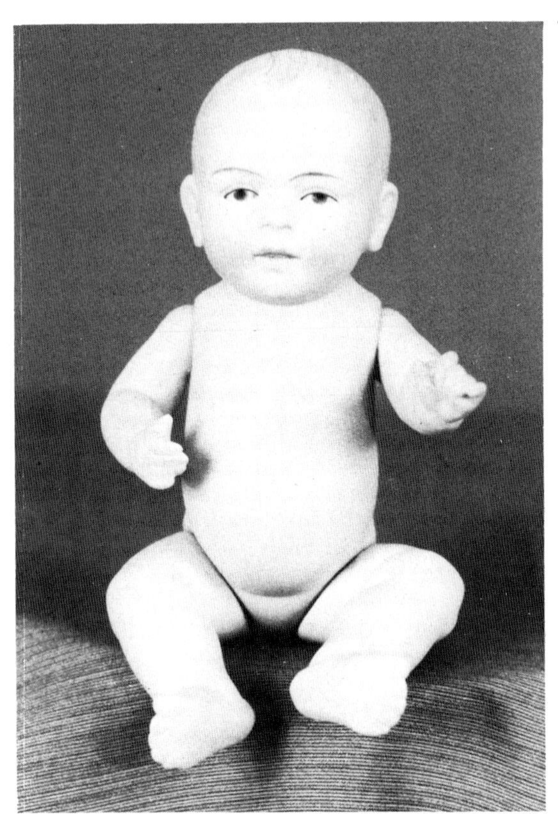

Illustration 497. With the great popularity of the regular line of character babies, it is not surprising that an experienced maker of all-bisques like Kestner would turn to making all-bisque babies as well. Judging from the numbers around, they were also very popular models. However, it seems as though the larger models of the all-bisque variety were more likely luxury items. They were made up to about 14in (35.6cm) in length; these larger models often have the same marked J.D.K. heads that were used on the composition bodies. This 6½in (16.5cm) baby is a marvel of detail for a toy. It is marked only "3" on the arms and legs, but there seems little doubt that he is a Kestner product. He has a very serene and pleasant face with a button nose which has molded nostrils that flare out and red nose dots. He also has an open/closed mouth with a protruding upper lip, a slight white space between lips, a molded lower lip, and a chin dimple. There are two darker red lip lines for accent. He has long blonde one-stroke eyebrows over his gray painted eyes with both red accent eye line and black lid line. The bisque is smooth and of excellent quality with slightly rosy cheeks. His blonde painted hair is lightly stroked with a molded forelock and small curls above each ear and at the nape of the neck. His ears are large and very detailed. His torso starts with a short fat neck, molded breasts, and even has a collarbone indentation. His stomach protrudes slightly and shows a molded navel. At the back he has molded shoulder blades, waist, and buttocks with dimples. Each arm is molded differently, one is bent inward, the other outward; they show fat rolls, elbows, wrists, dimples, and knuckles; the thumbs are free. His legs also are chubby with dimples, individual toes, and fat creases at the ankles. *H & J Foulke.*

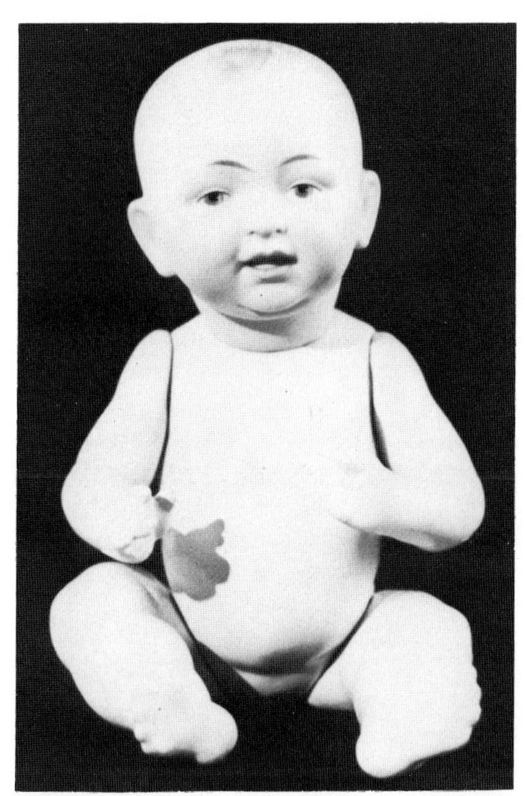

Illustration 498. This larger Kestner baby is from a different mold and has, in addition, a swivel neck. His eyes, which are in molded sockets, are very nicely painted with blue irises, black pupils and lid line, and a white dot highlight. His mouth is in the open/closed style with molded tongue and full painted lips. The detail around his smiling mouth is excellent. His "first" chin is pointed with a dimple. His eyebrows are one wavy stroke each. This face is like the model used with the composition bodies. *H & J Foulke.*

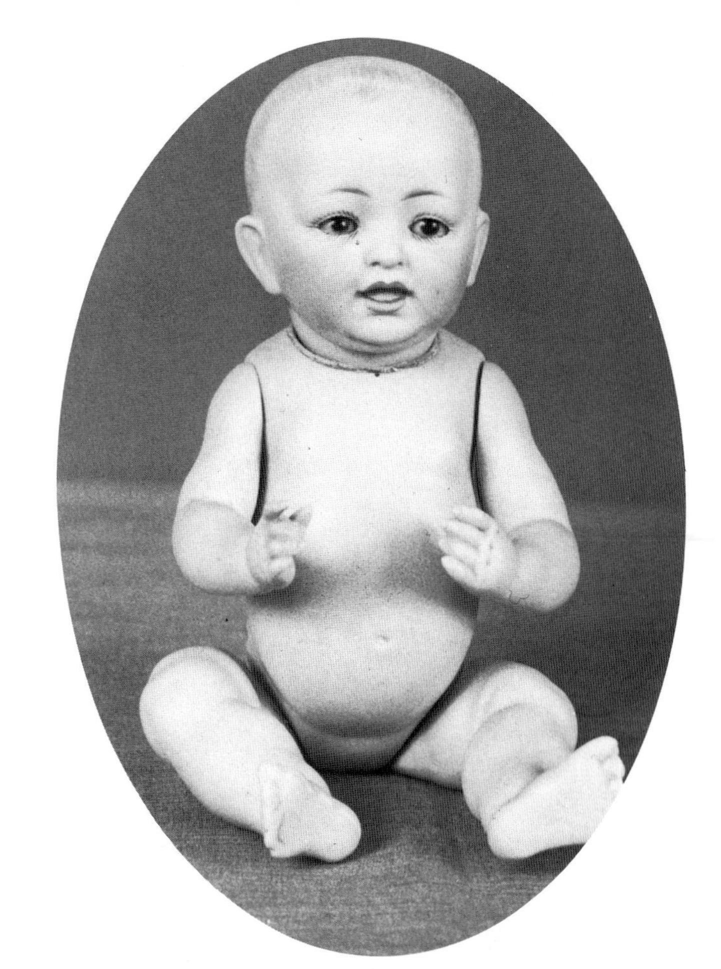

Illustration 499. This 9in (22.9cm) all-bisque baby has "J.D.K." incised on the back of his head, which is the same mold as the bald-head babies on the composition bodies. His neck swivels in a leather-lined socket. His brown sleep eyes are surrounded by a lot of painted lashes and his eyebrows are multi-stroked. His open/closed mouth has very full painted lips. His hands are especially well modeled with individual fingers, even fingernails and knuckles. His thumbs are free as is the forefinger of his right hand. Also the illlustration shows to advantage the molding detail in his very realistic legs with fat rolls, creases, crevices, and dimples. *Richard Wright Collection.*

All-bisque babies which appear to be Kestners are shown in the *1914 Marshall Field & Company Doll Catalog.* They are described as having "painted hair and eyes, fine bisque tinted; rubber cord jointed limbs." At 4-1/2in (11.5cm) they were $1.50 per dozen; superior grade at 4-1/4in (10.9cm) were $1.64 per dozen; 6-1/2in (16.5cm) were $4.00 per dozen; 7in (17.8cm) were $8.00 per dozen; and 10in (25.4cm) were $16.00 per dozen. Notice the increase in cost as the size gets larger. Admittedly these are wholesale prices before the 40 to 50 percent trade discount, but still these would be about their retail prices.

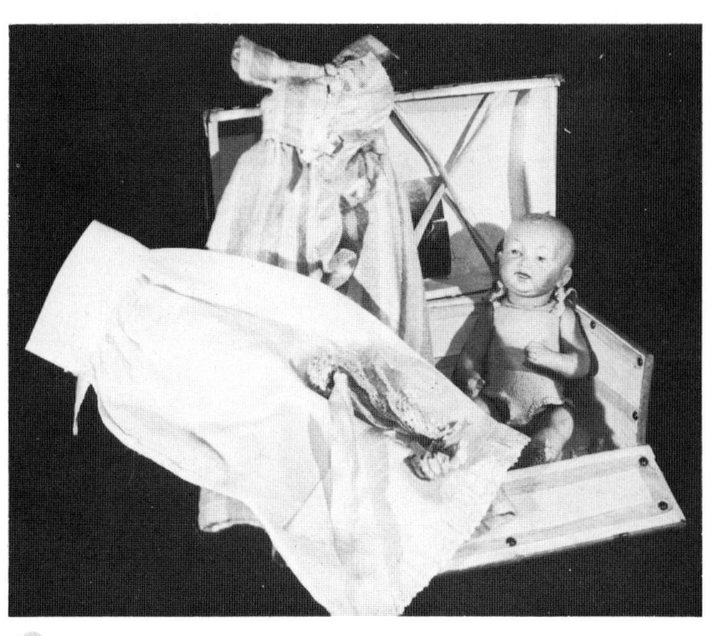

Illustration 500. This 9-1/2in (24.2cm) Kestner baby with swivel neck has all kid-lined parts. His gray painted eyes with white highlight and black lid line are in molded eye sockets. All of the detail and finishing on him is excellent. He is in his original dome-topped trunk with an F.A.O. Schwarz label and a wardrobe of original clothes including a cape with hood, a christening dress, a bonnet, booties, a shirt, diaper, sacque, three petticoats, a flannel dress, and a nightgown. Also included is an old photograph of the doll wearing his christening outfit. Truly an exciting find for any doll collector! *H & J Foulke.*

Illustration 501. It is hard to know exactly where to place some of these dolls when organizing a book. Although this baby is not all-bisque, she does have the 178 face used on the all-bisque character doll. It is such a sweet doll, just 7-1/2in (19.1cm) tall. She has tiny brown eyes with long painted lashes, blonde eyebrows in the Kestner style, a blonde mohair wig and a cute pug nose. Her open/closed mouth has two painted lower teeth. She is on a bent-limb composition baby body. *Jimmy and Faye Rodolfos Collection.*

Illustration 502. Here is a lovely large 7-1/2in (19.1cm) example of mold 178, an all-bisque toddler with swivel neck. He is a fabulous doll, with excellent quality smooth bisque. Some examples of this mold have been reported with the Kestner crown label on their stomach. *Jan Foulke Collection.*

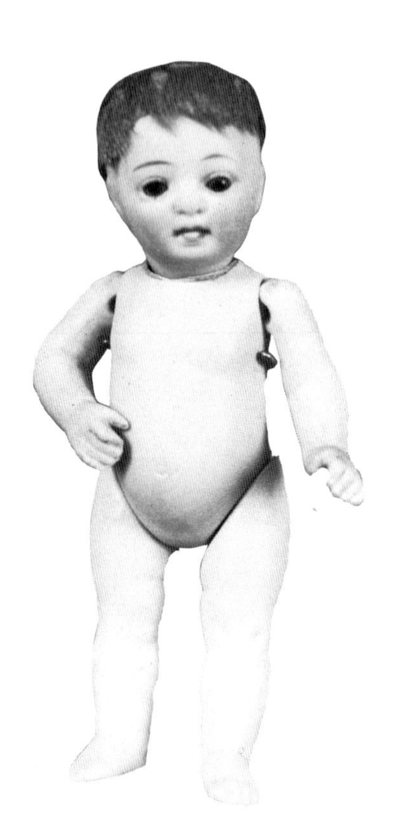

Illustration 503. Again, it is difficult to categorize some of these dolls, and this one is actually a toddler, not a baby, yet it has the same face, mold 178, which was used on the baby shown in *Illustration 501.* This boy is incised "178" on his head; "178 4/o" on his legs; "4/o" on his arms. He is 5½in (14.0cm) tall with a plaster dome covered by a skin wig. Both of these characteristics are typically Kestner. His tiny brown eyes sleep; his eyelashes are painted; his eyebrows are one stroke. He has the open/closed mouth with a molded-tooth effect. His neck swivels in a kid-lined socket, and his arms are bent in the manner of known Kestner babies. His legs are nicely modeled with fat creases, ridges, and dimpled knees. Kestner excelled in molding legs! His bare feet are unusual, as most little toddlers had molded low socks and shoes. *Richard Wright Collection.*

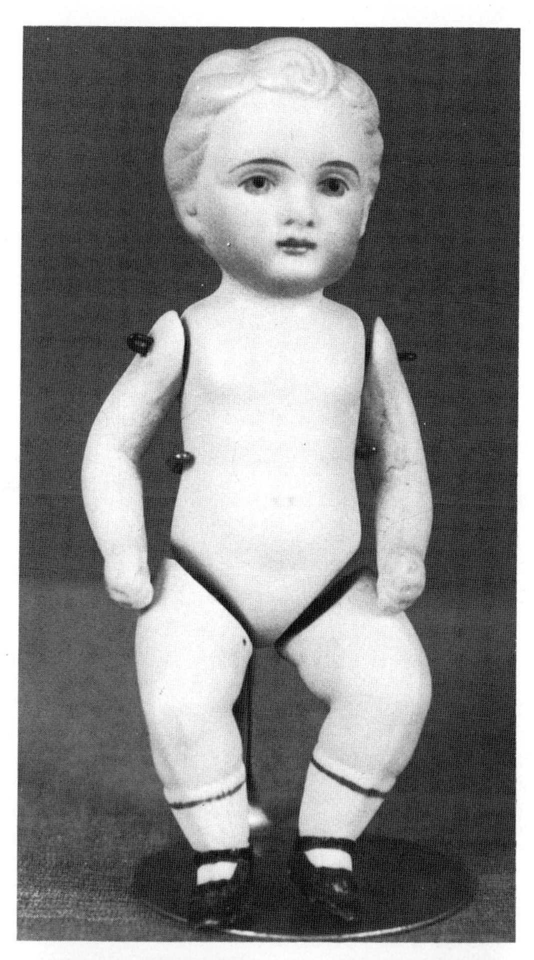

Illustration 504. Collectors tend to attribute these early babies to Kestner, but there is no real proof and they could just as easily have been made by another quality factory, such as Kling & Co. It is felt that the babies with the molded shoes are perhaps the earliest, probably dating to perhaps even a few years before 1900, which certainly precedes the bent-limb composition body by at least ten years. This particular one is 4¼in (10.9cm) long. He has the molded white ribbed stockings with magenta bands and black one-strap slippers with molded bows. His torso and leg modeling are good; his arms are quite mediocre but with clenched fists. Kestner all-bisques often have poor arms compared to their other body parts. His pouty face is very appealing. The bisque is pale with rosy cheeks. His molded blonde hair has exquisite detail of curls and comb marks. He has molded eye sockets with blue painted eyes, black pupils and lid linings as well as red accent lines. His mouth is closed with lips painted in the Kestner style. There are no marks on these early dolls and they are usually wire-strung, a method which was used both early and late in joining all-bisque parts. This baby has also been found with bare feet. *Richard Wright Collection.*

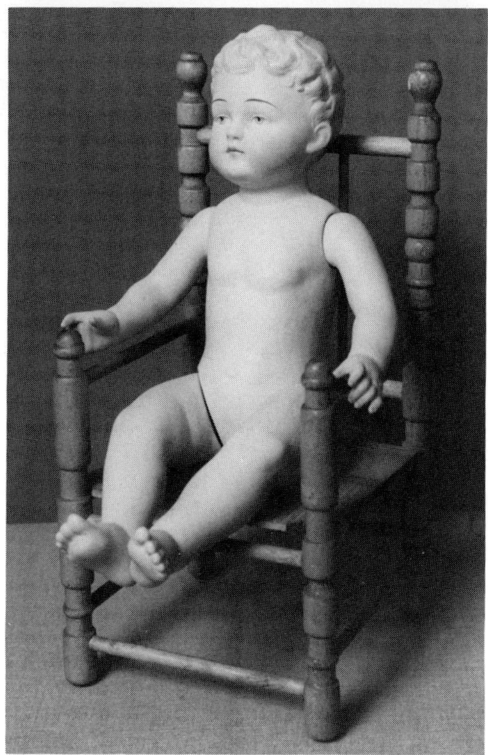

Illustration 505. This unusual 13in (33cm) all-bisque baby is fine early quality. He is brother to the small doll in *Illustration 506* in a rarely-found extra large size. His molded curly hair is blonde; painted eyes are blue. The whole doll is a work of art when the modeling is studied. The toes and fingers are particularly well done; even the soles of the feet are detailed. Genevieve Angione calls this baby the "13-mold infant," as she traced that many mold parts to the complete doll. *Joanna Ott Collection.*

Illustration 506. This 5in (12.7cm) baby is a different mold, but appears to be a product of the same factory, another likely Kestner doll. He has short blonde molded hair with a curl at the forehead, combed back at the sides and softly curled in the back. His cheeks are puffy; his nose is short and round. His eyebrows and eye decoration are the same as that on the doll in *Illustration 504,* but his irises are not quite as large. His lips are closed, but there is no shading line. He has molded breasts and a round protruding stomach. His bent legs have rounded knees, calves, upper fat rolls in back and molded toes with outlined nails. One hand is clenched; the other is open with fingers all together. Again, this doll is not marked. His bisque is very fine and pale. There are a good many of these molded blonde-haired babies around and they make a fascinating study as there are many variations in hair style, hands, feet, and other aspects of modeling. They are usually always of good quality with excellent workmanship, and it would be nice to know for sure whom to credit with their manufacture. *H & J Foulke.*

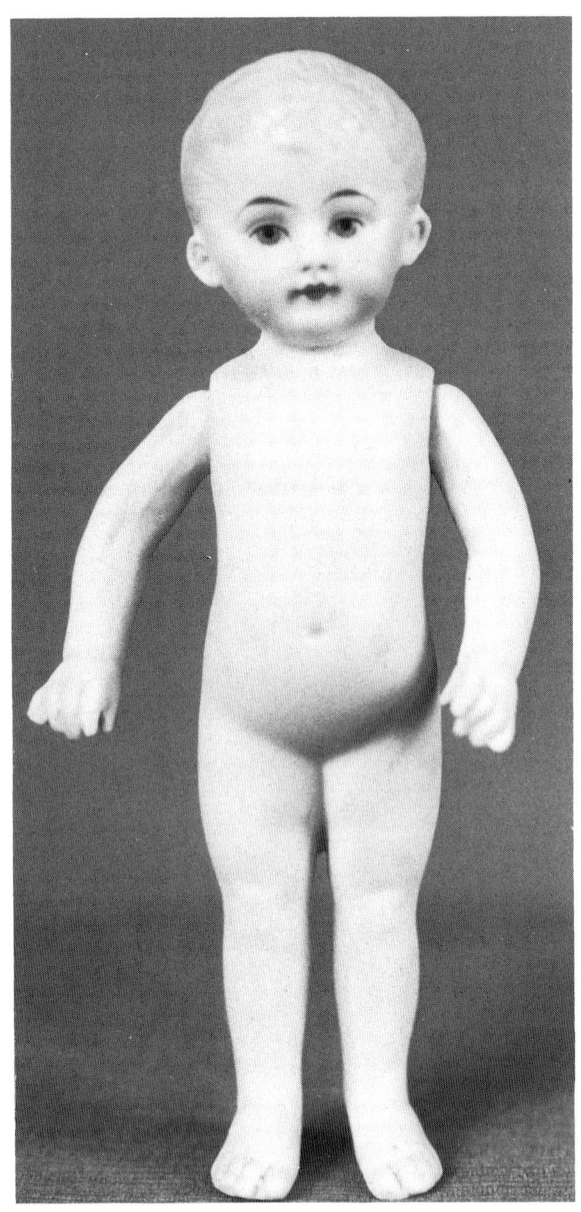

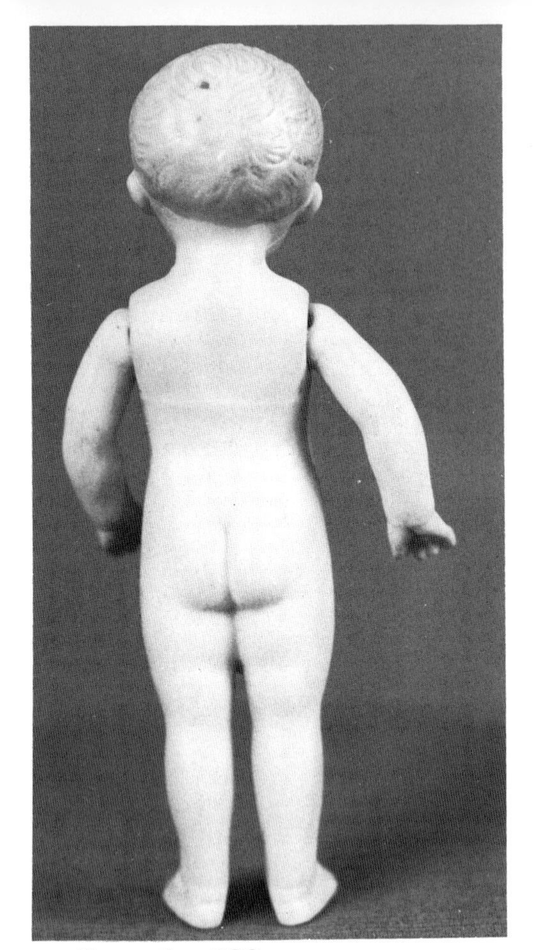

Illustration 508.

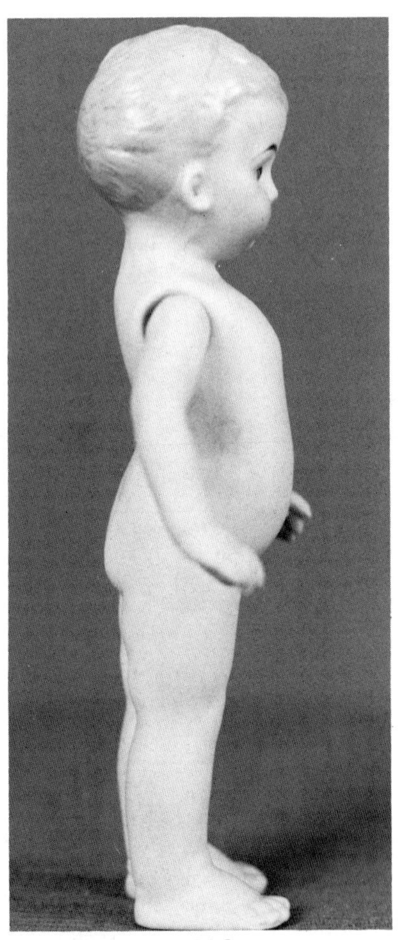

Illustration 509.

Illustrations 507, 508, and 509. Although not a baby, this child falls into the molded blonde hair group. We have also seen a standing boy with stiff arms and legs, a lying immobile baby, and a piano baby type doll with molded clothes, all with this same type of hair. These all appear to be from the same manufacturer. Interestingly enough similar babies and a little boy are pictured in the Kestner catalog of the 1930s, but the photograph is so small and muddy that it is impossible to determine if they are of this fine quality. This little boy is 6½in (16.5cm) tall and unmarked. His yellow molded hair has lovely curls, swirls and comb marks. His face is very pleasant with thin cheeks and a tiny closed heart-shaped mouth with upturned upper lip. His ears stick out but not unattractively; his nose is small and round. His blue-gray painted eyes are accented with a red eye line as well as corner eye dots. He has the shape of a young child: round protruding stomach and sway back. *Illustrations 508 and 509* show the detail in his modeling. He has fairly large feet with individual and detailed toes. His jointed shoulders are loop strung; his arms are barely bent at the elbow with wrist creases and nicely molded fingers with free thumbs. His fine quality and features make him a very probable Kestner product. *H & J Foulke.*

Dolls from the Circa 1930 Catalog

Quite a few times in this book we have referred to the Kestner catalog which is reprinted in *Die Deutsche Puppenindustrie 1814-1940* by Georgine Anka and Ursula Gauder. Unfortunately, the catalog is undated, but it appears to be from about 1930, and they purport it to be the last catalog which Kestner published before the firm was taken over by Kämmer & Reinhardt in 1932. There is an exciting page of all-bisque, mostly character dolls, many of which had never before been definitely recorded as Kestner dolls. Among those not pictured in this section are a black boy with boxing gloves on, a white boy boxer, a very mad boy with clenched fists and molded clothes, *Max* and *Moritz* comic characters with molded clothes, a boy with molded clothes in a rabbit costume, little girls with molded hair and bows, a set of frozen action characters very similar to those made by Gebrüder Heubach, small babies, naughty boys, and a large group of little girls with wigs, possibly some of which had glass eyes.

Illustration 510. Several of these *Kewpie*-type characters were offered. This 4½in (11.5cm) girl is number 520 in the catalog, but there is no mark on her. She has a one-piece head, torso, and legs with painted stockings and shoes which are not molded, just simply painted on. Her shoulders are jointed, and she has the starfish *Kewpie* hands. She has three clumps of molded and painted hair, and large side-glancing eyes with a white enamel highlight and thick upper painted lashes. Her uplifted eyebrows are painted lines. A button nose and a molded watermelon mouth with definite upper and lower lips complete her face. She has two holes at the side of her head for inserting a hair bow, but these are covered by her headpiece. She is in her original French provincial outfit. *Richard Wright Collection*.

Illustration 511. This is a larger version of catalog number 520 without the holes for the hair bow. He is 5in (12.7cm) tall, and again completely unmarked. He is also wearing what appear to be his original clothes. *Richard Wright Collection*.

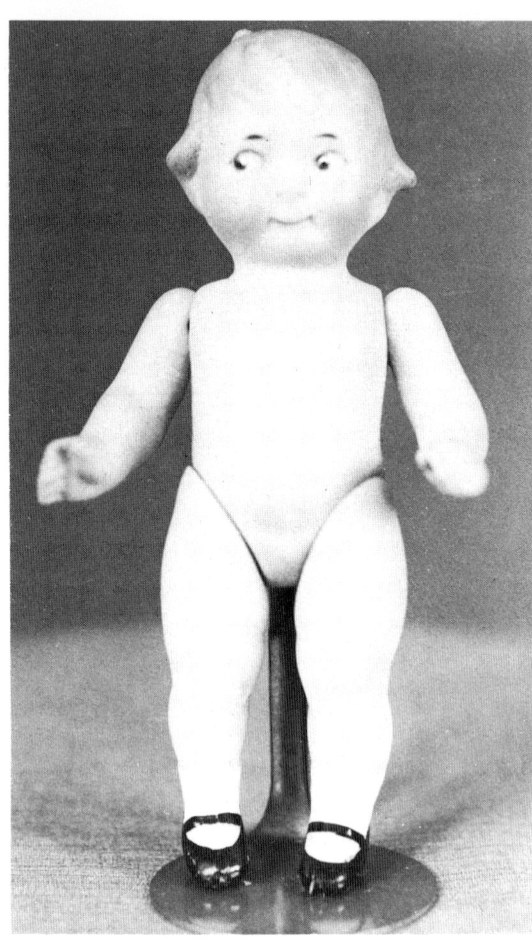

Illustration 512. This 5½in (16.5cm) character looks like the *Peterkin* doll designed for Horsman in 1910 by Helen Trowbridge. At this point, there is no way of knowing how long Kestner had been making this version of the doll or whether or not it was an authorized one. Horsman commissioned Fulper Pottery Company of Flemington, New Jersey, to make an all-bisque *Peterkin* in about 1919. "Peterkin" has a sweet face with a faintly molded watermelon mouth, fat cheeks with red circles of blush, a button nose with red nostril dots, brown side-glancing eyes with a white enameled highlight, and small brown eyebrows. His blonde hair is nicely molded with wings above each ear, a cowlick off center on his right side, and two clumps of hair coming well down on his forehead. His bisque is of a very nice quality with good modeling on his torso and delineation of breasts, shoulder blades, spine, and buttocks. His thighs have fat rolls; his knees are dimpled. His footwear consists of molded low white socks and black one-strap shoes. He has no marks except on his legs which are numbered "179," the standard legs which Kestner used on many of the small all-bisque characters. His catalog number is 192. *Richard Wright Collection*.

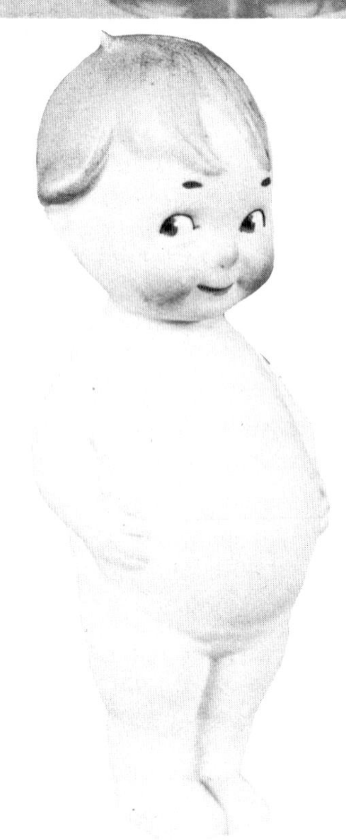

Illustration 513. This 6in (15.2cm) "Peterkin" is molded all in one piece wearing white Dr. Dentons with molded buttons up the back. His hands are clutching his stout tummy. (The jointed version is not this chubby.) There are nicely molded fabric creases at his shoulders, hips, and knees. He is unmarked, but his catalog number is 195. This doll also came with jointed shoulders as catalog number 198. *Richard Wright Collection*.

Illustration 514. This 4½in (11.5cm) boy is outfitted in probably what is Western garb, with a tan hat and shirt with a red neck piece. His gun belt holds a very large blue pistol and loops for holding bullets across the front. On his left side is a yellow bag. His blue eyes are intaglio style with only the irises and a black lid line painted. Only his top lip is painted, but it has darker red shading under it. There is a faintly molded ridge on his forehead over which the curving brown eyebrows are painted. The molded hair under his cap is light brown. His catalog number is 514. Number 511 is somewhat smaller with his hand on his holster. *H & J Foulke*.

Illustration 515. We are not sure that this is the Kestner doll which is shown in the catalog, but it is very similar to it. The black bisque has molded kinky hair and features, with brown painted eyes and a red painted mouth, and nose and eye dots. His shoulders are jointed, but his legs are fixed. He wears a gold molded necklace and a blue skirt molded to show fabric creases. His torso and arms are marked "771.3½." His catalog number is 6V. This was apparently a series of dolls, as we have seen others with different clothes. *H & J Foulke*.

Illustration 516. A rare pair of dolls is this 4¾in (12cm) set of *Max* and *Moritz*, well-known and popular German comic characters. They are all-bisque with swivel necks, jointed shoulders and hips, and painted shoes. Their faces are very well modeled with exceptional detail around the mouth, an important feature for establishing the character. *Moritz* is the doll with reddish hair styled with a topknot; *Max* has the dark brown hair with bangs. *Max*'s fingers are particularly well-done. These dolls are incised "187." In the catalog of the 1930s a *Max* and *Moritz* pair is shown with molded clothes; their catalog numbers are 186a (*Moritz*) and 186b (*Max*). *Coleman Collection. Photograph by Elizabeth Ann Coleman*.

Molded Boots

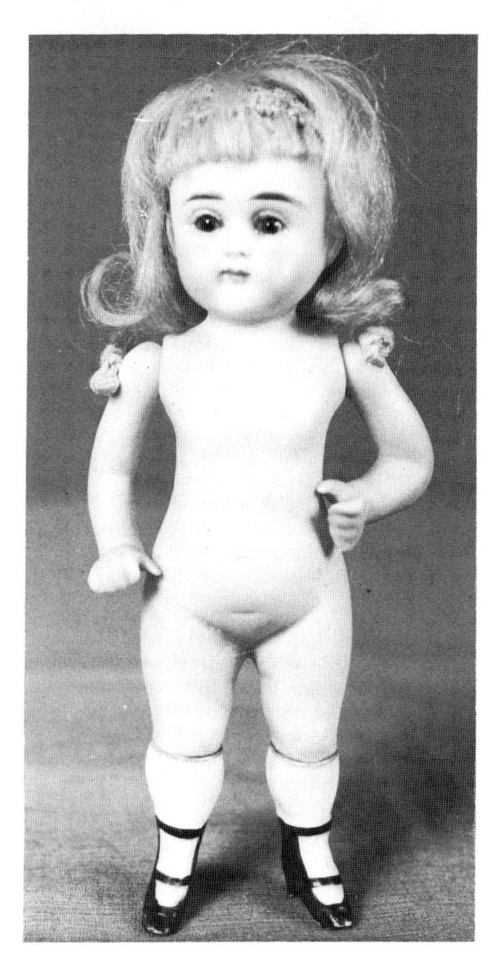

Illustration 517. We know that Kestner used the molded two-strap bootines on their dolls from the 1899 Butler Brothers wholesale catalog which advertised "Hollow Real Bisque Dolls Made by the celebrated Kestner and the very best in the market." These dolls were 3⅛in (7.9cm) with free legs (these are unjointed or stiff legs), jointed arms, mohair wig, natural features, natural glass eyes and fancy painted shoes and stockings. One dozen could be purchased for 88¢. The doll shown in the line drawing in the catalog has a round face, and is possibly the doll illustrated here. Although she is 6in (15.2cm) tall, we have also seen this face in a variety of sizes from 3in (7.6cm) on up with both painted and glass eyes. There are other faces also which usually have the black bootines, and one must be very cautious about assigning these dolls to Kestner, however, since we have also had marked Kling dolls with the same footwear. She has a slightly turned face, and her head has a very small crown opening to allow setting of the glass eyes. Her blonde eyebrows are glossy with flat undersides and much feathering along the top. Her cobalt blue eyes (a color often found with these dolls) are quite vivid with a molded upper eyelid, long painted lashes, and red eye dots. Her nostril dots and mouth with upturned upper lip are light orange; her cheeks are lightly blushed; her bisque is pale and of excellent quality. Her torso modeling is very good with breasts, navel, and buttocks. Her thighs are ample with knees being indicated with two dimples. Her footwear is extremely well-done with white molded hose with vertical ridges, a painted blue line at the very top of the molded hose ridge. The bootines have heels, rounded toes and two straps; the soles and heel bottoms are yellow. Her peg-jointed arms have cupped hands with free thumbs and one is bent more at the elbow than the other. Her original pale blonde mohair wig is on a good cloth cap. She is a sweet and very desirable doll. Unfortunately, she is entirely unmarked. *H & J Foulke*.

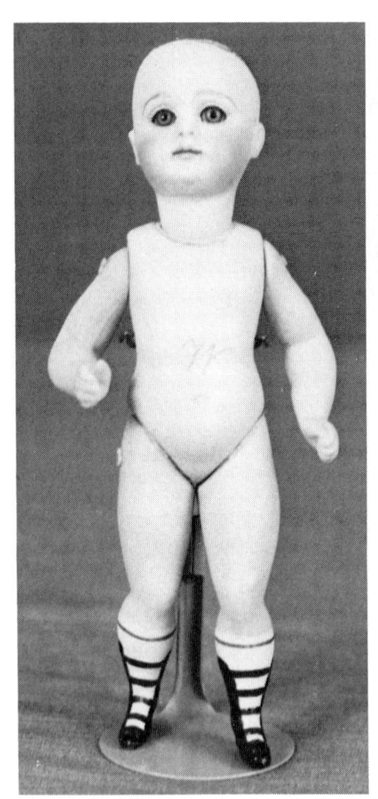

Illustration 519.

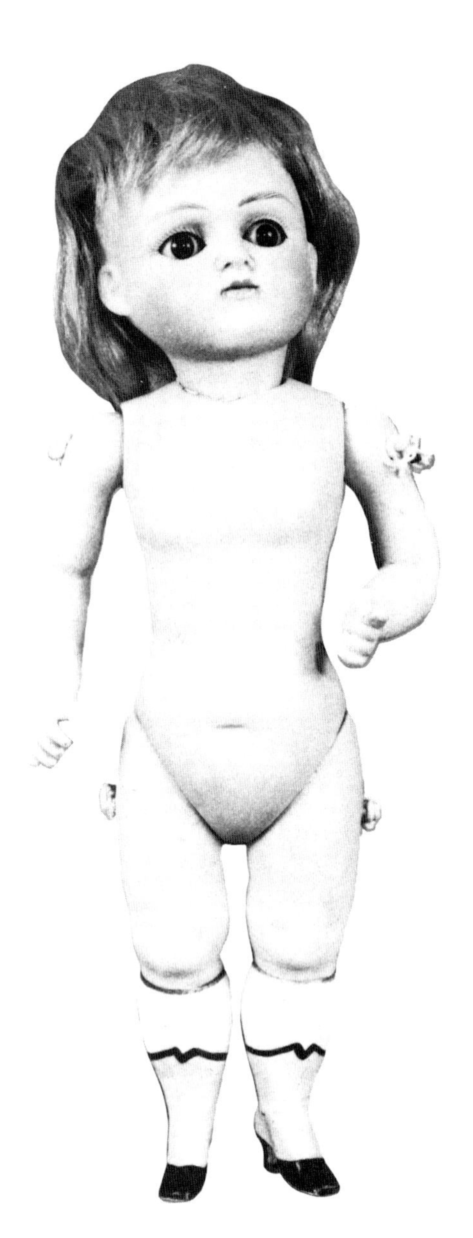

Illustration 518. There is no proof that Kestner is the manufacturer of this doll, yet she is generally attributed to that maker by collectors. If we attribute the closed-mouth pouty dolls on composition bodies to him, then this doll must follow also as she was evidently made by the same hand. This is a most beautiful, haunting face of lovely pale bisque with soft natural color. She is incised "1½" on her head and torso. She has blue sleep eyes with molded upper lids, lightly painted lashes, and long blonde curved eyebrows made from many individually painted strokes. Her pouty rose-colored mouth has two high peaks on the upper lip and a darker red line to separate the closed lips. Her footwear consists of white ribbed hose with magenta bands just below the top ridge, black bootines with four straps and molded pompons and heels. She has one cupped hand and one clenched fist. A small study of dolls of this type indicates that there is no definite pattern as to whether the hands are clenched or cupped. All types of variations occur. The neck and hip joints are kid-lined; shoulders and hips are pegged in the early style. She has her original plaster dome. This doll is considered one of the earlier types and probably dates from the 1880s. It is interesting to note that Kestner still had one doll with this type of footwear in his catalog of the 1930s. *H & J Foulke.*

Illustration 519. This lovely child is another variation of the doll in *Illustration 520.* She is called "The Wrestler" by collectors because her arms extend so far from her body and her thighs and calves are very generous. She has an open mouth with teeth; some versions have two square cut upper and one lower tooth. She also has pierced ears, an unusual feature for Kestner. Her molded boots are black with molded side insets, painted laces and heels, but she can be found with yellow boots. Her white ribbed hose have bands. She is incised "102." Some models have a plaster dome. *Betty Harms Collection.*

Illustration 520. This 11in (27.9cm) all-bisque is another outstanding doll. She has molded sockets and arms like the doll shown in *Illustration 517* with the cupped hands with free thumbs. Her torso is longer than on many of the all-bisque dolls, her waist more shapely. She has molded breasts, navel, and rounded stomach. Her knees are very rounded with two dimples on each. Her footwear consists of pale green boots with a molded black band, black toes and high heels. Her white stockings have vertical ribbing and contrasting upper bands at the top edge. Her round face has chubby cheeks, a closed mouth with a darker red lip line, and light blonde striated eyebrows. Her brown sleep eyes are surrounded by lightly painted lashes. Her blonde mohair wig is original. She has leather-lined sockets and pegged joints. Many aspects point to her as a likely Kestner product. *H & J Foulke.*

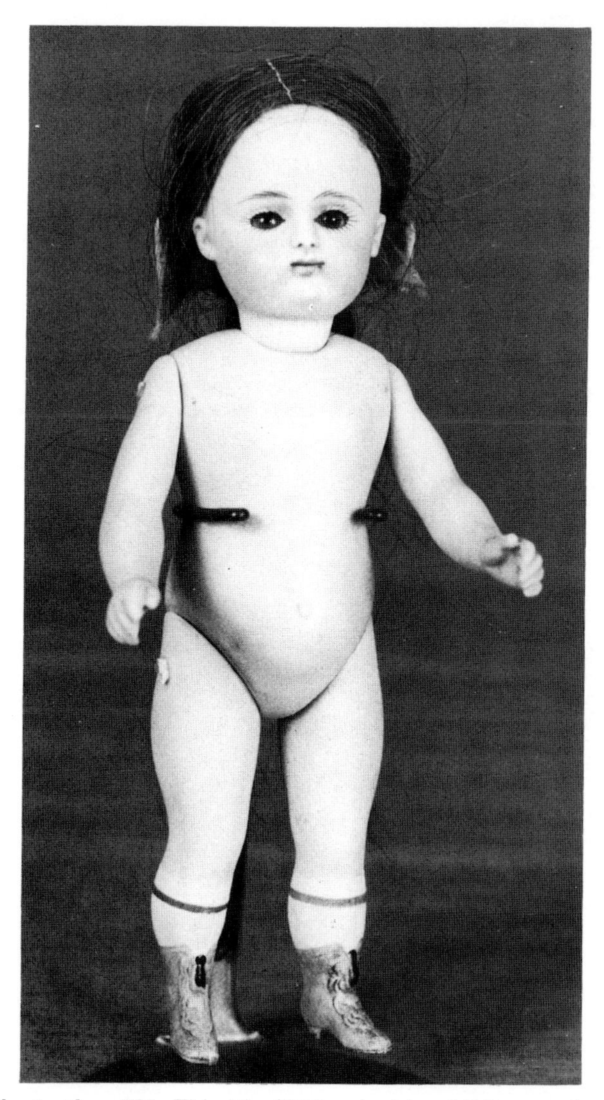

Illustration 521. This 7in (17.8cm) girl exhibits another type of footwear often found with these lovely closed-mouth all-bisques. Her footwear consists of gray molded boots with tiny heels and decorative molding on the fronts as well as black tassels. Her plain white hose have magenta bands just below the top ridge. She is definitely a relative of the doll shown in *Illustrations 518 and 519.* They both have molding detail in common as well as the straight pegged joints. This doll has both hands cupped. She has a lovely pouty face with tiny sleep eyes, blonde striated eyebrows, and long painted lashes. Her pouty mouth has a thin upper lip with upturned corners and a thick lower one; a red accent line separates her lips. Other examples of this doll have been found with blue boots. *Jane Alton Collection.*

Illustrations 522 and 523. This 7¾in (19.8cm) little girl again has the same torso, arm, and leg modeling as the doll shown in *Illustration 521.* Her footwear is different, however, and consists of gold hose with scalloped tops, heavy vertical ribs, and lighter crosshatches. Her face is lovely with a peachy complexion and blushed cheeks. Her light brown eyebrows are striated; her brown sleep eyes have long painted lashes. Her mouth is open showing four upper teeth and rosy lips, the upper with definitely upturned ends. Her ears are tiny. Over her plaster dome she has her original blonde mohair wig. A 10in (25.4cm) version of this doll has a plaster dome and head marked "4 made in Germany" in the Kestner manner. She has white instead of gold hose. *H & J Foulke.*

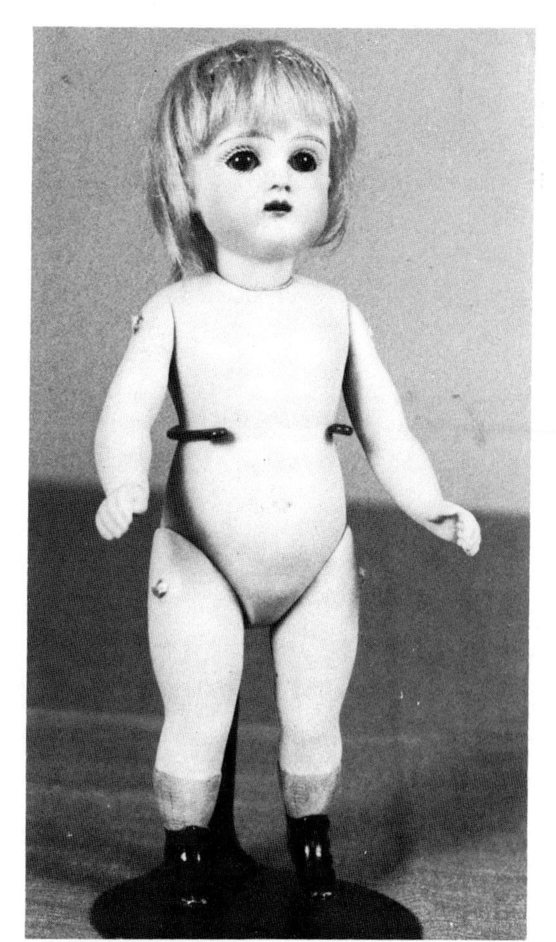

Illustration 522.

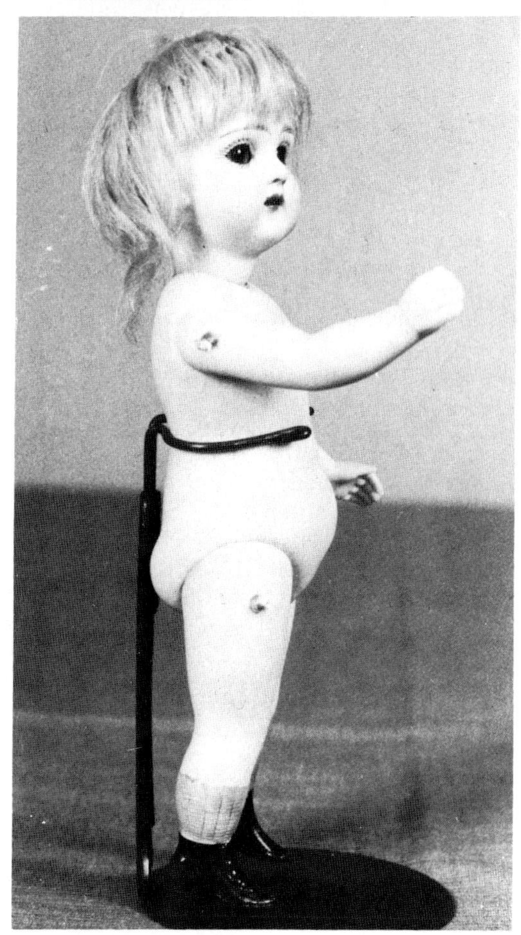

Illustration 523.

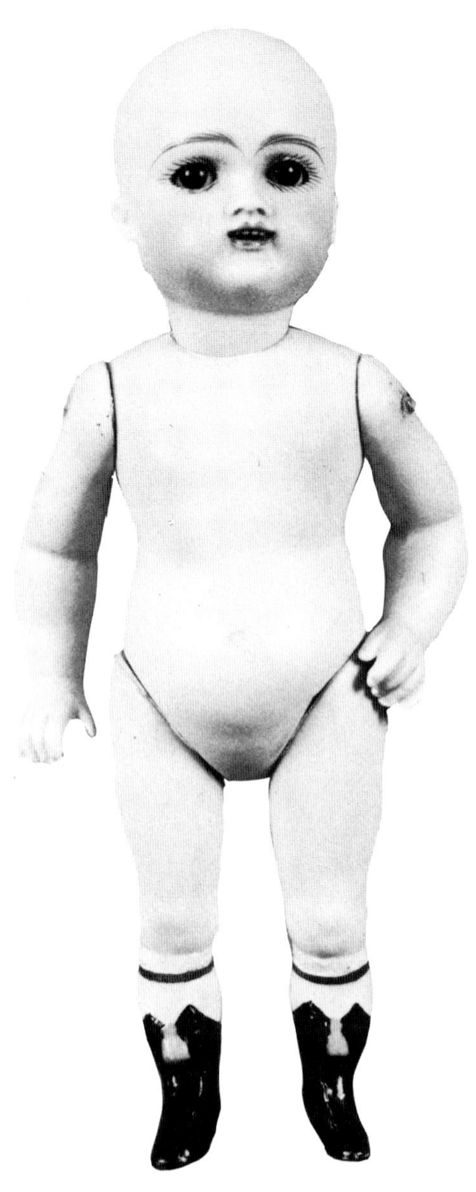

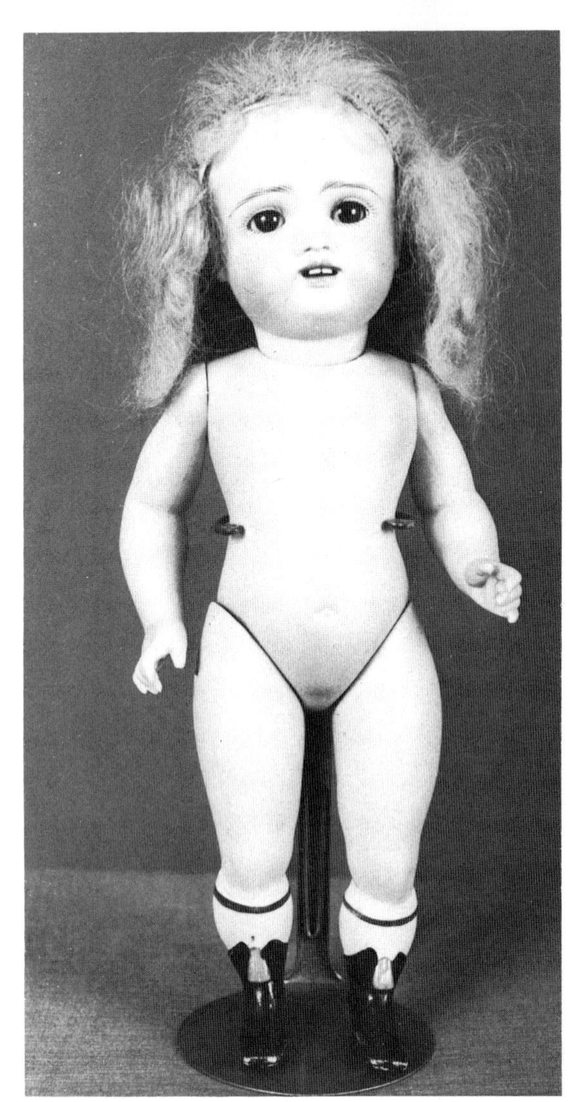

Illustration 525.

Illustration 524. This 9in (22.9cm) all-bisque is again from the same family but with different footwear. Her white hose have scalloped edges and magenta bands; her heeled boots are black with molded blue tassels. Her bisque is lovely and she has special rosy tinting on her molded breasts and dimpled knees. Her face is the same as the doll in *Illustrations 522 and 523* with a quite protruding upper lip. Notable also are her stout neck and tiny ears. *Joanna Ott Collection*.

Illustrations 525 and 526. Here is a larger 11in (27.9cm) version of the dolls shown in *Illustrations 521, 522, 523, and 524*. Her molding is lovely and she has special rosy shading on her breasts, stomach, knees, elbows, and tops of her hands. Her long blonde striated eyebrows match her replaced blonde mohair wig and contrast with her brown sleep eyes and painted lashes. Her open mouth shows two of the early style molded-in square-cut upper teeth. *Richard Wright Collection*.

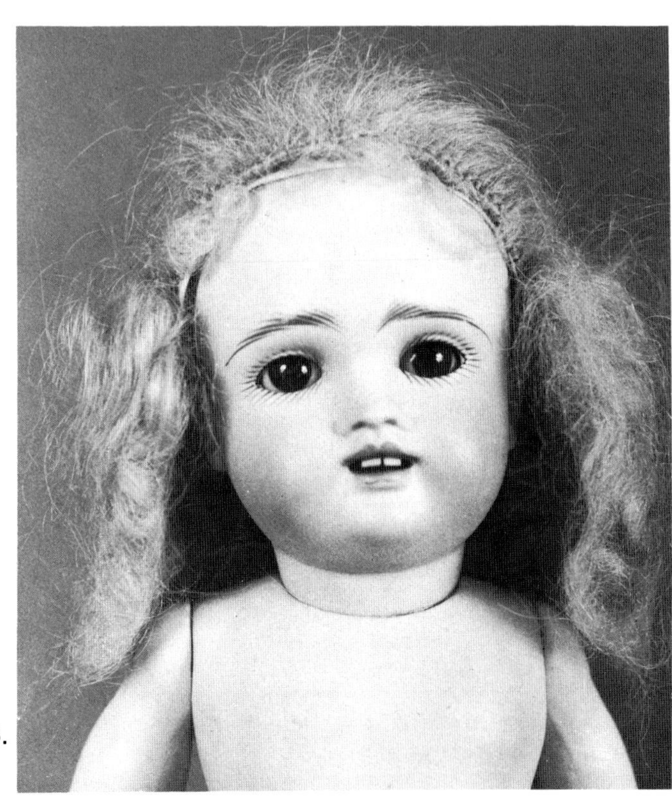

Illustration 526.

Bare Feet

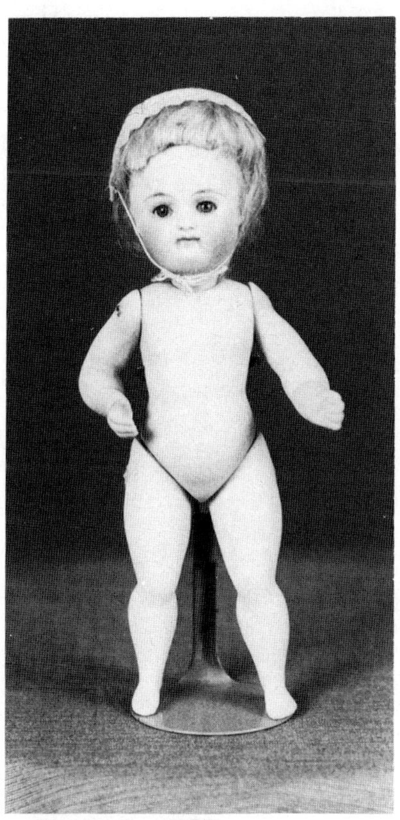

Illustration 527.

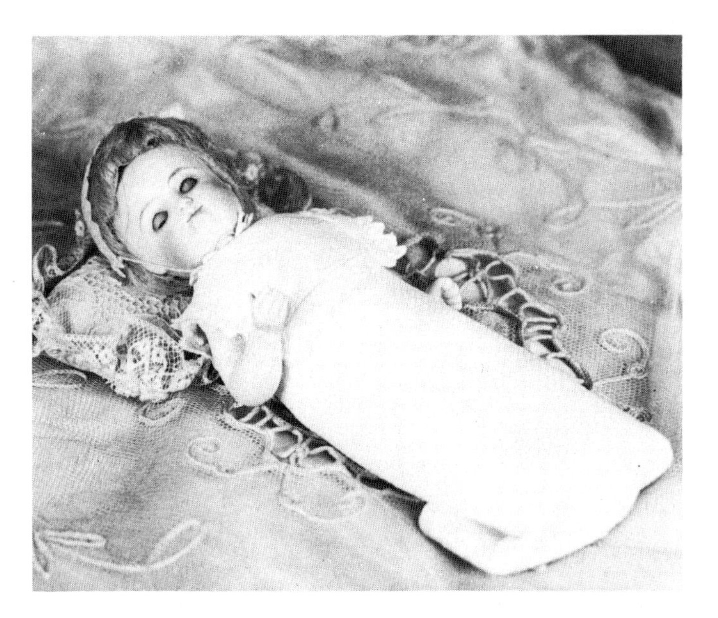

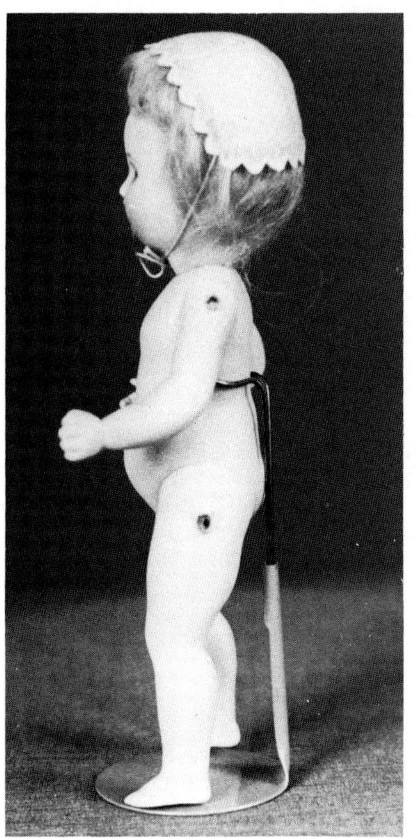

Illustration 529.

Illustrations 527, 528, and 529. Again there is no proof that Kestner made these all-bisques, but they certainly have characteristics in common with the closed-mouth dolls on composition bodies, and if one attributes the latter, then these must be by the same maker also. Also, it is very possible that a quality factory, such as Kling & Co. produced these dolls. This is probably the type of doll described in the doll column of *Harper's Bazar* in 1881 as a "tiny doll entirely of bisque with long natural blonde hair, eyes that open and close, and jointed limbs is a favorite with little girls who do not think size everything; these cost from 65¢ upward." She is certainly a marvelous little doll and is definitely a relative of *Illustrations 518 and 519* and *Illustration 521*. These early all-bisque dolls present a wide area of study, and here we can only touch the tip of the iceberg by suggesting that they are of Kestner manufacture. This little 6¼in (15.9cm) baby has outstanding modeling detail with molded breasts, waist, navel, rounded stomach, shoulder blades, and buttocks with two dimples. Her thighs, knees, and calves are nicely rounded. Of course, the bare feet are her rarest feature. Each of her arms are differently modeled with the left hand clenched and the right one cupped. Her flat joints are pegged; her neck swivels. She has the tiny gray sleep eyes, a color associated with Kestner, with lightly painted lashes. Her eyebrows are four light brown striated strokes. Her nose is rounded with red nostril dots. Her closed mouth has a protruding upper lip with a darker red lip line matching her red nostril dots. Her original short blonde mohair wig covers a plaster dome. Her only marking is an incised "1" on the back of her neck. She is wearing completely original baby clothes. She has a hemmed white lawn diaper, a shift, bib, pinning blanket, and cap of white waffle pique with embroidered trim. It seems strange to us to see a doll with straight limbs used as a baby, but at this early date there were probably no bent-limb bisque babies. Even those which we consider early types were probably not before 1890. *Jan Foulke Collection*.

Illustration 530. This 8½in (21.6cm) girl is a larger version of the baby in *Illustrations 527, 528, and 529*. She has her original plaster dome and blonde mohair wig, but her lovely white organdy dress with aqua stripes is a replacement. She has lovely bisque which has more of a complexion tint than that of the baby, hence her cheeks do not appear as rosy as the baby's. Her tiny brown sleep eyes are almond-shaped and are surrounded by finely painted lashes. Her long blonde eyebrows are striated with five strokes each. Her face is squared off at the bottom, but her cheeks are chubby and her "first" chin is rounded. She has the familiar protruding upper lip with a full bottom lip and darker red lip line. She is incised "111" on the back of her head. She is perhaps like the dolls advertised in Ridley's Fashion Magazine in 1886: "Small bisque bathing dolls, finest quality with moving eyes, head to turn, jointed arms and limbs with long flowing hair at $1 and up." Bathing dolls simply meant dolls which could be put into water. *Jan Foulke Collection*.

Illustration 531. This 11½in (29.2cm) girl is one of the largest of this type. She is a very heavy doll and would have been very difficult for a child to play with. Her torso, arm, and leg molding is excellent with many more crevices, muscles, fat rolls, and dimples than the smaller baby had. Her hands are particularly good with individually modeled fingers and knuckles; the molding in the palm of her cupped hand is unbelievably detailed. Her toes are also very well-done and are even rounded on the bottom of her foot. She is wearing her original dress of bright rose silk with black velvet ribbon and tatted lace trimming. Her cork pate is covered by her original pale blonde mohair wig in the Rembrandt style so often used by Kestner. She has the same facial features as the smaller dolls with light striated eyebrows and dark brown sleep eyes. She has no marking. *Jan Foulke Collection*.

Molded Clothes

There are dolls with molded clothes in the Kestner catalog of about 1930, and we know from an ad in Butler Brothers wholesale catalog of 1899 that Kestner made dolls with molded clothes at that time. The ad was for Kestner costume dolls, boys and girls in bloomer and knickerbocker costumes, each wearing a bonnet or cap. They came in a 2½in (6.4cm) size with free arms and legs (these were not movable, but molded separately) at 39¢ per dozen. The 3 ¾in (9.6cm) size was 77¢ per dozen. The line drawing accompanying the ad indicates quite an attractive doll. It is almost certain that more than one type of a molded clothes doll was made, but at this point, we are not sure just what.

Illustration 532. I have only seen two of these 6½in (16.5cm) all-bisque girls, but both times I had the feeling that they were probably Kestners. The face of the little girl's swivel head is so like many of the "Candy Store" dolls. She has the molded eye sockets, the black painted lids and pupils, the one-stroke eyebrows and the tiny closed mouth with upturned top lip. She also has the blonde mohair wig often found on Kestner dolls. Her jointed arms have nicely detailed hands. Her molded white dress shows fabric folds and has a molded and decorated yoke. Her footwear consists of the ubiquitous white ribbed stockings with blue bands and black Mary Janes with molded bow ties. Her legs are immobile and molded as though she is walking forward. The fine bisque and workmanship make Kestner a good possibility as maker of this lovely doll. *H & J Foulke*.

Illustration 533. Here is another superb item which is possibly a Kestner. She is 4½in (11.5cm) tall in a sitting position. Her features are beautifully modeled with the eyes and mouth of the Kestner type. Her molded shift is trimmed with painted blue piping and raised dots. The modeling of her hands and toes is quite detailed. She has a very long wig over her solid dome head. *Richard Wright Collection.*

Kestner Mold Numbers

All heads are socket type unless noted. All-Bisque dolls are not included.
*These mold numbers are listed by the Ciesliks, but I have never located any examples of them.

Mold #	Type	Date/Comments
120-127*	Dolly face	
128	Dolly face, open or closed mouth	
129	Dolly face, open mouth	
130*	Dolly face	
131*	Dolly face	
132	Dolly face, open mouth	9in (22.9cm) size only known
133	Dolly face, open mouth	7in (17.8cm) size only known
134	Dolly face, open mouth	Brown bisque
135*	Dolly face	
136*	Dolly face, closed mouth	also a H.S. & Co. dolly
137-140*	Dolly face	
141	Dolly face, open mouth	
142	Dolly face, open mouth	
143	Dolly face, open mouth	pre-1897 character-type
144	Dolly face, open mouth	
145	Dolly face, s/head, o.m.	1897
146	Dolly face, open mouth	
147	Dolly face, s/head, o.m.	
148	Dolly face, s/head, o.m.	
149	Dolly face, open mouth	
151*	Dolly face, open mouth	
152	Dolly face, open mouth	
153*	Dolly face, closed mouth	1887
154	Dolly face, s/head, o.m.	probably made 30 years
155	Dolly face, open mouth	small sizes only
156	Dolly face, open mouth	1898
159	Dolly face, s/head, o.m.	
160	Dolly face, open mouth	
161	Dolly face, open mouth	
162	Adult face, open mouth	compo lady body
164	Dolly face, open mouth	
166	Dolly face, s/head, o.m.	1898
167	Dolly face, open mouth	some real lashes
168	Dolly face, open mouth	1900
169	Dolly face, closed mouth	1892
170	Dolly face, open mouth	1903 walking body
171	Dolly face, open mouth	some real lashes
172	Adult face, s/head, c.m.	1910 Gibson Girl body
173	Dolly face, open mouth	
174	Dolly face, open mouth	1909 Wonder Doll Set
177	Character, molded hair	1909 painted eyes
178	Character child	1909 painted or glass
179	Character child	1909 two different faces
180	Character child	1909 painted or glass
181	Character child	1909 painted eyes
182	Character child	1909 painted or glass
183	Character child	1909 painted or glass
184	Character child	1909 painted or glass
185	Character child	1909 painted or glass
186	Character child	1909 painted or glass
187	Character child	1909 painted eyes
189	Character child	1909 glass eyes
190	Character child	1909 painted eyes
192?	Dolly face, o.m. and c.m.	1892 for K & R
195	Dolly face, s/head, o.m.	1910 real brows, lashes
196	Dolly face, o.m.	1910 real brows, lashes
199*	Dolly face, s/head	possibly for H. Handwerck
206	Character child	1909 glass eyes
208	Character child	1909 painted or glass
209	Character baby, s/head	1910 solid dome, o/c mouth
210	Character baby, s/head	1910 See Cieslik, p. 150, #894
211	Character baby	1910 open or o/c mouth

212	Character child	1910 glass eyes
214	Dolly face	1912
215	Dolly face	1912 real eyebrows
216*		
218*	Character baby	1912 sleep eyes, o/c mouth C.P.
219*	Character	1912 for C.P.
220	Character	1914 toddler body
221	Googly	1913
226	Character baby	1912
234	Character baby, s/head	1914
235	Character baby, s/head	1914
236*	Character baby	1913 sleep eyes, o.m.
237	Character baby Hilda	1914 Reg. #1070, bald head also
238	Character	# not verified
239	Character baby	1914
241	Character child	1914 open mouth
242*	Character	1914 sleep eyes, o.m.
243	Oriental baby	1914 molded hair version also
245	Character baby Hilda	1914 Reg. #1070, bald & black also
246*	Character	1915 sleep eyes, o.m.
247	Character baby	1915
249	Character child	open mouth
250	Dolly face	Walküre for K & H
254*	Character	1916
255	Infant screaming	1921 marked OIC
257	Character baby	1916 made to 1930s
260	Character child	1916 made to 1930s
262	Character baby	1916 for C.P.
263	Character baby	1916 for C.P.
264	Character child	1916 for C.P.
270*		for C.P.
272	Infant Siegfried	1925
279*	Infant	1925 Century
281	Mama, s/head	1925 Century
282*		1920 Walküre for K & H
292*	Character	1930 Walküre for K & H
680	Character baby	for Kley & Hahn

BIBLIOGRAPHY

Angione, Genevieve. *All-Bisque and Half-Bisque Dolls.* Camden, N.J.: Thomas Nelson & Sons, 1972.

Angione, Genevieve and Judith Whorton, *All Dolls Are Collectible.* N.Y.: Crown Publishers, Inc., 1977.

Anka, Georgine and Ursula Gauder. *Die Deutsche Puppenindustrie 1815-1940.* Stuttgart, Germany: Verlag Puppen and Spielzag, 1978.

Bachmann, Manfred. *Dolls — The Wide World Over.* N.Y.: Crown Publishers, Inc., 1973.

1890 Butler Brothers Wholesale Catalogue. Maryland: Carter Craft House.

Cieslik, Jürgen and Marianne. "The Kestner Story," *Doll Reader,* Volume X, Issue 2, February/March 1982.

__________. *German Doll Encyclopedia 1800-1939.* Cumberland, Maryland: Hobby House Press, Inc., 1985.

Coleman, Dorothy S., Elizabeth A., and Evelyn J. *The Collector's Book of Dolls' Clothes.* N.Y.: Crown Publishers, Inc., 1975.

__________. *The Collector's Encyclopedia of Dolls.* N.Y.: Crown Publishers, 1968.

__________. *The Collector's Encyclopedia of Dolls, Vol. II.* N.Y.: Crown Publishers, Inc., 1986.

__________. *The Age of Dolls.* Washington, D.C.: Coleman, 1965.

"How Dolls Are Made," *Doll Collectors Manual 1967.* Warner, N.H., Mayflower Press, 1967.

Johl, Janet Pagter. *The Fascinating Story of Dolls.* N.Y.: H. L. Lindquist, 1941.

Long, Ida and Ernest. *A Catalog of Dolls.* Calif.: Long's Americana, 1978.

1914 Marshall Field & Company Doll Catalog. Maryland: Hobby House Press, Inc., 1980.

Nobel, John. *A Treasury of Beautiful Dolls.* N.Y.: Hawthorne Books, Inc., 1971.

Schroeder, Jr., Joseph J. *The Wonderful World of Toys, Games & Dolls 1860-1930.* Illinois: Digest Books, Inc., 1979.

Westbrook & Ehrhardt. *Encyclopedia of American Collector Dolls.* Kansas City: Heart of America Press, 1975.

INDEX

MARKS AND MOLD NUMBERS: